SPACE-AGE ACRONYMS
ABBREVIATIONS AND DESIGNATIONS

SPACE-AGE ACRONYMS
ABBREVIATIONS AND DESIGNATIONS

By
Reta C. Moser

With a Foreword by
Bill M. Woods
Executive Director
Special Libraries Association

PLENUM PRESS
NEW YORK
1964

Library of Congress Catalog Card Number 64-20744

©1964 Plenum Press
A Division of Consultants Bureau Enterprises, Inc.
227 West 17th Street
New York, N.Y. 10011

No part of this publication may be reproduced in any
form without written permission from the publisher

Printed in the United States of America

FOREWORD

Apparently there aren't enough words in the English language (or any language) to express adequately all the ideas and concepts which come into existence and about which there is need to communicate.

Specialists in one field search for an effective medium for communication among themselves and with specialists in other fields, with managements in business, government, and education, and with interested and disinterested laymen. Whenever the language of scholarship or of day-to-day communication fails, an acceptable alternative is created.

There was then a time before life became so complicated, and before high-speed computers and space travel made man even more conscious of time, how to make the most efficient use of it, and how to extend the effectiveness of such use.

It wasn't that complete words were going out of style, but that vocabulary long in use had to be supplemented by something more widely understood. Obviously abbreviations, acronyms, and nicknames were needed—so they were created to fill the need.

During the 1930's it became popular to designate emergency agencies of government in this word shorthand. The space age just as quickly and even more dramatically has increased the need for greater speed and efficiency in communication. Persons in and out of the field must be kept informed. The acronym has been adopted in wholesale style. Some would suggest that the acronymic vocabulary is now so large that a monster has been bred, and that in multiple meanings communicative value has been lost.

Such is the reason for a dictionary-directory intended to keep space-age specialists as well as librarians, journalists, and others informed. It is intended to provide a map or chart through space-age acronyms and abbreviations—in fact, more than 10,000 acronyms and more than 17,000 definitions—and at the same time to help effect some standardization in terminology.

"A is for apple!"

Bill M. Woods
Executive Director
Special Libraries Association

PREFACE

How important is knowing the definition of an acronym? It is as important as the time lost searching for or the consequence of not finding it.

The purpose of this book is to offer to those in the space and defense industry a central acronymic reference source, one that can be updated, and one that can be used as a standard in the business. Great effort has been made to go to the source for the correct, original definition. For instance, MMRBM is often defined as Mobile Medium Range Ballistic Missile. The MMRBM is an Air Force project and, according to the Air Force official abbreviation manual, MMRBM is Mobile Mid-Range Ballistic Missile.

The acronyms contained in this book were created by the men and women of the space age to save time and space and eliminate unnecessary repetition and wordage. The sophistication and complexity of today's projects, systems, and equipment has necessitated descriptive, sentence-length nomenclature that is not only long but burdensome. The use of an acronym can reduce this nomenclature to a few, easy-to-remember letters without loss of its communicative value.

Acronymic appeal and acceptance has been so universal within the industry that now the easy-to-remember acronym is getting hard to remember with the thousands in daily use and their even more numerous multiple meanings. In order to maintain the communicative value of the here-to-stay acronym, we have "Space-Age Acronyms—Abbreviations and Designations."

Reta C. Moser
Venice, California

ACKNOWLEDGMENT

A book of this nature cannot be compiled without the help of many people. I want to thank all the technical librarians (it would take an appendix to mention them all individually) throughout the country who supplied me with acronyms and abbreviations peculiar to their companies.

I am deeply indebted for the information and assistance given by the military information officers, specifically Captain Barbara M. Boyd, of the Marine Corps Representative Armed Forces Information Office; Chief Journalist Joseph J. Brazan, of the Eleventh Naval District Public Information Office, Los Angeles Branch; Colonel Dan Biondi, of the United States Army Office Chief of Information, Los Angeles Branch; and Lt. Colonel William G. Thompson, of the Air Force Office of Information of the Office of the Secretary of the Air Force, Los Angeles Branch.

June Ruggles, of the National Aeronautics and Space Administration Western Operations Office, Public Information Division, was of particular assistance with the NASA acronyms.

Without the help of all the technical librarians and the information personnel, this book would not have been possible.

CONTENTS

ACRONYMS, ABBREVIATIONS 1

MISSILE, ROCKET, PROBE DESIGNATION SYSTEM 413

AIRCRAFT DESIGNATION SYSTEM 416

SHIP DESIGNATIONS 418

COMMUNICATION ELECTRONIC EQUIPMENT
    DESIGNATION SYSTEM 423
  Set or Equipment Designation System 423
  Component Designation System 425

A	- Absolute
	- Acceleration
	- Air Force
	- Airplane
	- Amphibian
	- Angstrom
	- Area
	- Argon
	- Arm
	- Army
	- Asbestos
	- Attack
A-	- Atomic
AA	- Absolute Altitude
	- Air to Air
	- Antiaircraft
	- Arithmetic Average
	- Armature Accelerator
	- Arrival Angle
	- Audible Alarm Panel
A/A	- Air Abort
	- Airborne Alert
	- Any Acceptable
AAA	- Active Acquisition Aid
	- Anti-Air ATABE
	- Antiaircraft Artillery
AA-A	- Antiaircraft Assistant
AAAA	- Army Aviation Association of America
AAAIS	- Advanced Army Airborne Indicating System
	- Antiaircraft Artillery Information Service
AAAOC	- Antiaircraft Artillery Operations Center
AAAS	- American Academy of Arts and Sciences
	- American Association for the Advancement of Science

■ AAB

AAB	- Aircraft Accident Board
	- Antiaircraft Balloon
AABY	- As Amended By
AAC	- Aeronautical Advisory Council
	- Aeronautical Approach Chart
	- Air Approach Control
	- Alaskan Air Command
	- Antiaircraft Cannon
	- Antiaircraft Common
	- Automatic Approach Control
AACB	- Aeronautics and Astronautics Coordinating Board
AACC	- American Automatic Control Council
	- Automatic Approach Control Coupler
AACS	- Airways and Air Communications Service
AACSM	- Airways and Air Communications Service Manual
AACSR	- Airways and Air Communications Service Regulation
AAD	- Admission and Disposition
	- Army Air Defense
AADA	- Advanced Air Depot Area
	- Antiaircraft Defended Area
	- Army Air Defense Area
AADC	- American Air Defense Command
	- Army Air Defense Command
AADCP	- Army Air Defense Command Post
AADIS	- Army Air Defense Information Service
AADOO	- Army Air Defense Operations Office
AADP	- Antiaircraft Defended Point
AADS	- Army Air Defense System
AAE	- American Association of Engineers
	- Army Aviation Engineers
AAEC	- Attitude Axis Emergency Control
AAEE	- Aircraft and Armament Experimental Establishment (British)
AAF	- Army Airfield
AAFB	- Auxiliary Air Force Base
AAFC	- Antiaircraft Fire Control
AAFCE	- Allied Air Forces, Central Europe (NATO)
AAFES	- Army and Air Force Exchange Service
AAFNE	- Allied Air Forces, Northern Europe (NATO)
AAFPS	- Army and Air Force Postal Service
AAFSE	- Allied Air Forces, Southern Europe (NATO)
AAG	- Air Adjutant General
AAGMC	- Antiaircraft Guided Missile Center
AAGMS	- Antiaircraft Guided Missile School
AAGR	- Air-to-Air Gunnery Range
AAHA	- Awaiting Action of Higher Authority

AAI	- Air-to-Air Identification
	- Angle of Attack Indicator
	- Asociacion Argentina Interplanetary
AAIA	- Army Area Analysis Intelligence Agency
AAIC	- Allied Air Intelligence Center
AAIE	- American Association of Industrial Editors
	- American Association of Industrial Engineers
AAL	- Ames Aeronautical Laboratory
AALO	- Antiaircraft Liaison Officer
AAM	- Air-to-Air Missile
	- Antiaircraft Missile
AAMG	- Antiaircraft Machine Gun
AAML	- Arctic Aeromedical Laboratory
AAMM	- Anti-Antimissile Missile
AAO	- Anti-Air Output Program
	- Astronaut Activities Office
AAOC	- Antiaircraft Operations Center
AAOD	- Army Aviation Operating Detachment
AAP	- Aerodynamics Advisory Panel
	- Anti-Air Processing Program
AAR	- Aircraft Accident Record
	- Army Area Representative
AARR	- Argonne Advanced Research Reactor
AAS	- Advanced Antenna System
	- Alerting Automatic Telling Status
	- American Astronautical Society
	- American Astronomical Society
	- Arnold Air Society
AASA	- Administrative Assistant to Secretary of the Army
AASB	- American Association of Small Business
AASD	- Antiaircraft Self-Destroying
AASE	- Association for Applied Solar Energy
AASL	- Associated Aero Science Laboratories
AASM	- Association of American Steel Manufacturers
AASR	- Airport and Airways Surveillance Radar
AAT	- Antiaircraft Talker
AATC	- Antiaircraft Training Center
	- Army Aviation Training Command
AATCAN	- Army Air Traffic Control and Navigation System
AATOC	- Airhead Air Traffic Coordination Center
AATRI	- Army Air Traffic Regulation and Identification
AAT&TC	- Antiaircraft Training and Test Center
AAU	- Auxiliary Air Units
AAUN	- American Association for the United Nations
AAUTC	- Army Aviation Unit Training Command

■ AAVCS

AAVCS	- Automatic Aircraft Vectoring Control System
AAV	- Airborne Assault Vehicle
AAVN	- Army Aviation
AAVS	- Automatic Aircraft Vectoring System (AAVCS preferred)
AAW	- Anti-Air Warfare
AAWEX	- Anti-Air Warfare Exercise
AB	- Able-Bodied
	- Adapter Booster
	- Aeronautical Board
	- Afterburner
	- Air Base
	- Air Bearing
	- Air Blast
	- Airborne
	- Anchor Bolt
A/B	- Air Base
	- Airborne
A-B	- Allen-Bradley
ABA	- Airborne Alert
AB/A	- Airborne Alert
ABAD	- Air Battle Analysis Division
ABAR	- Alternate Battery Acquisition Radar
ABATU	- Advanced Base Aviation Training Unit
ABB	- Automatic Back Bias
ABC	- Advanced Biomedical Capsule
	- American-British-Canadian
	- Atomic, Biological, and Chemical
	- Automatic Bandwidth Control
	- Automatic Bass Compensation
	- Automatic Bias Compensation
	- Automatic Brightness Control
ABCB	- Air Blast Circuit Breaker
ABCCTC	- Advanced Base Combat Communications Training Center
ABCD	- Advanced Base Construction Depot
ABD	- Advanced Base Depot
	- Advanced Base Dock
	- Airborne Ballistics Division
ABDACOM	- Advanced Base Depot Area Command
ABDL	- Automatic Binary Data Link
ABE	- Airborne Bombing Evaluation
ABEC	- Annular Bearing Engineering Committee
ABES	- Aerospace Business Environment Simulator
ABETS	- Airborne Beacon Test Set
ABF	- Aircraft Battle Force
ABG	- Air Base Group
	- Aural Bearing Generator

AC ■

ABGP	- Air Base Group
ABINF	- Airborne Infantry
ABIOL	- Advanced Base Initial Outfitting List
ABL	- Adaption Binary Load Program
	- Alleghany Ballistics Laboratory
ABLE	- Activity Balance Line Evaluation
	- Autonetics Base Line Equipment
ABM	- Abeam
	- Anti-Ballistic Missile
	- Automated Batch Mixing
ABMA	- Army Ballistic Missile Agency
ABMD	- Air Ballistic Missile Division
ABMEWS	- Anti-Ballistic Missile Early Warning System
ABN	- Airborne
ABO	- Astable Blocking Oscillator
ABOSS	- Advanced Bombardment System
ABPA	- Advanced Base Personnel Administration
ABPG	- Advanced Base Proving Ground
ABPO	- Advanced Base Personnel Officer
ABPU	- Advanced Base Personnel Unit
ABR	- Amphibian Boat Reconnaissance
ABRB	- Advanced Base Receiving Barracks
ABRD	- Advanced Base Repair Depot
	- Advanced Base Reshipment Depot
ABRES	- Advanced Ballistic Re-Entry System
ABRL	- Army Ballistic Research Laboratories
ABS	- Air Base Simulator
	- Air Base Squadron
	- Acid and Base Washed and Silanized
	- Air Break Switch
	- American Bureau of Shipping
ABSAP	- Airborne Search and Attack Plotter
ABSD	- Advanced Base Sectional Dock
	- Advanced Base Supply Depot
ABT	- Air Bearing Table
	- Air Blast Transformer
ABTF	- Airborne Task Force
ABTSS	- Airborne Transponder Subsystem
ABTU	- Advanced Base Torpedo Unit
	- Advanced Base Training Unit
	- Air Bombers Training Unit
ABWG	- Air Base Wing
AC	- Absolute Ceiling
	- Access Control
	- Accumulator
	- Aerodynamic Center

■ AC

AC	- Air Command
	- Aircraft
	- Aircraft Control
	- Airdrome Control (British)
	- Alaskan Command
	- Allis Chalmers
	- Alternating Current
	- Altocumulus
	- Amphibious Corps
	- Assigned Contractor
	- Associate Contractor
	- Audio Center
	- Autocollimator
	- Automatic Computer
	- Auxiliary Console
	- Aviation Cadet
A/C	- Aircraft
	- Associate Contractor
	- Autocollimator
A&C	- Arithmetic and Controls
ACA	- Adjacent Channel Attenuation
	- Advance Change Authorization
	- American Crystallographic Association
	- American Communications Association
	- Associate Contractor Administration
	- Australian Council for Aeronautics
	- Automatic Circuit Analyzer
	- Awaiting Combat Assignment
ACAC	- Allied Container Advisory Committee
ACAF	- Amphibious Corps, Atlantic Fleet
ACAL	- Associated Canadian Aircraft, Limited
ACAMR	- Associate Committee on Aviation Medical Research
ACAN	- Army Command and Administrative Network
ACAP	- Analysis of Critical Actions Program
ACAR	- Aero Car
ACAS	- Analytical Chemistry and Applied Spectroscopy
	- Atlantic Coast Air Service
ACB	- Air Circuit Breaker
	- Amphibious Construction Battalion
ACBWS	- Automatic Chemical Biological Warning System
ACC	- Acceptance
	- Accessory
	- Accumulator
	- Administrative Committee on Coordination
	- Air Center Commander
	- Air Control Center

6

ACC	- Air Coordinating Committee
	- Allied Control Commission
	- Area Control Center
	- Army Chemical Center
	- Astronomical Great Circle Course
ACCA	- Ad Hoc Crypto-Coordination Agency
ACCCE	- Association of Consulting Chemists and Chemical Engineers
ACCESS	- Aircraft Communication Control and Electronic Signaling System
ACCHAN	- Allied Command, Channel
ACCID	- Approach to Command Control Implementation and Design
ACCRY	- Accessory
ACCS	- Aerospace Command and Control System
ACCW	- Alternating Current Continuous Wave
ACD	- Antenna Control Display
ACDA	- Arms Control and Disarmament Agency
	- Aviation Combat Development Agency
ACDC	- Army Combat Developments Command
AC-DC	- Alternating Current—Direct Current
ACDUTRA	- Active Duty for Training
ACE	- Acceptance Checkout Equipment
	- Aerospace Control Environment
	- Allied Command, Europe
	- American Council on Education
	- Attitude Control Electronics
	- Automatic Checkout Equipment
	- Automatic Circuit Exchange
	- Aviation Construction Engineers
ACEL	- Air Crew Equipment Laboratory
ACE/LACE	- Air Cycle Engine/Liquid Air Cycle Engine
ACEORP	- Automotive and Construction Equipment Overhaul and Repair Plant
ACEPD	- Automotive and Construction Equipment Parts Depot
ACERP	- Advanced Communications Electronics Requirements Plan
ACES	- Air Collection Engine System
	- Assurance Control Economics System
	- Automatic Checkout and Evaluation System
	- Automatic Control Evaluation Simulator
ACE-S/C	- Acceptance Checkout Equipment—Spacecraft
ACF	- Alternate Communications Facility
	- Area Computing Facilities
ACFAS	- Association Canadiene-Française pour l'Avancement des Sciences
ACFT	- Aircraft
ACG	- Airborne Coordinating Group
ACGM	- Aircraft Carrier General Memorandum

■ ACI

ACI	- Air Combat Information
	- Air Combat Intelligence
	- Air Controlled Interception
	- Attitude Controls Indicator
ACIC	- Aeronautical Chart and Information Center
	- Auxiliary Combat Information Center
ACICO	- Assistant Combat Information Center Officer
ACIL	- American Council of Independent Laboratories, Inc.
ACIO	- Aeronautical Chart and Information Office
	- Air Combat Intelligence Office
	- Air Combat Intelligence Officer
ACIR	- Aviation Crash Injury Research
ACIS	- Aircraft Crew Interphone System
ACL	- Add and Carry Logical Word
	- Aeronautical Computer Laboratory
	- Allowable Cabin Load
	- Automatic Carrier Landing
ACLANT	- Allied Command, Atlantic
ACLS	- Automatic Carrier Landing System
ACLV	- Accrued Leave
ACM	- Active Countermeasure
	- Association for Computing Machinery
	- Audio Center Module
	- Authorized Controlled Material
	- Auxiliary Mine Layer
ACME	- Attitude Control and Maneuvering Electronics
	- Association of Consulting Management Engineers
ACM-GAMM	- Association for Computing Machinery—German Association for Applied Mathematics and Mechanics
ACMS	- Army Command Management System
ACN	- Advance Change Notice
	- American Council on NATO
	- Automatic Celestial Navigation
A&CN	- Approval and Clearance Notice
ACNA	- Advisory Council on Naval Affairs
ACNO	- Assistant Chief of Naval Operations
ACO	- Administrative Contracting Office
	- Administrative Contracting Officer
	- Air Control Officer
	- Aircraft Control Operator
	- Area Clearance Officer
	- Assembly and Checkout
	- Authorized Contracting Officer
A&CO	- Assembly and Checkout
ACOE	- Acceptance Checkout Equipment (ACE preferred)
ACofS	- Assistant Chief of Staff

8

ACOG	- Aircraft On Ground
ACOM	- Automatic Coding Machine
ACO/MGE	- Acceptance and Checkout—Maintenance Ground Equipment
ACORN	- Automatic Checkout and Recording Equipment
ACP	- Action Current, Potential
	- Aerospace Computer Program
	- Air Control Point
	- Aircraft Performance
	- Allied Communication Publication
	- Alternate Command Post
	- Audio Center Equipment
	- Auxiliary Check Point
	- Auxiliary Command Post
	- Azimuth Change Pulses
ACPATT	- All Commands Process Attached
ACPD	- Army Control Program Directive
ACPF	- Amphibious Corps, Pacific Fleet
ACPM	- Attitude Control Propulsion Motors
ACPM-1	- Aerospace Computer Program Model 1
ACQ	- Acquisition
ACQT	- Aviation Cadet Qualification Test
ACR	- Accumulator Register
	- Advanced Capabilities Radar
	- Aerial Combat Reconnaissance
	- Air Control Room
	- Air Cooled Compact Reactor
	- Aircraft Control Room
	- Airfield Control Radar
	- Approach Control Radar
	- Alaskan Communications Region
	- Associate Contractor
ACRC	- Audio Center Receiver
ACRD	- Airfield and Carrier Requirements Department
	- Army Chief of Research and Development
ACRE	- Automatic Checkout and Readiness Equipment
ACRES	- Airborne Communication Relay Station
ACS	- Adaptive Control System
	- Admiralty Computing Service (British)
	- Aircraft Control and Surveillance
	- Aircraft Control System
	- Alaskan Communication System
	- Alternating Current, Synchronous
	- American Ceramic Society
	- American Chemical Society
	- Armament Control System
	- Assistant Chief of Staff

■ ACS

ACS	- Attitude Control System
	- Automatic Control System
A/CS	- Assistant Chief of Staff
ACSB	- Apollo Crew Systems Branch
ACSC	- Air Command and Staff College
	- Armament Control System Checkout
ACSD	- Army Communication Service Division
ACSF	- Aircraft Storage Facility
	- Attack Carrier Striking Force
ACSI	- Assistant Chief of Staff, Intelligence
ACSIL	- Admiralty Centre for Scientific Information and Liaison (British)
ACSIM	- Arms Control Simulation
ACSM	- Apollo Command and Service Module
ACSP	- AC Spark Plug
ACSS	- Air Command and Staff School
	- Analog Computer Subsystem
ACT	- Air Control Team
	- Automatic Code Translation
	- Aviation Classification Test
ACT I	- Algebraic Compiler and Translator I
ACT III	- Algebraic Compiler and Translator III
ACTER	- Anti-Countermeasure Trainer
ACTG	- Advance Carrier Training Group
ACTION	- Automatic COMPOOL Generation
ACTM	- Audio Center Transmitter
ACTOR	- Askania Cine-Theodolite Optical-Tracking Range
ACTRAC	- Accurate Tracking
ACTREP	- Activities Report
ACTS	- Arc Current Time Simulator
ACTSECDEF	-Acting Secretary of Defense
ACU	- Acceleration Compensation Unit
ACV	- Air Cushion Vehicle
	- Alarm Check Valve
	- Alternating Current, Volt
ACW	- Airborne Collision Warning
	- Aircraft Control and Warning
	- Alternating Continuous Wave
A-CW	- Aerophysics—Curtiss-Wright
AC&W	- Air Communications and Weather
ACWO	- Aircraft Control and Warning Officer
ACWRON	- Aircraft Control and Warning Squadron
ACWS	- Aircraft Control and Warning System
ACWSS	- Aircraft Control and Warning System Stations
AD	- Active Duty
	- Actuator Drive

AD	- Air Defense
	- Air Depot
	- Air Distance
	- Air Division
	- Apollo Development
	- Automatic Detection
	- Average Deviation
A/D	- Air Depot
	- Airdrome (British)
	- Analog to Digital
A-D	- Analog to Digital
A&D	- Assembly and Disassembly
ADA	- Action Data Automation
	- Aerojet Differential Analyzer
	- Air Defense Area
	- Air Defense Artillery (ADAR preferred)
	- Angular Differentiating Accelerometer
	- Atomic Development Authority
	- Automatic Data Acquisition
ADAC	- Air Defense Artillery Complex
ADACC	- Automatic Data Acquisition and Computer Complex
ADACP	- Air Defense Artillery Command Post
ADAD	- Air Defense Artillery Director
ADALINE	- Adaptive Linear Classification Machine
ADAM	- Advanced Data Management System
ADAO	- Air Defense Artillery Officer
ADAPS	- Automatic Display and Plotting System
ADAPSO	- Association of Data Processing Service Organizations (American and Canadian)
ADAR	- Advanced Design Array Radar
	- Air Defense Artillery
ADAS	- Auxiliary Data Annotation Set
ADA-S	- Action Data Automation—Small
ADAT	- Automatic Data Accumulator and Transfer
ADAVAL	- Advise Availability
ADC	- Airborne Digital Computer
	- Air Data Computer
	- Air Defense Command
	- Air Development Center
	- Alaskan Defense Command
	- Analog-to-Digital Converter
	- Antenna Dish Control
	- Automatic Drive Control
ADCC	- Air Defense Command Computer
	- Air Defense Control Center
ADCH	- Air Defense Command Headquarters

■ ADCM

ADCM	- Air Defense Command Manual
ADCN	- Advance Drawing Change Notice
ADCO	- Aerophysics Development Corporation
ADCOC	- Area Damage Control Center
ADCOM	- Administrative Command
	- Air Defense Command
ADCON	- Advise All Concerned
	- Analog-to-Digital Converter
ADCOP	- Area Damage Control Party
ADCR	- Aerospace Division Commitment Record
	- Air Defense Command Regulation
ADD	- Aerospace Digital Development
	- Air Defense Division
	- Arming Decision Device
ADDA	- Air Defense Defended Area
ADDAC	- Analog Data Distributor and Computer
ADDAR	- Automatic Digital Data Acquistion and Recording
ADDAS	- Automatic Digital Data Assembly System
ADDC	- Air Defense Direction Center
ADDL	- Aircraft Dummy Deck Landing
ADDP	- Air Defense Defended Point
ADDRESOR	-Analog-to-Digital Data-Reduction System for Oceanographic Research
ADDS	- Automatic Data Digitizing System
ADDU	- Additional Duty
ADE	- After Delivery Economies
	- Atomic Defense Engineering
ADEC	- Aiken Dahlgren Electronic Calculator
ADEE	- Air Defense Electronic Environment
ADEP	- Air Depot
	- Authorized Direct Expenditure Plan
ADEPT	- Air Force Depot Equipment Performance Tester
	- Automatic Data Extractor and Plotting Table
ADES	- Air Defense Engineering Service
	- Automatic Digital Encoding System
ADES II	- Automatic Digital Encoding System II
ADF	- Air Defense Force
	- Automatic Direction Finder
	- Automatic Direction Finding
	- Auxiliary Detonating Fuze
ADFA	- Automatic Direction Finder Approach
ADFC	- Air Defense Filter Center
ADFHQ	- Air Defense Force Headquarters
ADFR	- Automatic Direction Finder, Remote-Controlled
ADG	- Air Defense Group
ADGRU	- Advisory Group

ADHCA	- Advise Headquarters of Completed Action
ADI	- Air Defense Intercept
	- American Documentation Institute
	- Attitude-Direction Indicator
ADIL	- Air Defense Identification Line
ADINTELCEN	-Advanced Intelligence Center
ADIOS	- Automatic Digital Input-Output System
ADIPM	- Air Defense Inspector Provost Marshall
ADIS	- Air Defense Integrated System
	- Automatic Data Interchange System
ADIT	- Analog-Digital Integrating Translator
ADIV	- Air Division
ADIZ	- Air Defense Identification Zone
ADL	- Acceptable Defect Level
	- Aids Distribution List
	- Arthur D. Little, Inc.
	- Atmospheric Devices Laboratory
	- Automatic Data Link
ADLO	- Air Defense Liaison Office
ADLOG	- Advance Logistical Command
ADLS	- Air Dispatch Letter Service
ADM	- Add Magnitude
	- Air Defense Missile
	- Atomic Demolition Munition
	- Automated Data Management
ADMA	- Aviation Distributors and Manufacturers Association
ADMB	- Air Defense Missile Base
ADMINO	- Administrative Officer
ADMO	- Air Defense Management Office
ADMS	- Asynchronous Data Multiplexer Synchronizer
ADMSG	- Advise by Message
ADMT	- Agena Detailed Maneuver Table
ADN	- Accession Designation Number
	- Acide Desoxyribonucleique (French for DNA)
ADO	- Advanced Development Objective
	- Audio Decade Oscillator
ADOC	- Air Defense Operations Center
ADONIS	- Automatic Digital On-Line Instrumentation System
ADOS	- Astronautical Defensive-Offensive System
ADP	- Ammonium Dihydrogen Phosphate
	- Automatic Data Processing
	- Automatic Destruct Program
ADPB	- Air Defense Planning Board
ADPC	- Automatic Data Processing Center
ADPE	- Automatic Data Processing Equipment
ADPS	- Automatic Data Processing System

■ **ADPSO**

ADPSO	- Association of Data Processing Service Organizations
ADR	- Address
	- Advisory Route
	- Aircraft Direction Room
	- Air Defense Requirement
	- Austin Data Recorder
ADRAC	- Automatic Digital Recording and Control
ADRG	- Automatic Data Routing Group
ADRN	- Advance Drawing Release Notice
ADRSA	- Assistant Data Recording System Analyst
ADRT	- Analog Data Recorder Transcriber
ADS	- Address Display Subsystem
	- Air Defense Sector
	- American Defense Society
	- Aviation Depot Squadron
ADSAT	- Anomalous Dispersion Spherical Array Target
ADSEC	- Air Defense Systems Engineering Committee
ADSG	- Atomic Defense and Space Group
ADSHIPDA	- Advise Shipping Data
ADSHIPDAT	- Advise Shipping Date
ADSID	- Air Defense Systems Integration Division (now AFSC)
ADSOC	- Administrative Support Operations Center
ADSOD	- Air Defense Systems Operation Division
ADSS	- Aircraft Damage Sensing System
ADSTADIS	- Advise Status and/or Disposition
ADSTKOH	- Advise Stock on Hand
ADSUP	- Automated Data Systems Uniform Practices
ADT	- Accelerated Development Test
	- Advanced Design Team
	- Atomic Damage Template
	- Automatic Data Translator
ADTIC	- Arctic, Desert, Tropic Information Center
ADU	- Angular Display Unit
ADUK	- Air Defense System, United Kingdom
ADVHED	- Advanced Headquarters
ADVSCOL	- Advanced Schools
ADVST	- Advanced Stoppage
ADW	- Air Defense Warning
	- Air Defense Wing
ADWKP	- Air Defense Warning Key Point
ADX	- Add Index Register
	- Automatic Data Exchange
AE	- Airborne Equipment
	- Air Escape
	- Apollo Engineering
	- Architecture and Engineering
	- Arithmetic Element

AERO

A-E	- Architect-Engineer
A&E	- Aircraft and Engines
	- Architecture and Engineering
	- Armament and Electronics
	- Assembly and Erection
	- Assembly and Equipment
	- Azimuth and Elevation
AEA	- Aircraft Electronics Association
	- American Economic Association
AEAO	- Airborne Emergency Action Officer
AEB	- Aft Equipment Bay
	- Apollo Engineering Bulletin
AEC	- American Engineering Council
	- Army Education Center
	- Army Engineer Center
	- Atomic Energy Commission
AECL	- Atomic Energy Commission, Limited (Canada)
AEC-NASA	- Atomic Energy Commission—National Aeronautics and Space Administration
AECOM	- Army Electronics Command (USAECOM preferred)
AED	- Aerospace Electrical Division
	- Agena Ephemeris Data
AEDC	- Arnold Engineering Development Center
AEDD	- Air Engineering Development Division
AEE	- Absolute Essential Equipment
AEEC	- Airline Electronic Engineering Committee
AEEL	- Aeronautical Electronic and Electrical Laboratory
AEF	- Airborne Equipment Failure
	- Aviation Engineer Force
AEG	- Active Element Group
AEI	- Associated Electrical Industries, Limited (British)
	- Azimuth Error Indicator
AEL	- Aeronautical Engine Laboratory
AEN	- Adaption Error Note
AEO	- Air Engineer Officer
AEOP	- Amend Existing Orders Pertaining To
AEP	- Atomic Electrical Project
AEPG	- Army Electronics Proving Ground
AER	- Abbreviated Effectiveness Report
	- After Engine Room
	- Airman Effectiveness Report
AERE	- Atomic Energy Research Establishment (British)
AERIS	- Airborne Electronic Ranging Instrumentation System
AERNO	- Aeronautical Equipment Reference Number
AERO	- Aeroballistics
	- Aerographer

- **AERO**

    AERO   - Aeronautics
              - Azimuth, Elevation, and Range Overtake
    AERO-A   - Aeroballistics—Aerodynamics Analysis
    AERO-D   - Aeroballistics—Dynamics Analysis
    AERO-DIR - Aeroballistics—Director
    AERO-E   - Aeroballistics—Experimental Aerodynamics
    AERO-F   - Aeroballistics-Flight Evaluation
    AEROFLOT - Main Administration of the Civil Air Fleet (USSR)
    AERO-G   - Aeroballistics-Aerophysics and Astrophysics
    AERO-P   - Aeroballistics-Future Projects
    AERO-PCA - Aeroballistics-Program Coordination and Administration
    AEROPNET - Air Operational Network
    AERO-PS   - Aeroballistics-Project Staff
    AEROSPACE - Aeronautics and Space
    AERO-TS   - Aeroballistics-Technical and Scientific Staff
    AES   - Aerospace Electrical Society
              - Aircraft Electrical Society
              - Aircraft Engineering Squadron
              - American Electrochemical Society
              - American Electroplasters' Society
              - Artificial Earth Satellite
              - Atomic Electric Power Plant (USSR)
    AESOP   - Artificial Earth Satellite Observation Program
    AET   - Actual Exposure Time
              - Approximate Exposure Time
    AEV   - Aerothermodynamic Elastic Vehicle
    AEW   - Airborne Early Warning
    AEWA   - Airborne Early Warning Aircraft
    AEWC   - Airborne Early Warning and Control
    AEW&C   - Airborne Early Warning and Control
    AEWRON   - Airborne Early Warning Squadron
    AEWTU   - Airborne Early Warning Training Unit
    AEWW   - Airborne Early Warning Wing
    AF   - Air Force
              - Audio Frequency
              - Augmented Final Fade
              - Automatic Fault Finding
              - Automatic Following
              - Auxiliary Field
    A&F   - Arming and Fuzing
    AFA   - Air Force Association
              - Army Flight Activity
    AFAAEC   - Air Force Academy and Aircrew Examining Center
    AFAC   - Air Force Armament Center
    AFAC&IC   - Air Force Aeronautical Chart and Information Center (ACIC preferred)

AFAFC	- Air Force Accounting and Finance Center
A-FA-H	- Aniline-Furfuryl Alcohol-Hydrazine
AFAITC	- Armed Forces Air Intelligence Training Center
AFAK	- Armed Forces Assistance in Korea
AFAPS	- Air Force Air Pictorial Service
AFASD	- Air Force Aeronautical Systems Division (ASD of AFSC preferred)
AFASE	- Association for Applied Solar Energy
AFATR	- Air Force Atlantic Test Range
AFB	- Air Force Base
	- Air Force Bulletin
	- Antifriction Bearing
AFBM	- Air Force Ballistic Missile
AFBMA	- Antifriction Bearing Manufacturing Association
AFBMC	- Air Force Ballistic Missile Center
AFBMD	- Air Force Ballistic Missile Division (now BSD)
AFBMD-FO	- Air Force Ballistic Missile Division—Field Operations (obsolete)
AFBMDIR	- Air Force Ballistic Missile Division Installation Regulation (obsolete)
AFBSD	- Air Force Ballistic Systems Division (BSD of AFSC preferred)
AFC	- Air Force Council
	- Area Frequency Coordinator
	- Automatic Frequency Control
AFCA	- Armed Forces Communications Association
AFCAL	- Association Francaise de Calcul
AFCC	- Air Force Combat Command
	- Air Force Communications Center
AFCCDD	- Air Force Command Control Development Division
AFCE	- Automatic Flight Control Equipment
AFCEA	- Armed Forces Communications and Electronics Association
AFCENT	- Allied Forces in Central Europe
AFCO	- Air Force Contracting Officer
	- All Fuel Cutoff
AFCOA	- Air Force Chief of Operations Analysis
AFCON	- Air Force Contractor
	- Air Force Controlled
AFCP	- Air Force Command Post
AFCPL	- Air Force Computer Program Library
AFCRC	- Air Force Cambridge Research Center
AFCRL	- Air Force Cambridge Research Laboratories
AFCS	- Adaptive Flight Control System
	- Air Force Chief of Staff
	- Air Force Coding System
	- Air Force Communications Service
	- Armament and Flight Control System

■ AFCS

AFCS	- Automatic Fire Control System
	- Automatic Flight Control System
AFCSA	- Air Force Scientific Advisory Board
AFCSL	- Air Force Communications Security Letter
AFCSM	- Air Force Communications Security Manual
AFCSS	- Air Force Communications Support System
AFD	- Active (discrete pellet) Functional Device
	- Air Force Depot
	- Assistant Flight Director
AFDAP	- Air Force Director of Development and Planning
AFDATACOM	- Air Force Data Communications
AFDB	- Auxiliary Floating Drydock, Big
AFDDC-NS	- Assistant Deputy Chief of Staff, Development, Nuclear Systems
AFDIER	- Air Force Departmental Industrial Equipment Reserve
AFDIS	- Air Force Director of Inspection Services
AFDL	- Auxiliary Floating Drydock, Little
AFDM	- Auxiliary Floating Drydock, Medium
AFDO	- Assistant Flight Director Office
AFDPP	- Air Force Director of Personnel Planning
AFDR	- Air Force Directorate of Requirements
AFDRQ	- Air Force Director of Requirements
AFDRT	- Air Force Director of Research and Technology
AFDSI	- Air Force Director of Special Investigations
AFEB	- Armed Forces Epidemiological Board
AFES	- Armed Forces Examining Station
AFESD	- Air Force Electronic Systems Division (ESD of AFSC preferred)
AFETR	- Air Force Eastern Test Range (formerly AFMTC and AMR)
AFF	- Army Field Force
AFFC	- Air Force Finance Center (AFAFC preferred)
AFFE	- Armed Forces, Far East
AFFTC	- Air Force Flight Test Center
AFFTD	- Air Force Foreign Technology Division (FTD of AFSC preferred)
AFG	- Arbitrary Function Generator
AFGU	- Aerial Free Gunnery Unit
AFHC	- Air Force Headquarters Command
AFHQ	- Air Force Headquarters
AFI	- Armed Forces Institute (USAFI preferred)
AFIC	- Air Force Intelligence Center
AFIED	- Armed Forces Information and Education Center
AFIG	- Air Force Inspector General
AFIP	- Air Force Information Program
	- Armed Forces Institute of Pathology
AFIPS	- American Federation of Information Processing Societies (formerly NJCC)

AFIRO	- Air Force Installation Regional Office
AFIS	- Armed Forces Information School
AFIT	- Air Force Institute of Technology
AFJPO	- Air Force Joint Project Office
AFKT	- Air Force Knowledge Test
AFL	- Above Field Level
	- Air Force Letter
AFLC	- Air Force Logistics Command (formerly AMC)
AFLCM	- Air Force Logistics Command Manual
AFLCR	- Air Force Logistics Command Regulation
AFLD	- Airfield
AFM	- Air Force Manual
	- Antifriction Metal
	- Automatic Fault Finding and Maintenance
A&FM	- Aerodynamics and Flight Mechanics
AFMA	- Armed Forces Management Association
AFMD	- Air Force Missile Division
AFMDC	- Air Force Missile Development Center
AFMED	- Allied Forces, Mediterranean
AFML	- Armed Forces Medical Library (obsolete)
AFMSS	- Air Force Material Supply and Services
AFMTC	- Air Force Missile Test Center (now AFETR)
AFN	- Armed Forces Network
AFNA	- Air Force-Navy (AN preferred)
AFNETR	- Air Force Nuclear Engineering Test Reactor
AFNRD	- Air Force National Range Division
AFNS	- Air Force-Navy Standard
AF-NS	- Air Force-Navy Standard
AFO	- Accounting and Finance Office
AFOAR	- Air Force Office of Aerospace Research (OAR preferred)
AFOAS	- Air Force Office of Aerospace Sciences
AFOCE	- Air Force Office of Civil Engineering
AFOIC	- Air Force Officer in Charge
AFOS	- Air Force Operational Service
AFOSR	- Air Force Office of Scientific Research
AFOTC	- Air Force Operational Test Center
AFP	- Air Force Pamphlet
	- Authority for Purchase
AFPAC	- Air Force, Pacific (PACAF preferred)
AFPC	- Air Force Procurement Circular
	- Armed Forces Policy Council
AFPI	- Air Force Procurement Instruction
AFPO	- Air Force Purchasing Office
AFPP	- Air Force Procurement Procedure

■ AFPR

AFPR	- Air Force Plant Representative
	- Air Force Procurement Request
	- Air Force Project Representative
AFPRO	- Air Force Plant Representative Office
AFPS	- Armed Forces Press Service
AFPTRC	- Air Force Personnel and Training Research Center
AFQC	- Air Force Quality Control
AFQCR	- Air Force Quality Control Respresentative
AFQT	- Armed Forces Qualification Test
AFR	- Acceptance Failure Rate
	- Access Function Register
	- Air Force Regulation
AFRC	- Air Force Records Center
AFRCE	- Air Force Regional Civil Engineer
AFRD	- Air Force Research Division (see AFRTD)
AFRM	- Advanced Flight Research Model
	- Air Frame
AFROIC	- Air Force Resident Officer in Charge
AFRPC	- Air Force Reserve Policy Committee
AFRR	- Air Force Resident Representative
AFRS	- Armed Forces Radio Service
AFRTD	- Air Force Research and Technology Division (R&TD of AFSC preferred)
AFRTS	- Armed Forces Radio and Television Service
AFS	- Air Force Specialty
	- Air Force Station
	- Air Force Stock
	- Air Force Supplied
	- Air Force Supply
	- Arm and Fuze System
A/FS	- Arm and Fuze System
AFSA	- Armed Forces Security Agency
AFSAB	- Air Force Scientific Advisory Board
AFSAW	- Air Force Special Activities Wing
AFSB	- Air Force Support Base
AFSC	- Air Force Speciality Code
	- Air Force Supply Catalog
	- Air Force Systems Command (formerly ARDC)
	- Armed Forces Staff College
	- Automatic Flight Stabilization and Control System
AFSCC	- Air Force Satellite Control Center
AFSD	- Air Force Stock Data
AFSF	- Air Force Stock Fund
AFSIE	- Air Force Standard Items and Equipment
AFSMAAG	- Air Force Section, Military Assistance Advisory Group

AFSN	- Air Force Serial Number
	- Air Force Service Number
	- Air Force Stock Number
AFSP	- Air Force Spare
AFSR	- Argonne Fast Source Reactor
AFSS	- Air Force Security Service
	- Air Force Service Statement
AFSSC	- Armed Forces Supply Support Center
AFSSD	- Air Force Space Systems Division (SSD of AFSC preferred)
AFSSO	- Air Force Special Security Officer
AFSWC	- Air Force Special Weapons Center
AFSWP	- Armed Forces Special Weapons Project
AFT	- Acceptance Functional Test
	- Air Freight Terminal
	- Automatic Fine Tuning
AFTAC	- Air Force Technical Applications Center
AF/TAT	- Air Force Technical Acceptance Team
AFTB	- Air Force Test Base
AFTC	- Airborne Flight Training Command
AFTO	- Air Force Technical Order
AFTRC	- Air Force Training Command
AFTRCC	- Aerospace Flight Test Radio Coordinating Council
AFTTH	- Air Force Technical Training Headquarters
AFUS	- Armed Forces of the United States
AFV	- Armored Fighting Vehicle (British)
AFVA	- Air Force Visual Aid
AFWL	- Armed Forces Writers League
AFWS	- Air Force Weapon System
	- Air Force Weapon Supply
AFWST	- Armed Forces Women's Selection Test
AFWTR	- Air Force Western Test Range (formerly PMR)
AFWW	- Air Force Weather Wing
AG	- Adjutant General
	- Air Group
	- Air to Ground
	- Anti-Gas
	- Armor Grating
	- Arresting Gear
A/G	- Air to Ground
A-G	- Aerojet-General Corporation
AGA	- Air Routes and Ground Aids
	- American Gas Association
AGACS	- Automatic Ground-to-Air Communications System
AGANI	- Apollo Guidance and Navigation Information
AGAP	- Attitude Gyro Accelerometer Package
AGARD	- Advisory Group for Aeronautical Research and Development

■ AGAVE

AGAVE	- Automatic Gimballed-Antenna Vectoring Equipment
AGC	- Adjutant General's Corps
	- Aerojet-General Corporation
	- Air-Ground Communications
	- Amphibious Command Ship
	- Apollo Guidance Computer
	- Automatic Gain Control
AGCA	- Automatic Ground-Controlled Approach
AGCC	- Air-Ground Communications Channel
AGCL	- Automatic Ground-Controlled Landing
AGCRS	- Army Gas-Cooled Reactor System
AGCS	- Automatic Ground Control Station
AGCT	- Army General Classification Test
AGCU	- Attitude Gyroscope Coupling Unit
AGDS	- American Gage Design Standard
AGE	- Aerospace Ground Equipment
	- Apollo Guidance Equipment
	- Automated Group Education
	- Automatic Guidance Electronics
AGEOP	- Aerospace Ground Equipment Out of Commission for Parts
AGEP	- Advisory Group on Electronic Parts
AGET	- Advisory Group on Electronic Tubes
AGF	- Automatic Guided Flight
	- Aviation Guided Flight
AGGD	- Apollo Guidance Ground Display
AGGDSSD	- Astro-Geodetic Geoid Data Station Spacing and Distribution
AGGR	- Air-to-Ground Gunnery Range
AGIC	- Air-to-Ground Information Center
AGL	- Above Ground Level
	- Airborne Gun Laying
AGLT	- Airborne Gun Laying for Turrets
AGM	- Air-to-Ground Missile (now ASM)
	- At Gage Marks
A&GM	- Artillery and Guided Missile School
AGMA	- American Gear Manufacturers Association
AGN	- Aerojet-General Nucleonics
	- Automatic Celestial Navigation
AGNIS	- Apollo Guidance and Navigation Industrial Support
AGO	- Adjutant General's Office (TAGO preferred)
AGOS	- Air-Ground Operations System
AGP	- Aircraft Grounded for Lack of Parts
AGPI	- Automatic Ground Position Indicator
AGRAVIC	- Unaffected by Gravitation, Weightless
AGREE	- Advisory Group on Reliability of Electronic Equipment
AGS	- Advanced Guidance System
	- Airborne Gunsight

AGS	- Alternating Gradient Synchrotron
	- Army General Staff
	- Atlantic Gateway
	- Automatic Gain Stabilization
AGT	- Aviation Gas Turbine
AGU	- American Geophysical Union
AGW	- Allowable Gross (take-off) Weight
AGZ	- Actual Ground Zero
AH	- Airfield Heliport
	- Anhydrous Hydrazine
A/H	- Air over Hydraulic
	- Alter Heading
AHC	- American Helicopter Company
AHGMR	- Ad Hoc Group on Missile Reliability
AHM	- Ampere-Hour Meter
AHP	- Air Horsepower
	- Allied Hydrographic Publication
AHS	- American Helicopter Society
	- Attack Heading Slot
AHSR	- Air Height Surveillance Radar
AHT	- Acoustic Homing Torpedo
AHV	- Accelerator, High Voltage
AHWT	- Ames Dimensional Hypersonic Wind Tunnel
AI	- Airborne Intercept
	- Airborne Interceptor
	- Aircraft Identification
	- Air Inspector
	- Air Installation
	- All Inertial
	- Altitude Index
	- Amplifier Input
	- Anti-Icing
	- Aptitude Index
	- Army Intelligence
	- Atomics International
	- Automatic Input
	- Azimuth Indicator
A/I	- Aptitude Index
AIA	- Aerospace Industries Association (formerly Aircraft)
	- American Institute of Aeronautics
	- American Institute of Architects
	- Association of Industrial Advertisers
AIAA	- Aerospace Industries Association of America
	- American Institute of Aeronautics and Astronautics (formerly ARS and IAS)
AIAOS	- Academic Instructor and Allied Officer School

■ **AIAS**

AIAS	- Army Institute of Advanced Studies
AIB	- American Institute of Banking
AIBS	- American Institute of Biological Sciences
AIC	- Advanced Intelligence Center
	- Aircraft Identification Control
	- Aircraft in Commission
	- Air Intelligence Command
	- Air Interception Control
	- Ammunition Identification Code
AICBM	- Anti-Intercontinental Ballistic Missile
AICC	- Air Intercept Control Common
AICE	- American Institute of Chemical Engineers
	- American Institute of Consulting Engineers
AICHE	- American Institute of Chemical Engineers
AICMA	- Association International des Constructeurs et Matériel Aerospatial
AICO	- Action Information Control Officer
AICP	- Atomic Incident Control Plan
AICPOA	- Advanced Intelligence Center, Pacific Ocean Area
AID	- Agency for International Development
	- Aircraft Identification Determination
	- Atomics International Division
AIDA	- Automatic Instrumented Diving Assembly
AIDE	- Adapted Identification Decision Equipment
	- Aerospace Installation Diagnostic Equipment
	- Automatic Instrumented Decision Equipment
AIDS	- Action Information Display System
	- Air Force Intelligence Data Handling System
A/E	- Airborne Interception Equipment
AIEE	- American Institute of Electrical Engineers (now IEEE)
AIETA	- Airborne Infrared Equipment for Target Analysis
AIEU	- Armament and Instrument Experimental Unit
AIF	- Air Intelligence Force
	- Atomic Industrial Forum
AIFS	- Advanced Instruction Flying School
AIG	- Address Indicating Group
	- All Inertial Guidance
AIGS	- All Inertial Guidance System
AII	- Acceptance Inspection Instructions
	- Apollo Implementing Instructions
	- Army Intelligence Interpreter
AIIE	- American Institute of Industrial Engineers
AIL	- Aileron
	- Airborne Instruments Laboratory
	- Air Intelligence Liaison
AILAS	- Automatic Instrument Landing Approach System

AILS	- Advanced Integrated Landing System (FAA)
	- Automatic Instrument Landing System
AIM	- Alarm Indication Monitor
	- Assistant Industrial Manager
	- Aviation Incentive Movement
AIMACO	- Air Materiel Command Computer
AIMME	- American Institute of Mining and Metallurgical Engineers
AIMMPE	- American Institute of Mining, Metallurgical and Petroleum Engineers
AIMS	- Advanced Intercontinental Missile System
	- Army Integrated Meteorological System
AINO	- Assistant Inspector of Naval Ordnance
AINSMAT	- Assistant Inspector of Naval Material
AINTSEC	- Air Intelligence Section
AIO	- Air Installation Officer
AIP	- Aeronautical Information Publication
	- American Institute of Physics
	- Auto-Igniting Propellant
AIR	- Accelerated Item Reduction
	- Air Arming Impact Rocket
	- Airborne Intercept Radar
	- American Institute of Research
	- Assembly Inspection Record
AIRARMUNIT	- Aircraft Armament Unit
AIRBM	- Anti-Intermediate Range Ballistic Missile
AIRCENT	- Allied Air Forces in Central Europe
AIRCOM	- Aerospace Communications Complex
	- Airways Communications System
AIRCOMNET	- Air Force Communications Network
AIRDEVRON	- Air Development Squadron
AIREDIV	- Aircraft Repair Division
AIRELO	- Air Electrical Officer
AIRENGPROPACCOVERHAUL	- Airplane Engine, Propeller, and Accessory Overhaul
AIREO	- Air Engineer Officer
AIRFMPAC	- Aircraft, Fleet Marine Force, Pacific
AIRLO	- Air Liaison Officer
AIRMG	- Aircraft Machine Gunner
AIROPNET	- Air Operational Network
AIRTRAINRON	- Air Training Squadron
AIRWEANET	- Air Weather Network
AIS	- Aeronautical Information Service
	- Ascension Island Station
	- Attitude Indicating System
AISC	- American Institute of Steel Construction
AISI	- American Iron and Steel Institute

■ AISS

AISS	- Air Intelligence Service Squadron
AIT	- Advanced Individual Training
	- Auto-Ignition Temperature
	- Automatic Information Test
AITA	- Air Industries and Transport Association
AITI	- Aero Industries Technical Institute
AJ	- Anti-Jam
	- Assembly Jig
A-J	- Anti-Jam
AJCC	- Alternate Joint Communications Center
AJD	- Anti-Jamming Display
AL	- Air Launch
	- Air Lock
	- Allegheny Ludlum
A/L	- Air Launch
	- Air Lift
	- Ammunition Loading
ALA	- American Library Association
	- Army Launch Area
ALAACS	- Alaskan Airways and Air Communications Service
ALANF	- Army Land Forces
ALARM	- Automatic Light Aircraft Readiness Monitor
ALARR	- Air Launched Air Rocket Recovery
ALBI	- Air-Launched Ballistic Intercept
ALBM	- Air-Launched Ballistic Missile
ALBUS	- All Bureaus
ALC	- Air Lines Circuit
	- Automatic Landing Control
	- Automatic Level Control
	- Automatic Load Control
ALCH	- Approach Light Contact Height
ALCO	- Airlift Coordinating Office
ALCOA	- Aluminum Company of America
ALCOM	- Alaskan Command
	- Algebraic Compiler
	- All Commands
ALCOP	- Alternate Command Post
ALCORCEN	- Air Logistics Coordination Center
ALCU	- Altocumulus
ALD	- Acoustic Locating Device
	- Analog Line Driver
	- At a Later Date
ALDC	- Automatic Load and Drive Control
ALDRI	- Automatic Low Data Rate Input
ALEMS	- Apollo Lunar Excursion Module Sensors

ALERT	- Automated Linguistic Extraction and Retrieval Technique
	- Automatic Logging Electronic Reporting and Telemetering
ALF	- Auxiliary Landing Field
ALFA	- Air Lubricated Free Attitude Trainer
	- Air Lubricated Free Axis
ALFCE	- Allied Land Forces, Central Europe (NATO)
ALFSE	- Allied Land Forces, Southern Europe (NATO)
ALG	- Advanced Landing Ground
ALGM	- Air-Launched Guided Missile
ALGOL	- Algorithmic Languate (formerly IAL)
ALIAD	- Alaskan Integrated Air Defense
ALIADS	- Alaskan Integrated Air Defense System
ALIAS	- Algebraic Logic Investigation of Apollo Systems
ALICE	- Adiabatic Low-Energy Injection and Capture Experiment
	- Alaskan Integrated Communications Exchange
ALIM	- Air-Launched Interceptor Missile
ALLNAVSTAS	- All Naval Stations
ALLS	- Apollo Lunar Logistic Support
ALMAJCOM	- All Major Commands
ALN	- Accounting Line Number
ALNICO	- Aluminum, Nickle, Cobalt
ALO	- Air Liaison Officer
	- Automatic Lock-On
ALOC	- Air Line of Communication
ALOO	- Albuquerque Operations Office
ALOXCON	- Aluminum-Oxide Electrolytic Capacitor
ALP	- Allied Logistics Publication
	- Automated Learning Process
ALPA	- Air Line Pilots Association
ALPB	- Aircraft Logistics Planning Board
ALPR	- Argonne Low Power Reactor
ALPS	- Advanced Liquid Propulsion System
ALPURCOMS	- All-Purpose Communication System
ALRI	- Airborne Long-Range Radar Input
ALRS	- Altitude Report Status
ALS	- Accumulator Left Shift
	- Aircraft Landing System
	- Alerting and Status
ALSEAFRON	- Alaskan Sea Frontier
ALSEC	- Alaskan Sector
ALSOR	- Air-Launched Sounding Rocket
ALSS	- Apollo Logistic Support System
ALST	- Alaska Standard Time
	- Altostratus
ALSTACON	- All Stations, Continental United States

■ ALT

ALT	- Administrative Lead Time
	- Alternate
	- Altitude
ALTA	- Association of Local Transport Airlines
ALTAC	- Algebraic Transistorized Automatic Computer Translator
ALTARE	- Automatic Logic Testing and Recording Equipment
ALTRAN	- Algebraic Translator
ALTREC	- Automatic Life Testing and Recording of Electronic Components
ALTRV	- Altitude Reservation
ALU	- Arithmetic and Logical Unit
ALUSNA	- American Legation, United States Naval Attache
AM	- Advancement of Management
	- Aeromedical Monitor
	- Air Movements
	- Amplitude Modulation
	- Auxiliary Memory
	- Aviation Medicine
A/M	- Automatic/Manual
A&M	- Assembly and Maintenance
AMA	- Aerospace Medical Association
	- Air Materiel Area
	- American Management Association
	- American Microfilming Association
	- Automatic Message Accounting
AM-A	- Auxiliary Memory A
AMAL	- Aviation Medical Acceleration Laboratory
AMAT	- Airborne Moving Attack Target
AMATC	- Air Materiel Armament Test Center
AMB	- Airways Modernization Board
	- Amber
AM-B	- Auxiliary Memory B
AMBLADS	- Advise Method, Bill of Lading, Date Shipped
AMC	- Aerodynamic Maneuver Capability
	- Aircraft Manufacturers Council
	- Air Materiel Command (now AFLC)
	- Army Materiel Command (USAMC preferred)
	- Army Medical Center
	- Automatic Mixture Control
	- Aviation Material Change
AMCLO	- Air Materiel Command Liaison Officer (obsolete)
AMCM	- Air Materiel Command Manual (now AFLCM)
AMCR	- Air Materiel Command Regulation (now AFLCR)
AMCS	- Airborne Missile Control System
	- Army Mobilization Capabilities Study

AMD	- Aerospace Medical Division (Air Force)
	- Aircraft Maintenance Department
	- Air Movement Data
	- Air Movement Designator
AMDA	- Advance for Mutual Defense Assistance
AMDC	- Army Missile Development Center
AMDL	- Air Munitions Development Laboratory
AMDLEVAC	- Aeromedical Evacuation
AMDSB	- Amplitude Modulation Double Sideband
AMDSB/SC	- Amplitude Modulation Double Sideband, Suppressed Carrier
AME	- Angle Measuring Equipment
AME/COTAR	- Angle Measuring Equipment—Correlation Tracking and Ranging
AMEDS	- Army Medical Service (formerly Medical Corps)
AMF	- Air Materiel Force
	- American Machine and Foundry Company
AMFC	- Atlantic Missile Flight Center
AMFEA	- Air Materiel Force, European Area
AMFPA	- Air Materiel Force, Pacific Area
AMFSO	- Assistant Missile Flight Safety Officer
AMG	- Aircraft Machine Gunner
	- Allied Military Government
	- Angle of the Middle Gimbal
	- Automatic Magnetic Guidance
AMGOT	- Allied Military Government of Occupied Territory
AMI	- Advanced Manned Interceptor
	- American Military Institute
AMICOM	- Army Missile Command (USAMICOM preferred)
AMIS	- Air Movements Information Section
AMIT	- Ampex to IBM Tape
AML	- Aeronautical Materials Laboratory
AMM	- Anti-Missile Missile
AMMIP	- Aviation Materiel Management Improvement Program
AMMO	- Ammunition
AMMP	- Apollo Master Measurements Program
AMMSDO	- Anti-Missile Missile and Space Defense Office
AMMTR	- Anti-Missile Missile Test Range
AMO	- Aircraft Material Officer
AMOCOM	- Army Mobility Command (USAMOCOM preferred)
AMOO	- Aerospace Medicine Operations Office
AMOS	- Acoustic Meteorological Oceanographic Survey
	- Alternate Military Occupational Speciality
	- Automatic Meteorological Observation Station
AMP	- Adaption Mathematical Processor
	- Advanced Manned Penetrator (formerly LAMP)

- **AMP**

    AMP   - American Marine Products
              - Ampere
              - Amphenol
              - Amplidyne
    AMPCO   - Associated Missile Products Company
    AMPD   - Army Mobilization Program Directive
    AMPERE   - APL (Applied Physics Laboratory) Management Planning
                and Engineering Resources Evaluation
              - Group for the Study of Atoms and Molecules from Radio-
                Electric Research
    AMPHIBEX - Amphibious Exercise
    AMPLAS   - Apparatus Mounted in Plastic
    AMPR   - Aeronautical Manufacturers Planning Report
    AMPS   - Advanced Manned Penetration System
              - Automatic Message Processing System
    AMPSS   - Advanced Manned Precision Strike System
    AMR   - Advanced Material Request
              - Advance Material Release
              - Airborne Magnetic Recorder
              - Air Movements Recorder
              - Atlantic Missile Range (now AFETR)
    AMRA   - Army Materials Research Agency
    AMRAC   - Anti-Missile Research Advisory Council
    AMRAD   - ARPA Measurements Radar
    AMRC   - Army Mathematical Research Center
    AMRD-NASC - Army Missiles and Rockets Directorate—NATO Supply
                Center
    AMRI   - Association of Missile and Rocket Industries
    AMRL   - Aerospace Medical Research Laboratories (Air Force)
              - Army Medical Research Laboratory
    AMRO   - Atlantic Missile Range Operations
    AMRTS   - Atlantic Missile Range Telemetry Submodule
    AMS   - Acoustic Measurement System
              - Aeronautical Material Specification
              - American Mathematical Society
              - American Meteorological Society
              - Apollo Mission Simulator
              - Army Management Structure
              - Army Map Service
              - Attitude Maneuvering System
              - Authority for Material Substitution
    AMSAM   - Anti-Missile Surface-to-Air Missile
    AMSCN   - Advance Master Schedule Change Notice
    AMSL   - Above Mean Sea Level
    AMSS   - Advanced Manned Space Simulator
              - Automatic Master Sequence Selector

AMSSB	- Amplitude Modulation Single Sideband
AMSSB/SC	- Amplitude Modulation Single Sideband, Suppressed Carrier
AMT	- Air Movements Talker
AMTA	- Airborne Moving Target Attack
AMTC	- Army Missile Test Center
AMTEC	- American Metalworking Technology for the European Community
	- Automatic Time Element Compensator
AMTF	- Air Mobile Task Force
AMTI	- Airborne Moving Target Indicator
AMTK	- Amphibious Tank
AMTRAC	- Amphibious Tractor
AMU	- Air Mileage Unit
AMUCOM	- Army Munitions Command (USAMUCOM preferred)
AMVER	- Atlantic Merchant Vessel Report
AMW	- Angular Momentum Wheel
AN	- Air Force-Navy
	- Army-Navy
	- Academy of Sciences (USSR)
ANA	- Aerojet Network Analyzer
	- Air Force-Navy Aeronautical
	- AND to Accumulator
	- Army-Navy Aeronautical
ANAB	- Alameda Naval Air Base
ANACDUTRA	- Annual Active Duty for Training
ANAF	- Army-Navy-Air Force
ANB	- Air Navigation Board
	- Army-Navy-British
ANBS	- Army-Navy-British Standard
ANC	- Air Force-Navy-Department of Commerce, Civil Aeronautics Administration (obsolete)
	- Army-Navy-Department of Commerce, Civil Aeronautics Administration (obsolete)
ANCLAV	- Automatic Navigation Computer for Land and Amphibious Vehicles
ANCR	- Aircraft Not Combat Ready
AND	- Air Force-Navy Design
	- Air Navigation Board
	- Army-Navy Design
ANDAC	- Air Navigation Data Center
ANDB	- Air Navigation Development Board
ANDS	- Alphanumeric Displays
ANDUS	- Anglo-Dutch-United States
ANE	- Aeronautical and Navigation Electronics
ANEEG	- Army-Navy Electronics Evaluation Group
ANFA	- Aniline Furfuryl Alcohol

■ ANFE

ANFE	- Aircraft Not Fully Equipped
ANG	- Air Force-Navy-Army (combined service missile designation)
	- Air National Guard
ANGLICO	- Air and Naval Gunfire Liaison Company
AN/GRA	- Air Force-Navy Ground Radar
ANGUS	- Air National Guard of the United States
ANIP	- Army-Navy Instrumentation Program
ANJSB	- Arny-Navy Joint Specification Board
ANL	- Argonne National Laboratory
	- Automatic Noise Limiter
ANMB	- Army-Navy Munitions Board
ANNA	- Army-Navy-NASA-Air Force
ANO	- Above Named Officer
ANO	- Alphanumeric Output
ANORE	- Aircraft Not Operationally Ready due to Lack of Equipment
ANORM	- Aircraft Not Operationally Ready due to Lack of Maintenance (see AOCM)
ANORP	- Aircraft Not Operationally Ready due to lack of Parts (see AOCP)
ANOV	- Analysis of Variance
ANP	- Aircraft Nuclear Propulsion
ANPB	- Army-Navy Petroleum Board
ANPD	- Aircraft Nuclear Propulsion Department
ANPO	- Aircraft Nuclear Propulsion Office
ANPP	- Aircraft Nuclear Propulsion Program
	- Army Nuclear Power Program
ANPPF	- Aircraft Nuclear Power Plant Facility
ANPT	- Aeronautical National Taper Pipe Thread
ANR	- Alaskan NORAD Region
ANRAC	- Aids Navigation Radio Control
ANS	- American Nuclear Society
	- AND to Storage
ANSIA	- Army-Navy Shipping Information Agency
ANTAC	- Air Navigation and Tactical Control System
ANTC	- Air Navigation Technical Committee
ANTIS	- Anti-Missile Weapon System
ANTO	- Aeronautical Scientific and Technical Society (USSR)
ANTS	- Advanced Naval Training School
ANTU	- Air Navigation Training Unit
AO	- Access Opening
	- American Optical Company
	- Aviation Ordnance
AOA	- American Ordnance Association
AOB	- Advanced Operational Base
	- Air Order of Battle
AOBSR	- Air Observer

AOC	- Air Operations Center
	- Airport Operations Council
	- Army Ordnance Corps
	- Automatic Output Control
	- Automatic Overload Circuit
	- Aviation Officer Candidate
AOCM	- Aircraft Out of Commission for Maintenance (see ANORM)
AOCO	- Atomic Ordnance Cataloging Office
AOCP	- Aircraft Out of Commission for Parts (see ANORP)
AOD	- Apollo Operations Director
	- Assistant Operations Director
AODS	- All Ordnance Destruct System
AOEC	- Airways Operation Evaluation Center
AOF	- Air Objective Folder
AOG	- Airplane on Ground
AOGO	- Advanced Orbiting Geophysical Observatory
AOIL	- Aviation Oil
A-OK	- All OK
AOLM	- Apollo Orbiting Laboratory Module
AOMC	- Army Ordnance Missile Command (now USAMICOM)
AOND	- Administrative Office, Navy Department
AOO	- Aviation Ordnance Officer
AOOR	- Army Office of Ordnance Research
AOP	- Assembly and Operations Plan
AOPA	- Aircraft Owners and Pilots Association
AOQ	- Average Outgoing Quality
	- Aviation Officers Quarters
AOQL	- Average Outgoing Quality Limit
AOR	- Add One to the Right
	- Air Operations Room
AORG	- Army Operational Research Group
AORL	- Apollo Orbital Research Laboratory
AOS	- Add or Subtract
	- Analog Output Submodule
	- Announcement and Order Sheet
	- Atlantic Ocean Ship
AOSL	- Authorized Organizational Stockage List
AOSO	- Advanced Orbiting Solar Observatory
AOSP	- Automatic Operating and Scheduling Program
AOTC	- Aviation Officers Training Corps
AOTE	- Amphibious Operational Training Element
AOTOP	- Advent Orbital Test and Operation Plan
AOU	- Azimuth Orientation Unit
AP	- Accelerometer Package
	- Access Panel
	- Access Point

■ AP

AP	- Acid Proof
	- Acquisition Point
	- Aft Perpendicular
	- Aiming Point
	- Air Pilot
	- Airport
	- Air Position
	- Ammunition Point
	- Armor Piercing
	- Artificial Pupil
	- Atomic Powered
	- Autopilot
A/P	- Airplane
	- Allied Papers
	- Autopilot
APA	- American Psychological Association
	- Appropriation Purchases Account
APAC	- Airborne Parabolic Arc Computer
APADS	- Automatic Programmer and Data System
APAR	- Automatic Programming and Recording
APASTO	- ADCC (Air Defense Command Computer) Programming and System Training Office
APATS	- Automatic Programmer and Test System
APC	- Additional Planning Capability
	- Angular Position Counter
	- Approach Control
	- Area Positive Control
	- Armored Personnel Carrier
	- Armor-Piercing Capped
	- Automatic Phase Control
APCA	- Automatic Phono-Cardiac Analyzer
APCBC	- Armor-Piercing Capped, Ballistic Capped
AP&CC	- American Potash and Chemical Corporation
APCHE	- Automatic Programmed Checkout Equipment
APCI	- Armor-Piercing Capped Incendiary
APCIT	- Armor-Piercing Capped Incendiary with Tracer
APCN	- Assembly Page Change Notice
APCON	- Approach Control
APCR	- Apollo Program Control Room
APCS	- Air Photographic and Charting Service
APCT	- Armor-Piercing Capped with Tracer
APD	- Advanced Program Development
	- Air Procurement District
	- Analog-to-Pulse Duration
	- Angular Position Digitizer
	- Apollo Project Directive

APDA	- Atomic Power Development Associates
	- Auxiliary Pump Drive Assembly
APDL	- Aids Production and Distribution List
APE	- Automatic Positioning Equipment
APEL	- Aeronautical Photographic Experimental Laboratory
APERS	- Anti-Personnel
APEX	- ARCAS (All Purpose Rocket for Collecting Atmospheric Soundings) Piggyback Emulsion Experiment
APF	- Aircraft Parachute Flare
APG	- Aberdeen Proving Ground
	- Army Planning Group
APGC	- Air Proving Ground Center
APHFFF	- Ames Prototype Hypersonic Free-Flight Facility
API	- Air Position Indicator
	- American Petroleum Institute
	- Armor-Piercing Incendiary
APIC	- Automatic Power Input Controller
APIT	- Armor-Piercing Incendiary with Tracer
APK	- Accelerometer Package
APL	- Advanced Parts List
	- Advanced Procurement List
	- Applied Physics Laboratory (Johns Hopkins University)
	- Assembly Page Listing
	- Assembly Parts List
	- Automatic Phase Lock
	- Average Picture Level
APL/JHU	- Applied Physics Laboratory, Johns Hopkins University
APM	- Assembly Page Maintenance
APMD	- Army Projects Management Department
APO	- Accountable Property Office
	- Apogee
	- Apollo Project Office
	- Army (or Air Force) Post Office
	- Assembly Page Order
APOD	- Aerial Port of Debarkation
APOE	- Aerial Port of Embarkation
APOTA	- Automatic Positioning of Telemetering Antenna
APP	- Access Point PACE (Pre-Flight Acceptance Checkout Equipment)
	- Army Procurement Procedures
	- Auxiliary Power Plant
APPC	- Advance Planning Procedure Change
	- Automatic Power Plant Checker
APPLE	- Advanced Propulsion Packaged Liquid Engine
APPO	- Advanced Product Planning Operation
APPR	- Army Package Power Reactor

■ APR

APR	- Airman Performance Report
APRA	- Armed Forces Production Resources Agency
APREQ	- Approval Request
APRI	- Air Priority
APRIL	- Automatically Programmed Remote Indication Logging
APRL	- Army Prosthetic Research Laboratory
APRU	- Applied Psychology Research Unit
APS	- Accessory Power Supply
	- Airborne Power System
	- American Physical Society
	- Appearance Station
	- Applied Physics Staff
	- Applied Psychological Services
	- Armor-Piercing Sabot
	- Auxiliary Power Supply
	- Auxiliary Propulsion System
APT	- Airman Proficiency Test
	- Armor-Piercing with Tracer
	- Augmented Programmer Training
	- Automatic Programming Tool
	- Automatic Picture Transmission
APT III	- Automatic Programming Tool
APTI	- Actions per Time Interval
APTPDA	- Advance Payment of Travel Per Diem Authorized
APTS	- Automatic Picture Transmission Subsystem
APTT	- Apollo Parts Task Trainer
APU	- Accessory Power Unit
	- Airborne Power Unit
	- Audio Playback Unit
	- Auxiliary Power Unit
APWL	- Automatically Processed Wire List
APX III	- Automatic Programming System Extended
AQ	- Aircraft Quality
	- Apollo Qualification
	- Acquisition Message
AQGV	- Azimuth Quantized Gated Video
AQL	- Acceptable Quality Level
AQREC	- Army Quartermaster Research and Engineering Center
AQT	- Applicant Qualification Test
AR	- Acid Resisting
	- Acoustic Reflex
	- Acquisition Radar
	- Actual Range
	- Advanced Readiness
	- Aerial Refueling
	- Aeronautical Research

AR	- Aircraft Rocket
	- Amphibian Reconnaissance
	- Armament
	- Armored Reconnaissance
	- Army Regulations
	- Aspect Ratio
	- Assembly and Repair
A/R	- Action and/or Reply
	- Aerial Refueling
	- Armed Reconnaissance
	- At the Rate of
A&R	- Assemble and Recycle
	- Assembly and Repair
ARA	- Actual Range Angle
	- Aerial Refueling Area
	- Army Reactor Area
	- Auxiliary Recovery Antenna
ARAD	- Airborne Radar and Doppler
ARADCOM	- Army Air Defense Command
ARADMAC	- Army Aeronautical Depot Maintenance Center
ARB	- Aircraft Reactors Branch
	- Air Research Bureau
	- Army Reactors Branch
ARBOR	- Argonne Boiling Reactor
ARC	- Advanced Re-Entry Concepts
	- Aiken Relay Calculator
	- Air Reserve Center
	- Airworthiness Requirements Committee
	- Ames Research Center
	- Analysis of Registration and Collimation
	- Atlantic Research Corporation
	- Automatic Range Control
	- Average Response Computer
ARC-1	- Average Response Computer
ARCAS	- All Purpose Rocket for Collecting Atmospheric Soundings
	- Automatic Radar Chain Acquisition System
ARCO	- Aircraft Resources Control Office
ARCON	- Automatic Rudder Control
ARCP	- Aerial Refueling Control Point
ARCS	- Air Resupply and Communication Service
ARCT	- Aerial Refueling Control Time
ARD	- Advanced Research and Development
	- Air Research Division
	- Army Research and Development
	- Auxiliary Repair Dock
AR&D	- Air Research and Development

■ ARDA

ARDA	- Analog Recording Dynamic Analyzer
	- Astronautical Research and Development Agency
ARDC	- Air Research and Development Command (now AFSC)
ARDME	- Automatic Radar Data Measuring Equipment
ARE	- Aircraft Reactor Experiment
	- Apollo Reliability Engineering
AREE	- Apollo Reliability Engineering Electronics
ARENTS	- ARPA Environmental Test Satellite
AREP	- Ammunition Reliability Evaluation Program
ARF	- Aeronautical Research Foundation
	- Air Reserve Forces
	- Armour Research Foundation
	- Automatic Return Fire
ARFC	- Air Reserve Flying Center
ARFCOS	- Armed Forces Courier Service
ARFPC	- Air Reserve Forces Policy Committee
ARGMA	- Army Rocket and Guided Missile Agency
ARGO	- Advanced Research Geophysical Observatory
ARGUS	- Automatic Routine Generating and Updating System
ARIES	- Atmospheric Research Information Exchange Study
	- Authenic Reproduction of an Independent Earth Satellite
ARINC	- Aeronautical Radio, Incorporated
ARIP	- Automatic Rocket Impact Predictor
ARIS	- Advanced Range Instrumentation Ships
	- Atomic Reactors in Space
	- Attitude and Rate Indicating System
ARITH-MATIC	- Arithmetic Automatic Coding System
ARL	- Acceptable Reliability Level
	- Aeronautical Research Laboratories (OAR Lab)
	- Amateur Radio League
	- Astronautical Research Laboratory (OAR Lab at AF Academy)
ARLO	- Air Reconnaissance Liaison Officer
ARM	- Anti-Radar Missile
	- Apollo Requirements Manual
ARMC	- Automatic Repeat Request Mode Counter
ARMEL	- Armament and Electronics Equipment
ARMS	- Aerial Radiological Measurements and Survey Program
	- Automatic Radio Meteorological Station (USSR)
ARN	- Acide Ribonucleique (French for RNA)
	- Action and Reply Notice
ARNA	- Army with Navy
ARNG	- Army National Guard
ARNGUS	- Army National Guard of the United States
ARO	- Airborne Range Only
	- Air Research Organization
	- Army Research Office
	- Automatic Range Only

ARODS	- Airborne Radar Orbital Determination System
AROU	- Aviation Repair and Overhaul Unit
ARP	- Active Recording Program
	- Advanced Re-Entry Program
	- Aeronautical Recommended Practice
	- Airborne Radar Platform
	- Air Raid Precaution
	- Automatic Radio Direction Finder (USSR)
ARPA	- Advanced Research Projects Agency
ARPAT	- Advanced Research Projects Agency Terminal Defense System
ARQ	- Automatic Request
ARR	- Air Regional Representative
	- Antenna Rotation Rate
ARRC	- Air Reserve Records Center
ARRL	- Aeronautical Radio and Radar Laboratory
	- American Radio Relay League
ARS	- Accumulator Right Shift
	- Active Repeater Satellite
	- Advanced Reconnaissance System Satellite
	- Advanced Re-Entry System
	- Aerial Reconnaissance and Security
	- Aerial Refueling Squadron
	- Air Regulating Squadron
	- Air Rescue Service
	- American Rocket Society (now AIAA)
	- Atmosphere Revitalization Section
	- Attitude Reference System
ARSD	- Aviation Repair Supply Depot
ARSR	- Air Route Surveillance Radar
ARSTRIKE	- Army Strike
ART	- Area Responsibilities Transfer
	- Airborne Radiation Thermometer
	- Automatic Range Tracker
AR&T	- Advanced Research and Technology
ARTC	- Aircraft Research and Testing Committee
	- Air Route Traffic Control
ARTCC	- Air Route Traffic Control Center
ARTOC	- Army Tactical Operations Center
ARTRON	- Artifical Neuron
ARTS	- Advanced Radar Traffic System
	- Army Research Task Summary
ARTU	- Automatic Range Tracking Unit
ARTY	- Artillery
ARV	- Armored Recovery Vehicle
ARVN	- Army Vietnam

■ ARW

ARW	- Air Raid Warning
AS	- Academy of Sciences
	- Add-Subtract
	- Aeronautical Standard
	- Airscoop
	- Air Speed
	- Air Station
	- Air to Surface
	- Anti-Submarine
	- Area Surveillance
	- Army Security
A/S	- Air Speed
	- Air to Surface
	- Arm/Safe
	- As Stated
ASA	- Acoustical Society of America
	- Air Security Agency
	- American Standards Association
	- Army Security Agency
	- Assistant Secretary of the Army
	- Atomic Security Agency
	- Aviation Supply Annex
ASACS	- Airborne Surveillance and Control System
ASAE	- American Society of Aeronautical Engineers
ASAF/M	- Assistant Secretary of the Air Force, Materiel
ASA(FM)	- Assistant Secretary of the Army, Financial Management
ASAF/R&D	- Assistant Secretary of the Air Force, Research and Development
ASA(I&L)	- Assistant Secretary of the Army, Installations and Logistics
ASAMAT	- Assistant Secretary of the Army, Materiel
ASAP	- Anti-Submarine Attack Plotter
	- As Soon As Possible
ASA(R&D)	- Assistant Secretary of the Army, Research and Development
ASB	- Acoustical Standards Board
	- Air Safety Board
	- Air Surveillance Branch
ASC	- Aeronautical Systems Center
	- Air Service Command
	- Air Situation Coordinator
	- Air Support Control
	- Army Signal Corps
	- Automatic Sensitivity Control
	- Auxiliary Switch (breaker) Normally Closed
ASCAT	- Analog Self-Checking Automatic Tester

ASCC	- Air Standardization Coordinating Committee
	- Automatic Sequence-Controlled Calculator
ASCE	- American Society of Civil Engineers
ASCO	- Automatic Sustainer Cutoff
ASCOP	- Applied Science Corporation of Princeton
	- Automatic Telemetry Decommutation
ASCP	- Army Strategic Capabilities Plan
ASCS	- Automatic Stabilization and Control System
ASCU	- Air Support Control Unit
ASD	- Advanced Surveillance Drone
	- Aeronautical Systems Division (of AFSC)
	- Air Situation Display
	- Artillery Spotting Division
	- Assistant Secretary of Defense
	- Aviation Supply Depot
ASD/C	- Assistant Secretary of Defense/Comptroller
ASD/CD	- Assistant Secretary of Defense/Civil Defense
ASDD	- Apollo Signal Definition Document
ASDE	- Airport Surface Detection Equipment
	- Antenna Slave Data Equipment
ASDEC	- Applied System Development Evaluation Center
ASDEFORLANT	- Anti-Submarine Defense Force, Atlantic
ASDEFORPAC	- Anti-Submarine Defense Force, Pacific
ASDF	- Air Self-Defense Force
ASDIC	- Anti-Submarine Detection Investigation Committee
	- Armed Services Documents Intelligence Center
ASD/I&L	- Assistant Secretary of Defense/Installations and Logistics
ASD/ISA	- Assistant Secretary of Defense/International Security Affairs
ASD/M	- Assistant Secretary of Defense/Manpower
ASD/PA	- Assistant Secretary of Defense/Public Affairs
ASDR	- Airport Surface Detection Radar
ASE	- Aerospace Support Equipment
	- Airborne Search Equipment
	- Air Surveillance Evaluation
	- Automatic Stabilization Equipment
ASEA	- American Society of Engineers and Architects
ASEC	- American Standard Elevator Code
ASEE	- American Society for Engineering Education
ASESA	- Armed Services Electro-Standards Agency
ASESB	- Armed Services Explosives Safety Board
ASETC	- Armed Services Electron Tube Committee
ASF	- Alaskan Sea Frontier
	- American Steel Foundaries
ASFDO	- Anti-Submarine Fixed Defense Officer
ASFG	- Atmospheric Sound-Focusing Gain
ASFIR	- Active Swept Frequency Interferometer Radar

■ ASFTS

ASFTS	- Auxiliary Systems Function Test Stand
ASG	- Aeronautical Standards Group
	- Air Surveillance Group
ASGPD	- Attitude Set and Gimbal Position Display
ASGS	- Assistant Secretary General Staff
ASHACE	- American Society of Heating and Air Conditioning Engineers
ASHCAN	- Area Scheduling Conference Network
ASHRAE	- American Society of Heating, Refrigerating and Air Conditioning Engineers
ASHVE	- American Society of Heating and Ventilating Engineers
ASI	- Air Speed Indicator
	- Amended Shipping Instructions
	- Augmented Spark Igniter
	- Azimuth Speed Indicator
AS&I	- Arming, Safing, and Initiating
ASIS	- Abort Sensing and Implementation System
	- American Society for Industrial Security
ASK	- Amplitude Shift Keying
	- Askania Theodolite Camera
ASL	- Above Sea Level
	- Approved Source List
	- Authorized Stockage List
ASLE	- American Society of Lubrication Engineers
ASLIB	- Association of Special Libraries and Information Bureaus
ASLO	- American Society of Limnology and Oceanography
ASM	- Air-to-Surface Missile
	- American Society for Metals
	- Apollo Service Module
	- Apollo Systems Manual
	- Army System Manager
ASMCOM	- Army Supply and Maintenance Command (USASMCOM preferred)
ASME	- American Society of Mechanical Engineers
ASMI	- Aerodrome Surface Movement Indicator
	- Airfield Surface Movement Indicator
ASMPA	- Armed Services Medical Procurement Agency
ASMRO	- Armed Services Medical Regulating Office
ASMS	- Advanced Surface Missile System
ASN	- Army Serial Number
	- Average Sample Number
ASNE	- American Society of Naval Engineers
ASO	- Aeronautics Supply Office
	- Air Signal Officer
	- Air Surveillance Officer
	- Area Supply Officer
	- Auxiliary Switch (breaker) Normally Open
	- Aviation Supply Office

ASOC — Air Support Operations Center
ASOP — Army Strategic Objectives Plan
— Aviation Supply Office, Philadelphia
ASOS — Automatic Storm Observation Service
ASP — Advanced Study Project
— Aerospace Plane
— Airborne Support Platform
— Anti-Submarine Patrol
— Ammunition Supply Point
— Apollo Spacecraft Project
— Assemble Sequence Parameters
— Atmospheric Sounding Project
— Atmospheric Sounding Projectile
— Automatic Schedule Procedure
ASPB — Armed Services Petroleum Board
ASPC — Analysis of Spare Parts Change
ASPDE — Automatic Shaft Position Data Encoder
ASPEP — Association of Scientists and Professional Engineering Personnel
ASPI — Apollo Supplemental Procedural Information
ASPL — Assistant Sector Programming Leader
ASPM-1 — Aerospace Program Model 1
ASPO — Advanced Systems Project Office
— Apollo Spacecraft Project Office
ASPPA — Armed Services Petroleum Purchasing Agency
ASPPO — Armed Services Procurement Planning Office
ASPR — Armed Services Procurement Regulation
ASQC — American Society for Quality Control
ASR — Acceptance Summary Report
— Accumulators Shift Right
— Airport Surveillance Radar
— Air-Sea Rescue
— Air Search Radar
— Air Surveillance Radar
— American Society of Rocketry
— Automatic Send and Receive
— Available Supply Rate
— Aviation Safety Regulation
ASRE — American Society of Refrigerating Engineers
ASRG — Advanced Sciences Research Group
ASRL — Aerolastic and Structures Research Laboratory
ASROC — Anti-Submarine Rocket
ASRT — Air Support Radar Team
ASSE — American Society of Sanitary Engineers
ASSET — Aerothermodynamic/Elastic Structural Systems Environmental Test
— Air Surveillance Subsystem Evaluation and Training

■ ASSOTW

ASSOTW	- Airfield and Seaplane Stations of the World
ASST	- American Society for Steel Treating
ASSTL	- Assistant Sector System Training Leader
ASSY	- Assembly
AST	- Aerial Survey Team
	- Apollo Systems Test
	- Army Survey Team
	- Atlantic Standard Time
ASTE	- American Society of Tool Engineers
ASTEC	- Advanced Solar Turbo-Electric Concept
	- Anti-Submarine Technical Evaluation Center
ASTER	- Anti-Submarine Terrier
ASTIA	- Armed Services Technical Information Agency (now Defense Documentation Center)
ASTM	- American Society for Testing Materials
ASTMC	- ARS (American Rocket Society) Structures and Materials Committee
ASTME	- American Society of Tool and Manufacturing Engineers
ASTOR	- Anti-Submarine Torpedo Ordnance Rocket
ASTP	- Accelerated Service Test Program
ASTR	- Airborne Test Reactor
	- Aircraft Shield Test Reactor
	- Astrionics
ASTRA	- Automatic Strobe Tracking
ASTRAC	- Arizona Statistical Repetitive Analog Computer
ASTRO	- Air Space Travel Research Organization
	- Artificial Satellite Time and Radio Orbit
ASTS	- Advanced SAGE Tracking Study
ASTSECNAV	- Assistant Secretary of the Navy
ASTSECNAVFIN	- Assistant Secretary of the Navy, Financial Management
ASTSECNAVINSLOG	- Assistant Secretary of the Navy, Installations and Logistics
ASTSECNAVRESDEV	- Assistant Secretary of the Navy, Research and Development
ASTT	- Action Speed Tactical Trainer
ASTU	- Air Support Training Unit
	- Automatic Systems Test Unit
ASU	- Aircraft Scheduling Unit
	- Area Service Unit
ASV	- Aerothermodynamic Structural Vehicle
	- Airborne Radar for Detecting Surface Vessels
	- Air-to-Surface Vessel
	- Angle Stop Valve
ASW	- Air-to-Surface Warfare
	- Antiaircraft Switch

ASW	- Anti—Satellite Weapon
	- Anti-Submarine Warfare
ASWCR	- Airborne Surveillance Warning and Control Radar
ASWD	- Anti-Submarine Warfare Division
	- Army Special Weapons Depot
ASWDU	- Anti-Submarine Warfare Development Unit
ASWEPS	- Anti-Submarine Weapons Environmental Prediction System
ASWEX	- Anti-Submarine Warfare Exercise
ASWG	- American Steel and Wire Gage
ASWO	- Air Stations Weekly Orders
ASWORG	- Anti-Submarine Warfare Operational Research Group
ASWTC	- Anti-Submarine Warfare Training Center
ASWTU	- Anti-Submarine Warfare Training Unit
ASWU	- Anti-Submarine Warfare Unit
AT	- Accelerometer-Timer
	- Advanced Trainer
	- Aerial Torpedo
	- Air Temperature
	- Airtight
	- Ambient Temperature
	- Ampere Turn
	- Anti-Tank
	- Anti-Torpedo
	- Assembly Telling
	- Astronomical Time
	- Atomic Time
	- Attenuator
	- Automatic Telling
	- Automatic Typewriter
A/T	- Action Time
	- Ammunition Torque
	- Angle Tracker
A-T	- Adinine-Thymine
A&T	- Assembly and Test
ATA	- Action Assembly Table Program
	- Actual Time of Arrival
	- Air Transport Association
ATAA	- Air Transport Association of America
ATABE	- Automatic Target Assignment and Battery Evaluation
ATAD	- Air Technical Analysis Division
ATAF	- Allied Tactical Air Force
ATAPO	- ATABE Adaption Printout
ATAR	- Anti-Tank Aircraft Rocket
ATAW	- Advanced Tactical Assault Weapon
ATB	- Amphibious Training Base
	- Apollo Test Box
	- Arctic Test Branch

■ ATBM

ATBM	- Advanced Tactical Ballistic Missile
	- Anti-Tactical Ballistic Missile
	- Average Time Between Maintenance
ATC	- Airborne Test Conductor
	- Aircraft Technical Committee
	- Airport Traffic Control
	- Air Traffic Control
	- Air Training Command
	- Alert Transmit Console
	- Army Training Center
	- Assistant Test Conductor
	- Automatic Temperature Compensator
	- Automatic Throttle Control
	- Automatic Timing Control
ATCA	- Air Traffic Conference of America
	- Air Traffic Control Association
ATCC	- Air Traffic Control Center
	- Atlantic Division, Transport Control Center
ATCE	- Ablative Thrust Chamber Engine
ATCFAS	- Air Traffic Control Flight Advisory Service
ATCM	- Air Training Command Manual
ATC/NAV	- Air Traffic Control and Navigation Panel
ATCO	- Air Traffic Control Office
	- Air Traffic Coordinating Officer
ATCRBS	- Air Traffic Control Radar Beacon System
ATCS	- Air Traffic Communications Service
	- Air Traffic Communications Station
ATCSS	- Air Traffic Control Signaling System
ATD	- Actual Time of Departure
ATDESA	- Automatic Three-Dimensional Electronics Scanning Array
ATDS	- Airborne Tactical Data System
	- Automatic Telemetry Decommutation System
ATDU	- Aircraft Torpedo Development Unit
ATE	- Air-Turbo Exchanger
	- Altitude Transmitting Equipment
	- Automatic Test Equipment
ATECOM	- Army Test and Evaluation Command (USATECOM preferred)
ATERM	- Air Terminal
ATF	- Actual Time of Fall
	- Actuating Transfer Function
	- Automatic Target Finder
	- Automatic Terrain Following Radar
ATFA	- Atomic-Type Field Army
ATFC	- Air Task Force Commander
ATFOS	- Alignment and Test Facility for Optical Systems

ATR ■

ATG	- Advanced Technology Groups
	- Air to Ground
ATGM	- Anti-Tank Guided Missile
ATHODYD	- Aerothermodynamic Duct
ATI	- Air Technical Intelligence
	- A-Track Initiator
	- Average Total Inspection
ATIC	- Aerospace Technical Intelligence Center
	- Air Technical Intelligence Center
ATL	- Advanced Technology Laboratories
	- Aero-Thermal Laboratory
	- Atlantic
	- Automatic Telling
ATLAS	- Automatic Tape Load Audit System
ATLB	- Air Transport Licensing Board
ATM	- Air Traffic Management
	- Anti-Tank Missile
	- Atmosphere
ATMC	- Air Traffic Movement Control Center
ATMU	- Aircraft Torpedo Maintenance Unit
ATO	- Action Technical Order
	- Aircraft Transfer Order
	- Air Tactics Officer
	- Apollo Test and Operations
	- Area Traffic Officer
	- Assisted Take-Off
	- Atlantic Ocean
AT&O	- Apollo Test and Operations (ATO preferred)
ATOM	- Advanced Technology of Management
ATOMDEF	- Atomic Defense
ATOMDEV	- Atomic Device
ATORP	- Anti-Torpedo
	- Atomic Torpedo
ATOS	- Assisted Take-Off System
ATOT	- Angle Track on Target
ATP	- Acceptance Test Procedure
	- Action Table Print
	- Advanced Technical Payload
	- Alert Transmit Panel
	- Allied Tactical Publication
	- Army Training Plan
	- Assembly Test Program
ATR	- Actual Time of Refueling
	- Airborne Test Reactor
	- Aircraft Trouble Report
	- Air Traffic Regulation

- **ATR**

ATR	- Air Transport Radio
	- Antenna Transmit-Receive
	- Anti-Transmit-Receive
	- Anti-Transmitter-Receiver
	- Apollo Test Requirement
	- Assembly Test Recording
	- Attenuated Total Reflectance
	- Aviation Training Record
ATRAN	- Automatic Terrain Recognition and Navigation System
ATRC	- Air Training Command (ATC preferred)
	- Anti-Tracking Control
ATRS	- Assembly Test Recording System
ATRT	- Anti-Transmit-Receive Tube
ATS	- Accelerometer-Timer Switch
	- Acquisition and Tracking System
	- Advanced Technological Satellite
	- Aeronautical Training Society
	- Air to Ship
	- Air Traffic Services
	- Air Transport Service
	- Army Transport Service
	- Associated Technical Services, Incorporated
	- Associated Training Specialist
	- Astronomical Time Switch
	- Atlantic Test Site
	- Atlantic Tracking Ship
ATSS	- Acquisition and Tracking Subsystem
ATT	- All Thrust Termination
	- Army Training Test
AT&T	- American Telephone and Telegraph Company
ATTC	- Aviation Technical Training Center
ATTR	- All Thrust Terminate Relay
ATTS	- Automatic Telemetry Tracking Station
ATU	- Advanced Training Unit
	- Automatic Tracking Unit
ATUM	- Anti-Tank Nonmetallic
ATUR	- Apollo Test Unsatisfactory Report
ATW	- Aerospace Test Wing
ATWS	- Automatic Track While Scanning
AU	- Air to Underwater
	- Air University
	- Angstrom Unit
	- Arithmetic Unit
	- Army Unit
	- Assembler Unit
	- Astronomical Unit

A/U	- Air to Underwater
AUDAR	- Autodyne Detection and Ranging
AUDIT	- Automatic Unattended Detection Inspection Transmitter System
AUDREY	- Automatic Digit Recognizer
AUGU	- Augmenting Unit
AUI	- Associated Universities, Incorporated
AUL	- Air University Library
AUM	- Air-to-Underwater Missile
AUM-N	- Air-to-Underwater Missile, Nuclear
AUNT	- Automatic Universal Translator
AUP	- Ames Unitary Plan
AUSA	- Association of the United States Army
AUT	- Advanced Unit Training
AUTEC	- Atlantic Undersea Test and Evaluation Center
AUTHGR	- Authority Granted
AUTO CV	- Automatic Check Valve
AUTOLABS	- Automatic Low Altitude Bombing System
AUTOMAP	- Automatic Machining Program
AUTOMET	- Automatic Meteorological Corrections
AUTOPIC	- Automatic Personal Identification Code
AUTOPROMT	- Automatic Programming of Machine Tools
AUTOPSY	- Automatic Operating System
AUTOSPOT	- Automatic System for Positioning Tools
AUTOSYN	- Automatically Synchronous
AUV	- Armored Utility Vehicle
AUWT	- Ames Unitary Wind Tunnel
AUX	- Auxiliary
AUXTRAC	- Auxiliary Track
AV	- Audio-Visual
	- Average
	- AVIONICS
A-V	- Audio-Visual
AVC	- Automatic Volume Control
AVCS	- Advanced Videcon Camera System
AVE	- Aerospace Vehicle Equipment
	- Automatic Volume Expansion
AVGAS	- Aviation Gasoline
AVGH	- Average Grid Heading
AVH	- Average Heading
AVI	- Avoid Verbal Information
AVIONICS	- Aviation Electronics
AVL	- Approved Vendors List
AVLUB	- Aviation Lubricant
AVNL	- Automatic Video Noise Limiter
AVO	- Avoid Verbal Orders

■ AVOID

AVOID	- Accelerated View of Input Data
AVOIL	- Aviation Oil
AVRE	- Assault Vehicle, Royal Engineers
AVS	- Aircraft Vectoring System
AVSEP	- Audio Visual Superimposed Electrocardiogram Presentation
AVSS	- Apollo Vehicle Systems Section
AVT	- Air Velocity Transducer
	- Applications Vertical Test
AW	- Above Water
	- Acid Waste
	- Air to Water
	- Air Warning
	- Air Weapons
	- Arming Wire
	- Articles of War
	- Automatic Weapon
	- Automatic Word
A/W	- All Weather
AWA	- All Weather Attack
	- Aviation Writers Association
AWACS	- Airborne Warning and Control System
AWAR	- Air Weighted Average Resolution
AWASP	- Advanced Weapon Ammunition Support Point
AWC	- Air War College
	- Army War College (USAWC preferred)
AWCO	- Area Wage and Classification Office
AWCS	- Air Weapons and Control System
AWD	- Activation Working Group
	- Air Warfare Division
AWECOM	- Army Weapons Command (USAWECOM preferred)
AWG	- American Wire Gage
AWI	- Accommodation Weight Investigation
AWLS	- All-Weather Landing System
AWOC	- All-Weather Operations Committee
AWRE	- Atomic Weapons Research Establishment (British)
	- Australian Weapons Research Establishment
AWRS	- Airborne Weather Radar System
AWS	- Air Warning System
	- Air Weapon System
	- Air Weather Service
	- American War Standard
	- American Welding Society
AWSG	- Armstrong-Whitworth and Sperry Gyroscope
AXC	- Address to Index
AXD	- Auxiliary Drum

AXP   - Allied Exercise Publication
AXSIGCOM - Axis of Signal Communication
AXT   - Address to Index Register, True
          - Automatic Crosstell
AYI   - Angle of Yaw Indicator
AZ   - Azimuth
AZAR   - Adjustable Zero, Adjustable Range
AZEL   - Azimuth and Elevation
AZ-EL   - Azimuth and Elevation
AZON   - Azimuth Only
AZRAN   - Azimuth and Range
AZS   - Automatic Zero Set

B

B	- Ballistic
	- Bandwidth
	- Base
	- Beam
	- BIT
	- Bomber
	- Bonded
	- Boron
	- Braid
	- Breath
B-	- Boilerplate
BA	- Bell Aerosystems Company
	- Binary Add
	- Blind Approach
	- Bombing Altitude
	- Breathing Apparatus
	- Buffer Amplifier
	- Bundle Assembly
BAADS	- Bangor Air Defense Sector
BABS	- Blind Approach Beacon System
BAC	- Barometric Altitude Control
	- Bell Aerospace Corporation
	- Bendix Aviation Corporation
	- Binary Asymmetric Channel
	- Boeing Airplane Company (now TBC)
	- British Aircraft Corporation
	- Bureau of Air Commerce
BACAIC	- Boeing Airplane Company Algebraic Interpretive Computing System
BACP	- Business Advisory Committee on Procurement
BACS	- Boeing Applied Computing Service
BADA	- Base Air Depot Area
BADC	- Binary Asymmetric Dependent Channel

BADGE   - Base Air Defense Ground Environment System
BADIC   - Biological Analysis Detection Instrumentation and Control
BAE   - Beacon Antenna Equipment
BAFCOM   - Basic Armed Forces Communication Plan
BAGR   - Bureau of Aeronautics General Representative (obsolete)
BAGS   - Bullpup All-Weather Guidance System
BAI   - Boeing Airborne Instrumentation Equipment
BAIC   - Binary Asymmetric Independent Channel
BAL   - Base Authorization List
BALAST   - Balloon Astronomy
BALLWIN   Ballistic Winds
BALMI   - Ballistic Missile
BALOG   - Base Logistical Command
BALSPACON - Balance of Space-to-Space Control Agencies
BALUN   - Balanced to Unbalanced
BAMAGAT   - Block-a-Matic, Block-a-Gram, and Block-a-Text
BAMBI   - Ballistic Missile Boost Intercept
BAMG   - Browning Aircraft Machine Gun
BAMIRAC   - Ballistic Missile Radiation Analysis Center
BAMO   - Bureau of Aeronautics Material Officer (obsolete)
BAO   - Bureau of Air Operations
BAP   - Beacon Aircraft Position
BAPL   - Base Assembly Parts List
BAQ   - Bachelor Airmen Quarters
          - Basic Allowance for Quarters
BAR   - Barometer
          - Battery Acquisition Radar
          - Browning Automatic Rifle
          - Bureau of Aeronautics Representative (obsolete)
BARB   - British Angular Rate Bombsight
BARC   - Barge, Amphibious, Resupply, Cargo
BARLANT   - Barrier Force, Atlantic
BARN   - Bombing and Reconnaissance
BARO   - Barometer
BAROSWITCH - Barometric Switch
BARPAC   - Barrier Force, Pacific
BART   - Battle-Alert System
BAS   - Basic Air Speed
          - Basic Allowance for Subsistence
BASEC   - Base Section
BASEFOR   - Base Force
BASIC   - Battle Area Surveillance and Integrated Communications
BASICPAC - Battle Area Surveillance and Integrated Communications
            Processor and Computer
BASO   - Bureau of Aeronautics Shipment Order (obsolete)

■ BASOPS

BASOPS	- Base Operations
BAT	- Basic Air Temperature
	- Battalion Anti-Tank
	- Blind Approach Training
	- Buffer Area Table
BATC	- Boeing Atlantic Test Center
BATDIV	- Battleship Division
BATE	- Base Activation Test Equipment
	- Base Assembly and Test Equipment
BATFOR	- Battle Force
BATM	- Bureau of Air Traffic Management
BAWTR	- Babcock and Wilcox Test Reactor
BB	- Back to Back
	- Ball Bearing
	- Baseband
	- Battleship
	- Bomb Bay
	- Breadboard
	- Busy BIT
BBC	- British Broadcasting Corporation
	- Bromobenzylcyanide
BBES	- Bang-Bang Erection System
BBM	- Break-Before-Make
BBO	- Booster Burnout
BBRO	- Bomb Bay Ring Out
BC	- Back Connected
	- Ballistic Camera
	- Barge, Cargo
	- Barium Crown
	- Base Count
	- Battery Commander
	- Before Christ
	- Between Centers
	- Binary Code
	- Binary Counter
	- Broadcast
BCA	- Battery Control Area
BCB	- Battery Control Building
BCC	- Battery Control Center
	- Beacon Control Console
BCD	- Binary Coded Decimal
BCDP	- Battery Control Data Processor
BCE	- Beam Collimation Error
BCERPO	- Base Civil Engineer Real Property Office
BCFSK	- Binary Code Frequency Shift Keying

BCH	- Binary Coded Hollerith
BCI	- Base Line Configuration Identification
	- Battery Condition Indicator
	- Binary Coded Information
BCLU	- Base Construction Liaison Unit
BCN	- Beacon
BCO	- Base Contracting Officer
	- Battery Control Officer
	- Binary Coded Octal
	- Booster Engine Cutoff
BCOB	- Broken Clouds or Better
BCR	- Boro-Carbon Resistor
BCRT	- Binary Coded Range Time
BCRTS	- Binary Coded Range Time Signal
BCS	- Bardeen-Cooper-Schrieffer Theory of Superconductivity
	- British Computer Society
BCSD	- Model B Shift Driver
BCSO	- British Commonwealth Scientific Office
BCT	- Basic Combat Training
	- Battalion Combat Team
BD	- Base Detonating
	- Binary Decoder
	- Binary Divide
	- Bomb Disposal
	- Building Density
B/D	- Binary to Decimal
B-D	- Binary to Decimal
BDA	- Bermuda (remote site)
	- Bomb Damage Assessment
BDC	- Binary Decimal Counter
	- Boeing Development Center
BDCAA	- Booster Dynamic Condition at Abort
BDD	- Binary to Decimal Decoder
BDES	- Ball-Disc Erection System
BDF	- Base Detonating Fuze
BD-FT	- Board-Foot
BDG	- Beacon Data Generation
	- Bridge
BDH	- Bearing, Distance, Heading
BDHI	- Bearing, Distance, Heading Indicator
BDI	- Bearing Deviation Indicator
BDIS	- Battle Damage Information Script
BDL	- Battery Data Link
	- Beach Discharge Lighter
BDM	- Ballistic Defense Missile
	- Bomber Defense Missile

■ BDS

BDS	- Base Design Section
	- Bomb Damage Survey
	- Bomb Director System
BDSA	- Business and Defense Services Administration
BDSD	- Base Detonating Self-Destroying
BDS-OFI	- Base Design Section—Operational Facilities Installations
BDS-SFI	- Base Design Section—Support Facilities Installations
BDT	- Binary Deck to Binary Tape
BDU	- Battle Damage Umpire
BDXAV	- Bendix Aviation Corporation
BE	- Band Elimination
	- Base Ejection
	- Bombing Encyclopedia
B/E	- Bill of Exchange
BEA	- British European Airways
BEAC	- Boeing Engineering Analog Computer
BEBA	- Breeze Electron Ballistic Accelerometer
BECO	- Booster Engine Cutoff
BEF	- Buffered Emitter Follower
BEMA	- Business Equipment Manufacturers Association of America
BENELUX	- Belgium-Netherlands-Luxembourg
BENREP	- Big Ben Report
BENT	- Beginning Evening Nautical Twilight
BER	- BIT Error Rate
BESM	- High-Speed Electronic Computer (USSR)
BESS	- Binary Electromagnetic Signal Signature
BEST	- Blockhouse Equipment Switching Test
BET	- Best Estimate of Trajectory
BEV	- Billion Electron Volts
BEW	- Board of Economic Warfare
BF	- Back Faces
	- Back Feed
	- Base Fuze
	- Battleship Firing
	- Beat Frequency
	- Boiler Feed
	- Both Faces
	- Bottom Face
B/F	- Brought Forward
	- Buffer/Formatter
BFC	- Bureau of Foreign Commerce
BFCT	- Boiler Feed Compound Tank
BFE	- Beam Forming Electrode
	- Buyer Furnished Equipment
BFG	- B. F. Goodrich Company
BFL	- Back Focal Length

BFM	- Basic Field Manual
	- Branch of Full Minus
BFO	- Beat Frequency Oscillator
BFP	- Boiler Feed Pump
BFR	- Buffer
BFTC	- Boeing Flight Test Center
BFTU	- Boeing Field Test Unit
BFW	- Boiler Feed Water
BFZ	- Branch on Full Zero
BG	- Back Gear
	- Background
	- Battle Group
	- Beach Group
	- Blast Gauge
BGCG	- Battery Guidance Command Group
B/GE	- Burroughs/General Electric
BGLT	- Battle Group Landing Team
BGMP	- Hydrometeorological Forecasts Bureau (USSR)
BGR	- Bombing Gunnery Range
BGTS	- Boeing Gulf Test Section
BH	- Boiler Horsepower (BHP preferred)
	- Brake Horsepower (BHP preferred)
	- Brinell Hardness
B/H	- Blockhouse
BHN	- Brinell Hardness Number
BHO	- Branch Hydrographic Office
BHP	- Brake Horsepower
	- Boiler Horsepower
BHP-HR	- Brake Horsepower-Hour
BHR	- Biotechnical and Human Research
BHWT	- Boeing Hypersonic Wind Tunnel
BI	- Background Investigation
	- Barometric Indicator
	- Base and Increment
	- Base Ignition
	- Battery Inverter
	- Break-In Cycle
BIA	- Booster Insertion and Abort
BI-APS	- Battery Inverter Accessory Power Supply
BIAR	- Base Installation Action Request
BIB	- Baby Incendiary Bomb
	- Biographical Information Blank
BID	- Base Installation Department
BI-IPS	- Battery Inverter Instrument Power Supply
BILE	- Balanced Inductor Logic Element
BIM	- Basic Industrial Materials Program

■ BI-MM

BI-MM	- Base Installation—Minuteman
BIN	- Binary
	- BOMARC Interception
BINAC	- Binary Automatic Computer
BIO	- Branch Intelligence Office
BIODEF	- Biological Defense
BIOMED	- Biological-Medical
BIONICS	- Biology-Electronics
BIOPACK	- Biological Package
BIOR	- Business Input-Output Rerun Compiling System
BIOS	- Biological Investigation of Space
BIOWAR	- Biological Warfare
BIPAD	- Binary Pattern Detector
BIPCO	- Built-In-Place Components
BIRDIE	- Battery Integration and Radar Display Equipment
BIS	- Board of Inspection and Survey
	- British Interplanetary Society
BISEPS	- BOMARC Integration SAGE Evaluation Program Simulation
BISU	- Base Interface Surveillance Unit
BIT	- Binary Digit
	- Built-In Test
BITE	- Base Installation Test Equipment
BIZMAC	- Business Machine Computer
BJCEB	- British Joint Communications-Electronics Board
BJM	- Bluejacket's Manual
BJSM	- British Joint Services Mission
BK	- Barge Knockdown
BL	- Base Line
	- Benson-Lehner
	- Bomb Line
	- Bottom Layer
	- Breech Loading
	- Butt Line
	- Buttock Line
B/L	- Bill of Lading
BLADE	- Base Level Automation of Data Through Electronics
BLADES	- Bell Laboratories Automatic Design System
BLADING	- Bill of Lading
BLC	- Boundary Layer Control
BLCS	- Boundary Layer Control System
BLESSED	- Bell Little Electrodata Symbolic System for the Electrodata Michigan Bell Telephone
BLEU	- Blind Landing Experimental Unit
BLF	- Baryta Light Fling
BLH	- Baldwin-Lima-Hamilton
BLIS	- Bell Labs Interpretive System

BLLE	- Balanced-Line Logic Element
BLM	- Branch on Left Minus
BLO	- Backtell Lateraltell Output
	- Bombardment Liaison Officer
BL&P	- Blind Loaded and Plugged
BLS	- Bureau of Labor Statistics
BLT	- Battalion Landing Team
	- Blind Loaded with Tracer
BM	- Ballistic Missile
	- Bench Maintenance
	- Bench Mark
	- Bendix-Mishawaka
	- Binary Multiply
	- Breech Mechanism
B/M	- Bench Maintenance (BM preferred)
	- Bill of Materials
	- Buffer/Multiplexer
BMA	- British Medical Association
BMAG	- Body-Mounted Attitude Gyroscope
BMC	- Ballistic Missile Center
	- Ballistic Missiles Committee
BMCS	- Business Management Control System
BMCSRP	- Business Management Control System Research Project
BMCT	- Beginning Morning Civil Twilight
BMD	- Ballistic Missile Division (now BSD)
	- Base Maintenance Division
BMDC	- Ballistic Missile Defense Committee
BMD-FO	- Ballistic Missile Division—Field Office
BMDS	- Ballistic Missile Defense System
BME	- Bench Maintenance Equipment
BMEP	- Brake Mean Effective Pressure
BMETO	- Ballistic Missiles European Task Organization
BMEWS	- Ballistic Missile Early Warning System
BMG	- Body-Mounted (attitude) Gyroscope (BMAG preferred)
	- Browning Machine Gun
BMI	- Ballistic Missile Interceptor
	- Battelle Memorial Institute
BMM	- Ballistic Missile Manager
BMMC	- Ballistic Missile Management Complex
BMNT	- Beginning Morning Nautical Twilight
BMO	- Ballistic Missile Office
BMOM	- Base Maintenance and Operational Model
BMRS	- Ballistic Missile Re-Entry System
BMS	- Ballistic Missile Specification
BMT	- Before Morning Twilight
BMTD	- Ballistic Missile Terminal Defense

■ BMTS

BMTS	- Basic Military Training School
	- Bench Maintenance Test Set
BMU	- Beachmaster Unit
BMWS	- Ballistic Missile Weapon System
BN	- Battalion
	- Binary Number
B/N	- Beacon
B-N	- Barited-Nickel
BNF	- Bomb Nose Fuze
BNGS	- Bomb Navigation Guidance System
BNL	- Brookhaven National Laboratory
BNS	- Bomb Navigation System
BO	- Bail Out
	- Base Order
	- Beat Oscillator
	- Blackout
	- Blanket Order
	- Blocking Oscillator
	- Break-Out Cycle
B/O	- Back Order
BOA	- Broad Ocean Area
BOADS	- Boston Air Defense Sector
BOB	- BOMARC B Program
	- Bureau of the Budget
BOBHIP	- BOMARC B History Printout
BOC	- Base Operations Center
	- Blowout Cell
	- Blowout Coil
	- Body on Chassis
BOD	- Beneficial Occupancy Date
	- Biochemical Oxygen Demand
BODU	- Bureau of Ordnance Design Unit
BOI	- Basis of Issue
	- Break Off Inspection
BOLT	- Beam of Light Transistor
BOM	- Base Operations Manager
	- Bill of Materials
BOMARC	- Boeing-Michigan Air Research Conference
BOMREP	- Bombing Report
BOMRON	- Bombing Squadron
BONUS	- Boiling Nuclear Superheat Reactor
BOP	- Basic Operation Plan
	- Basic Overall Polarity
	- Binary Output Program
BOPO	- Browned-Off Passed-Over
BOP TEST	- Basic Overall Polarity Test

BPU ■

BOQ	- Bachelor Officers Quarters
BOR	- Board of Review
BOSCO	- BOMARC SAGE Compatibility Program
BOSS	- Bioastronautical Orbiting Space System (NASA)
	- BMEWS Operational Simulation System
	- Boeing Operational Supervisory System
	- Bomb Orbital Strategic System (Air Force)
BOSS-WEDGE	- Bomb Orbital Strategic System—Weapon Development Glid Entry
BOT	- BOMARC Training Program
BOU	- Boat Operating Unit
BOV	- Burnout Velocity
BP	- Back Pressure
	- Bailed Property
	- Bandpass
	- Base Percussion
	- Base Plate
	- Base Point
	- Bendix-Pacific
	- Between Perpendiculars
	- Bioscience Programs
	- Black Powder
	- Blood Pressure
	- Blueprint
	- Boilerplate
	- Boiling Point
	- Bonded Part
B/P	- Blueprint
	- Boilerplate
B&P	- Budgetary and Planning
BPA	- Basic Pressure Altitude
BPB	- Base Planning Board
BPC	- Back Pressure Control
BPD	- Buy per Drawing
BPF	- Bandpass Filter
BPG	- Break Pulse Generator
BPI	- Polytechnic Institute of Brooklyn
BPL	- Bell Propulsion Laboratory
BPMS	- Buy per Manufacturing Specification
BPO	- Basic Post Flight
	- BOMARC Prelaunch Output
BPPSP	- Block Point Plan Scheduling Procedure
BPR	- Bimonthly Progress Report
	- Bridge Plotting Room
BPS	- BITS per Second
BPU	- BOMARC Pickup

■ BR

BR	– Barrage Rocket
	– Basic Research
	– Brake Relay
	– Branch
	– Briefing Room
BRAMATEC	– Brain Mapping Technique
BRANE	– Bombing Radar Navigation Equipment
BRAS	– Ballistic Rocket Air Suppression
BRASO	– Branch, Aviation Supply Office
BRC	– Bio-Medical Recovery Capsule
BRD	– Booster Requirements Document
	– Bureau of Research and Development
BRDC	– Bureau of Research and Development Center
BREN	– Bare Reactor Equipment at Nevada Test Site
BRG	– Bearing
BRI	– Bomb Run Insert
BRIG	– BMEWS Raid Input Generator
BRINSMAT	– Branch Office, Inspector of Naval Material
BRK	– Break
BRKN	– Broken Arrow
BRL	– Ballistic Research Laboratories
	– Bomb Release Line
BRLESC	– Ballistic Research Laboratories Electronic Scientific Computer
BRLG	– Bomb, Radio, Longitudinal, Generator-Powered
BRL SYS	– Barrier Ready Light System
BRLT	– Ballistic Research Laboratories Transonic Range
BRM	– Branch on Right Minus
BROFICON	– Broadcast Figher Control
BROOM	– Ballistic Recovery of Orbiting Man
BRP	– Beacon Ranging Pulse
BRR	– Bridge Receiving Room
BR STD	– British Standard
BRTD	– Bright Radar Tube Display
BRU	– Base Records Unit
BRZN	– Bravo Zone
BS	– Battleship
	– Battleship Squadron
	– Binary Subtract
	– Blue Streak
	– Both Sides
B/S	– Bill of Sale
	– Blip Scan
	– Blue Streak
B&S	– Brown and Sharpe
BSB	– Baseband

BSC	- Binary Symmetric Channel
	- Boresight Camera
BSD	- Ballistic Systems Division (of AFSC)
BSDC	- Binary Symmetric Dependent Channel
BSDL	- Boresight Datum Line
BSE	- Base Support Equipment
	- Large Soviet Encyclopedia (USSR)
BSFC	- Brake Specific Fuel Consumption
BSGA	- Ballistic Systems Quality Administration
BSI	- Booster Situation Indicator
	- British Standards Institute
BSIC	- Binary Symmetric Independent Channel
BSIE	- Bio-Sciences Information Exchange
BSMV	- Bi-Stable Multi-Vibrator
BSO	- Battle Simulation Officer
BSR	- Backspace Register
	- Blip-Scan Radar
	- Bureau of Safety Regulations
BSRL	- Boeing Scientific Research Laboratory
BST	- Booster
	- Boresight Tower
	- Brief System Test
	- British Summer Time
B&SWG	- Brown and Sharpe Wire Gauge
BSWT	- Boeing Supersonic Wind Tunnel
BT	- Barge Training
	- Bathythermograph
BTC	- Basic Training Center
	- Booster Test Conductor
BTCC	- Basic Traffic Control Center
BTD	- Bomb Testing Device
BTDL	- Back Transient Diode Logic
BTE	- Battery Timing Equipment
	- Brake Thermal Efficiency
B-TELL	- Back Telling
BTF	- Ballistic Test Facility
	- Bomb Tail Fuze
BTI	- B-Track Initiator
	- Bureau of Technical Information (USSR)
BTL	- Backtell
	- Bell Telephone Laboratories
BTO	- Bombing Through Overcast
B to B	- Back to Back
BTRY	- Battery
BTS	- Boeing Test Support
	- B Table Simulator

■ BTT

BTT	- Beginning to Tape Test
BTU	- British Thermal Unit
BTWT	- Boeing Transonic Wind Tunnel
BU	- Base Unit
B/U	- Backup
BUA	- British United Airways
BUAER	- Bureau of Aeronautics (now BUWEPS)
BUCO	- Buildup Interceptor Control
BUDOCKS	- Bureau of Yards and Docks
BUF	- Buffer
BUG	- BOMARC Unintegrated Guidance
	- BOMARC-UNIVAC-GPA/35
BUIC	- Backup Interceptor Control
BUI SYS	- Barrier Up Indicator System
BUMED	- Bureau of Medicine and Surgery
BUNO	- Bureau Number
BUORD	- Bureau of Ordnance (now BUWEPS)
BUPERS	- Bureau of Personnel
BUPS	- Beacon, Ultra Portable S-Band
BUR	- Bureau
BUS	- BOMARC-UNIVAC-SAGE
BUSANDA	- Bureau of Supplies and Accounts
BUSHIPS	- Bureau of Ships
BUWEPINST	- Bureau of Naval Weapons Instruction
BUWEPS	- Bureau of Naval Weapons (formerly BUAER and BUORD)
BUWEPSTLO	- Bureau of Naval Weapons Technical Liaison Office
BV	- Balanced Voltage
	- Breakdown Voltage
BVD	- Beacon Video Digitizer
BW	- Ballistic Wind
	- Bandwidth
	- Basic Weight
	- Beam Width
	- Bendix-Westinghouse
	- Biological Warfare
	- Borg-Warner
	- Braided Wire Armor
B-W	- Bandwidth
B&W	- Babcock & Wilcox
	- Black and White
BWC	- Backward Wave Converter
BWG	- Birmingham Wire Gage
BWIA	- British West Indian Airways, Limited
BWL	- Belt Work Line
BWO	- Backward Wave Oscillator
BWP	- Basic War Plan

BWPA	- Backward Wave Power Amplifier
BWR	- Bandwidth Ratio
BWRWS	- Biological Warfare Rapid Warning System
BWT	- Boeing Wind Tunnel
BWV	- Back Water Valve

C

C	- Candle
	- Capacitance
	- Capacitor
	- Carbon
	- Cargo
	- Celsius
	- Center
	- Centigrade
	- Change
	- Closure
	- Confidential
	- Core
	- Cotton
	- Curie
	- Cycle
C-	- Configuration (see "S-" for Saturn Configuration)
CA	- Circuitry Adapter
	- Civil Affairs
	- Coast Artillery
	- Command Attitude
	- Consonant Amplification
	- Convening Authority
	- Cost Accounting
C&A	- Contract and Administration
CAA	- Civil Aeronautics Administration (now FAA)
CAAC	- Civil Aviation Administration of China
	- Civil Aviation Advisory Committee
CAAM	- Civil Aeronautics Administration Manual
CAAR	- Compressed Air Accumulatory Rocket
CAB	- Captured Air Bubble
	- Civil Aeronautics Board
	- Civil Aeronautics Bulletin
	- Civil Air Branch

CABH	– Cleared Altitude Block Height
CAC	– Canaveral Administration Complex (now KAC)
	– Caribbean Air Command
	– Change Analysis Commitment
	– Clear All Channels
	– Clear and Add Clock
	– Coast Artillery Corps
	– Combat Alert Center
	– Continental Air Command
	– Control and Coordination
CACM	– Central American Common Market
CACS	– Continental Airways and Communications Service
CAC&W	– Continental Aircraft Control and Warning
CAD	– Cartridge Actuated Device
	– Clear and Add
	– Communications Access Device
	– Compensated Avalanche Diode
	– Computer Access Device
	– Contract Administration Division
	– Copy and Add
CADC	– Cambridge Automatic Digital Computer
	– Central Air Data Computer
	– Continental Air Defense Command
CADCO	– Core and Drum Corrector
CADDAC	– Central Analog Data Distributing and Computing System
CADF	– Central Air Defense Force
	– Commutated Antenna Direction Finder
CADFISS	– Computation and Data Flow Integrated Subsystem Test
CAD-I	– Computer Access Device—Input
CADIC	– Chemical Analysis Detection Instrumentation Control
CADIN	– Continental Air Defense Integration North
CADIZ	– Canadian Air Defense Identification Zone
CADO	– Central Air Documents Office
CADPO	– Communications and Data Processing Operation
CADS	– Centralized Air Defense System
	– Continental Air Defense System (see CONAD)
CADSAME	– Call Signs and/or Address Group Remain Same
CADSS	– Combined Analog-Digital Systems Simulator
CADW	– Civil Air Defense Warning
CAE	– Continental Aviation and Engineering Corporation
CAEC	– Continental Aviation and Engineering Corporation
CAF	– Complete Assembly for Ferry
CAFB	– Chanute Air Force Base
CAFD	– Contact Analog Flight Display
CAFM	– Commercial Air Freight Movement
CAFS	– Checkout and Firing Subsystem

■ CAFSC

CAFSC	– Control Air Force Specialty Code
CAG	– Canopus Acquisition Gate
	– Carrier Air Group
	– Combined Arms Group
CAGC	– Coded Automatic Gain Control
CAGE	– Canadian Air Ground Environment
	– Common Air Ground Environment
	– Compiler and Assembler by General Electric
CAI	– Canadian Aeronautical Institute
	– Computer Analog Input
CAI/OP	– Computer Analog Input/Output
CAIRC	– Caribbean Air Command
CAL	– Caliber
	– Calibrate
	– Calorie
	– Command Authorization List
	– Continental Air Lines
	– Copy and Add Logical Word
	– Cornell Aeronautical Laboratory
	– Point Arguello, California (remote site)
CAL/CERT	– Calibration/Certification
CALI	– Cornell Aeronautical Laboratory, Incorporated
CALL	– Carat Assembled Logical Loader
CALST	– Cornell Aeronautical Laboratory Shock Tunnel
CALT	– Cleared Altitude
CALTEC	– California Institute of Technology
CAM	– Checkout and Automatic Monitoring
	– Civil Aeronautics Manual
	– Civil Air Movement
	– Clear and Add Magnitude
	– Commerical Air Movement
	– Consolidated Aircraft Maintenance
CAMA	– Centralized Automatic Message Accounting
CAMAL	– Continuously Airborne Missile Launcher and Low-Level Penetration Bomber
CA/MG	– Civil Affairs and Military Government
CAMP	– Computer Applications of Military Problems Users Group
	– Cost of Alternative Military Programs
CAMRON	– Consolidated Aircraft Maintenance Squadron
CAN	– Correlation Air Navigation
CANEL	– Connecticut Aircraft Nuclear Engine Laboratory
CANUKUS	– Canada-United Kingdom-United States
CANUS	– Canada-United States
CAO	– Collective Analysis Only
	– Committee on Amphibious Operations
CAOS	– Completely Automatic Operational System

CAP   - Card Assembly Program
            - Catapult and Arresting-Gear Pool
            - Civil Air Patrol
            - Combat Air Patrol
            - Component Acceptance Procedure
            - COMPOOL Assembler Program
            - Conflict Analysis Program
            - Console Action Processor
            - Current Assessment Plan
CAPCHE   - Component Automatic Programmed Checkout Equipment
CAPE   - Communication Automatic Processing Equipment
CAPPI   - Constant Altitude Plan-Position Indicator
CAPRI   - Coded Address Private Radio Intercom
CAPTAIN   - Carter's Adaptation Processor to Aid Interception
CAR   - Capital Assets Record
            - Civil Aeronautics Regulations
            - Contractural Action Request
CARDE   - Canadian Armament Research and Development Establishment
CARF   - Central Altitude Reservation Facility
CARP   - Computed Air Release Point
CARPAC   - Carriers, Pacific Fleet
CART   - Central Automatic Reliability Tester
CARTASKFOR - Carrier Task Force
CAS   - Calibrated Air Speed
            - Collision Avoidance System
            - Compare Accumulator with Storage
            - Complete Assembly for Strike
CASCU   - Commander Aircraft Support Control Unit
CASD   - Carrier Aircraft Service Division
CASDN   - Comite d'Action Scientifique de la Defense Nationale
CASE   - Computer Automated Support Equipment
CASF   - Composite Air Strike Forces
CASI   - Canadian Aeronautics and Space Institute
            - Combined Approach System Investigation
CAS/PWI   - Collision Avoidance System Proximity Warning Indicator
CASU   - Carrier Aircraft Service Unit
            - Combat Aircraft Service Unit
CAT   - Carburetor Air Temperature
            - Category
            - Celestial Atomic Trajectile
            - Central Air Transport
            - Centralized Automatic Testing
            - Civil Air Transport
            - Clean Air Turbulence
            - Closest Approach Time

■ **CAT**

CAT	– Component Acceptance Test
	– Computer of Average Transients
	– Control and Assessment Team
CATB	– Canadian Air Transport Board
CATC	– Canadian Air Transport Command
CAT-DAB	– Category Display Assignment BIT
CATE	– Current AFSC (formerly ARDC) Technical Efforts
CATOR	– Combined Air Transport Operations Room
CATS	– Category Switch
	– Civil Affairs Training School
	– Comprehensive Analytical Test System
CATU	– Combat Aircrew Training Unit
CAV	– Constant Angular Velocity
CAVU	– Ceiling and Visibility Unlimited
CB	– Center of Buoyancy
	– Chemical-Biological
	– Circuit Breaker
	– Citizens Band
	– Common Battery
	– Comparator Buffer
	– Construction Battalion
CB1	– Current BIT 1
CB2	– Current BIT 2
CBA	– C-Band Transponder Antenna
	– Cocoa Beach Apollo
CBAR	– Change Board Analysis Record
CBBU	– Construction Battalion Base Unit
CBC	– Change Board Commitment
	– Complete Blood Count
	– Construction Battalion Center
	– Contraband Control
CBCC	– Chemical-Biological Coordination Center
	– Common Bias, Common Control
CBCR	– Change Board Commitment Record
CBD	– Construction Battalion Detachment
CBI	– China-Burma-India
	– Complete Background Investigation
CB/L	– Commerical Bill of Lading
CBLS	– Carrier-Borne Air Liaison Section
CBM	– Continental Ballistic Missile
CBMU	– Current BIT Monitor Unit
CBO	– Coding Board Officer
CBR	– Chemical-Biological-Radiological
	– Current Balance Record
CBRA	– Conditionality Branch Address
CBRC	– Chemical, Biological, Radiological Center

CCA ■

CBRW	– Chemical-Biological-Radiological Warfare
CBS	– Controlled Blip Scan
CBSC	– Common Bias, Single Control
CBU	– Container, Bomb, Utility
CBX	– C-Band Transponder
CC	– Capsule Communicator
	– Central Control
	– Channel Command
	– Chemical Corps (CMLC preferred)
	– Chrysler Corporation
	– Circuit Closing
	– Cirrocumulus
	– Closing Coil
	– Color Code
	– Combat Command
	– Command Car
	– Command Center
	– Command Computer
	– Command Console
	– Command Control
	– Commercial Control
	– Common Carrier
	– Communication Center
	– Communications Control
	– Computer Calculator
	– Construction Contractor
	– Construction Corps
	– Control Center
	– Control Code
	– Coordinate Converter
	– Coordinates Computed
	– Cross Correlation
	– Cross Couple
	– Cubic Centimeter
C/C	– Change of Course
	– Command Center
	– Command Control
	– Coordinate Converter
	– Crew Chief
C&C	– Calibration and Certification
	– Calibration and Checkout
	– Command and Control
	– Communications and Control
CCA	– Carrier-Controlled Approach
	– Combat Center Active
	– Continental Control Area
	– Contract Change Authorization

71

■ CCB

CCB	- Change Control Board
	- Close Control Bombing
	- Committee Change Board
	- Configuration Control Board
	- Console-to-Computer Buffer
	- Contraband Control Base
	- Contract Change Board
CCBD	- Configuration Control Board Directive
	- Contract Change Board Directive
CCC	- Chief Cable Censor
	- Command Control Center
	- Command Control Console
	- Communications Control Console
	- Computer Communications Console
	- Computer Communications Converter
	- Coordinate Conversion Computer
CCD	- Central Command Decoder
	- Combat Center Director
	- Core Current Driver
CC/D	- Command Control/Destruct
CCDC	- Cape Cod Direction Center
CC/DC	- Combined Combat Center, Direction Center
CCDD	- Command and Control Development Division (now AFSC)
CCDO	- Combat Center Duty Officer
CCDS	- Command Control Destruct System
CCDSO	- Command Control Division System Office
CCE	- Command Control Equipment
CCF	- Central Computer Facilities
	- Combat Center Function
	- Communication Central Facility
CCFT	- Controlled Current Feedback Transformer
CCG	- Combat Center Group
	- Communications Change Group
	- Constant Current Generator
CCGD	- Commandant Coast Guard District
CCH	- Command Control Handover and Keying
CCIR	- Communications Change Initiation Request
CCIS	- Command Control Information System
CCITT	- Comite Consultatif Internationale—Telegraphie et Telephonie
CCK	- Convergence Current Register
CCL	- Communications Change Log
	- Conversion and Check Limit
CCM	- Commodity Class Manager
	- Constant Current Modulation
	- Controlled Carrier Modulation
	- Counter-Countermeasure

CD

CCMD	- Chrysler Corporation Missile Division
	- Continuous Current-Monitoring Device
CCMR	- Central Contract Management Region
CCMTA	- Cape Canaveral Missile Test Annex (now CKMTA)
CCMTC	- Cape Canaveral Missile Test Center (now CKMTC)
CCN	- Command Control and Keying
	- Contract Change Notification
CCO	- Crystal Controlled Oscillator
CCP	- Card Input-Preliminary Processing
	- Command Control Panel
	- Command Control Post
	- Communications Control Panel
	- Contract Change Proposal
CCPL	- Combat Center Programming Leader
CCR	- Change Commitment Record
	- Communications Change Request
	- Contract Change Record
	- Contract Change Request
CCS	- Cape Cod System
	- Central Computing Site
	- Change Commitment Summary
	- Change Control System
	- Combat Center Standby
	- Combined Chiefs of Staff
	- Command Control System
CC&S	- Central Computer and Sequencer
CCSD	- Chrysler Corporation Space Division
CCSI	- Combat Center Status Indicator
CCSU	- Computer Cross-Select Unit
CCSX	- Command Control System Exercise
CCT	- Constant Current Transformer
	- Control Center Tracking
	- Chrystal Controlled Transmitter
CC&T	- Combat Center and Crosstell
CCTS	- Combat Crew Training Squadron
CCTV	- Closed Circuit Television
CCU	- Camera Control Unit
	- Communication Control Unit
	- Communication Coupling Unit
	- Computer Control Unit
	- Coupling Control Unit
CCW	- Command Control and Weather
	- Counterclockwise
CD	- Call Director
	- Camouflage Detection
	- Capacitor-Diode

73

■ CD

CD		– Civil Defense
		– Close Doublet
		– Coastal Defense
		– Coefficient of Drag
		– Cold-Drawn
		– Confidential Document
		– Conning Director
		– Countdown
		– Crystal Driver
		– Cumulative Destruction
C/D		– Calls per Day
		– Chaff Delivery
		– Command Destruct
		– Countdown
C&D		– Communications and Data
		– Controls and Displays
CDA		– Command Data Acquisition
CDB		– Current Data BIT
CDBN		– Column-Digit Binary Network
CDC		– Call Direction Code
		– Combat Development Center
		– Combat Direction Central
		– Command and Data-Handling Console
		– Computer Development Center
		– Control Data Corporation
		– Count-Double Count
CDCM		– Coupling Display Manual Control—IMU (Inertial Measurement Unit)
CD/CN		– Command Control
CDCO		– Coupling Display Manual Control – Optics
CDCU		– Communications Digital Control Unit
CDD		– Combat Data Director
CDDP		– Console Digital Display Programmer
CDE		– Cornell Dubilier Electronics
CDEVC		– Computer Development Center
CDF		– Compare and Difference Full Words
CDFF		– Command Distributor Flip-Flop
CDG		– Capacitor Diode Gate
CDI		– Classified Defense Information
CDL		– Compare and Difference Left Half Words
		– Confidential Damage Level
CDM		– Circuit Directory Maintenance
		– Civil Defense Mobilization
		– Compare and Difference of Masked BITs
CDMB		– Civil and Defense Mobilization Board
CDO		– Command Duty Officer

C&E

CDOG	– Combat Developments Objective Guide
CDOH	– Coupling Display Optical Hand Controller
CDOM	– Chief Draftsman Office Memorandum
CDP	– Central Data Processor
	– Checkout Data Processor
	– Correlated Data Processor
CDPC	– Central Data Processing Computer
CDR	– Command Destruct Receiver
	– Compare and Difference Right Half Words
	– Critical Design Review
	– Critical Design Revision
	– Current Directional Relay
CDRD	– Computations and Data Reduction Division
CDRM	– Critical Design Review Meeting
CDS	– Central Distribution System
	– Communications and Data Subsystems
	– Compatible Duplex System
	– Comprehensive Display System
CDSC	– Coupling Display SCT Manual Control
CDSE	– Computer-Driven Simulation Environment
C&DSS	– Communications and Data Subsystem
CDT	– Central Daylight Time
	– Coincidence Detection Program
	– Command Destruct Transmitter
	– Contractor Development Testing
CDU	– Coastal Defense Unit
	– Coupling Data Unit
	– Coupling Display Unit
CDUM	– Coupling Display Unit—IMU (Inertial Measurement Unit)
CDUO	– Coupling Display Unit—Optics
CDW	– Civil Defense Warning
	– Computer Data Word
CDX	– Control-Differential Synchro Transmitter
CE	– Chief Engineer
	– Circular Error
	– Common Era
	– Communications-Electronics
	– Commutator End
	– Conditional Effectiveness
	– Construction Equipment
	– Controller Error
	– Corps of Engineers
C/E	– Corps of Engineers (CE preferred)
C-E	– Communications-Electronics
C&E	– Communications and Electronics

75

■ CEA

CEA	- Circular Error Average
	- Council of Economic Advisors
CEADS	- Central European Air Defense Sector
CEAPD	- Central Air Procurement District
CEBM	- Corona, Eddy Current, Beta Ray, Microwave
CEBMCO	- Corps of Engineers Ballistic Missile Construction Office
CEC	- Certification of Equipment Completion
	- Civil Engineering Corps
	- Consolidated Electrodynamics Corporation
	- Consolidated Electronics Corporation
CECM	- Composite Engineering Change Memorandum
CECO	- Chandler Evans Corporation
CECOS	- Civil Engineering Corps Officers
CECS	- Communications-Electronics Coordinating Section
CED	- Communications and Electronics Division
	- Communications-Electronics Doctrine
CEDAC	- Central Differential Analyzer Control
CEE	- Combat Emplacement Excavator
CEF	- Carrier Elimination Filter
CEG	- Combat Evaluation Group
CEI	- Communications-Electronics Instruction
CEIP	- Communications-Electronics Implementation Plan
CEIR	- Corporation for Economic and Industrial Research
CEL	- Celestial
	- Civil Engineering Laboratory
	- Combat Evaluation Launch
	- Crew Evaluation Launcher
CELESCOPE	- Celestial Telescope
CELLSCAN	- Cell (blood) Scanning System
CEMF	- Counter-Electromotive Force
CEMON	- Customer Engineering Monitor
CEMS	- Central Electronic Management System
CENTAG	- Central Army Group
CENTO	- Central Treaty Organization
CEO	- Chief Engineer's Office
	- Communications-Electronics Officer
CEP	- Circular Error Probable
	- Computed Ephemeris Position
	- Computer Entry Punch
CEPS	- Command Module Electrical Power System
CER	- Change Evaluation Request
	- Circle of Equal Probability
	- Complete Engineering Release
	- Crew Environment Requirements
CERB	- Coastal Engineering Research Board
CERMA	- Centre d'Etudes et de Recherches de Medecine Aeronautique

CERMET	- Ceramic to Metal
CERN	- Centre Europeen de Recherches Nucleaires (European Center for Nuclear Research)
CERO	- Coastal Engineering Research Office
CESA	- Canadian Engineering Standards Association
CESI	- Closed Entry Socket Insulation
CET	- Capsule Elapsed Time
	- Cumulative Elapsed Time
CETA	- Centre d'Etudes pour la Traduction Automatique
CETC	- Corps of Engineers Technical Committee
CETIS	- Centre de Traitement de l'Information Scientifique
CF	- Captive Flight
	- Carrier Frequency
	- Cathode Follower
	- Center Frequency
	- Center of Flotation
	- Centrifugal Force
	- Check Flight
	- Checking Fixture
	- Circuit Finder
	- Controlled Fragmentation
	- Conversion Factor
	- Counterfire
C/F	- Carry Forward
	- Center Frequency
CFA	- Combination Fabrication and Assembly
	- Crossed-Field Amplifier
CFAE	- Contractor Furnished Aeronautical Equipment
CFAR	- Constant False Alarm Rate
CFAW	- Commander Fleet Air Wing
CFC	- Central Fire Control
	- Contractor Furnished Container
CFD	- Contractor Functional Demonstration
	- Cumulative Frequency Distribution
CFE	- Contractor Furnished Equipment
CFF	- Change Film Frame
	- Critical Flicker Frequency
CFK	- Confidence Firing Kit
CFL	- Cold Flow Laboratory
	- Context Free Language
C/FLT	- Captive Flight
CFM	- Cathode Follower Mixer
	- Cubic Feet per Minute
CF-NRTS	- Central Facilities, National Reactor Test Station
CFP	- Contractor Furnished Property
	- Customer Furnished Part
	- Customer Furnished Property

■ **CFR**

CFR	- Carbon-Film Resistor
	- Contract Flight Rules
CFS	- Carrier Frequency Shift
	- Critical Flight Subsector
	- Cubic Feet per Second
CFSSB	- Central Flight Status Selection Board
CG	- Camera Gun
	- Cargo Glider
	- Center of Gravity
	- Centigram
	- Coast Guard
	- Coincidence Gate
	- Command Group
	- Commanding General
	- Cytosine Guanine
CGA	- Contrast Gate Amplifier
CGAIRDET	- Coast Guard Air Detachment
CGAIRFMFPAC	- Commanding General Air Fleet Marine Force, Pacific
CGAS	- Coast Guard Air Station
CGASC	- Cornell-Guggenheim Aviation Safety Center
CGDO	- Coast Guard District Office
CGI	- Cognizant Government Inspector
CGL	- Coast Guard League
CGR	- Central GEEIA Region
CGS	- Centimeter-Gram-Second
	- Coast and Geodetic Survey
C&GS	- Coast and Geodetic Survey
CGSC	- Coast Guard Supply Center
CGSE	- Centimeter-Gram-Second-Electrostatic
CGSM	- Centimeter-Gram-Second-Electromagnetic
CGSS	- Cryogenic Gas Storage System
CGT	- Current Gate Tube
CGTM	- Command Guided Tactical Missile
CH	- Case Harden
	- Compass Heading
	- Conductor Head
C/H	- Command Heading
CHABA	- Committee on Hearing and Bio-Acoustics
CHACOM	- Chain of Command
CHADS	- Chicago Air Defense Sector
CHAMPION	- Compatible Hardware and Milestone Program for Integrating Organizational Needs
CHANCOMTEE	- Channel Committee
CHARM	- CAA High Altitude Remote Monitoring
CHB	- Cargo Handling Battalion
	- Chain Home Beamed

CINCEASTLANT

CHGFA	- Costs Chargeable to Fund Authorization
CHINFO	- Chief of Information
CHO	- Choke Oil
CHOP	- Change of Operational Control
CHOPLIN	- Change My Operation Plan
CHORD	- Change My Operation Order
C-HR	- Candle-Hour
CHS	- Central Heading System
CHSKED	- Change Schedule
CHT	- Charactron Tube
	- Cylinder Head Temperature
CHU	- Centigrade Heat Unit
CI	- Card Input
	- Cast Iron
	- Center of Impact
	- Cost Inspector
	- Counter Intelligence
C/I	- Carrier-to-Interference
C&I	- Communications and Instrumentation
CIA	- Central Intelligence Agency
CIB	- Central Intelligence Board
	- Configuration Identification Board
CIC	- Chemical Institute of Canada
	- Combat Information Center
	- Combat Intercept Control
	- Commander in Chief (CINC preferred)
	- Command Input Coupler
	- Communications Instructor Console
	- Counter Intelligence Corps
	- Customer Identification Code
CICO	- Combat Information Center Officer
CID	- Cable Interconnect Diagram
	- Criminal Investigation Division
CIDPS	- Continental Intelligence Data Processing System
CIDS	- Chemical Information Data System
CIED	- Card Input Editor
CIF	- Central Information Facility
	- Central Instrumentation Facility
	- Cost, Insurance, and Freight
CIGIF	- Central Inertial Guidance Test Facility
CIIC	- Counter Intelligence Interrogation Center
CIL	- Certificate in Lieu of
CINC	- Commander in Chief
CINCAL	- Commander in Chief, Alaska
CINCARIB	- Commander in Chief, Caribbean
CINCEASTLANT	- Commander in Chief, Eastern Atlantic Area (NATO)

■ **CINCENT**

CINCENT - Commander in Chief, Allied Forces, Central Europe (NATO)
CINCHAN - Commander in Chief, Allied Forces, Channel (NATO)
CINCLANT - Commander in Chief, Atlantic Fleet (CINCLANTFLT preferred)
CINCLANTFLT - Commander in Chief, Atlantic Fleet
CINCMED - Commander in Chief, Allied Forces, Mediterranean (NATO)
CINCNORAD - Commander in Chief, North American Air Defense Command
CINCNORTH - Commander in Chief, Allied Forces, Northern Europe (NATO)
CINCONAD - Commander in Chief, Continental Air Defense Command
CINCPAC - Commander in Chief, Pacific Fleet (CINCPACFLT preferred)
CINCPACAF - Commander in Chief, Pacific Air Forces
CINCPACFLT - Commander in Chief, Pacific Fleet
CINCSAC - Commander in Chief, Strategic Air Command
CINCSOUTH - Commander in Chief, Allied Forces, Southern Europe (NATO)
CINCSPECOMME - Commander in Chief, Specified Command, Middle East
CINCUNC - Commander in Chief, United Nations Command
CINCUS - Commander in Chief, United States Fleet (now COMINCH)
CINCUSAFE - Commander in Chief, United States Air Forces, Europe
CINCUSAREUR - Commander in Chief, United States Army, Europe
CINCWESTLANT - Commander in Chief, Western Atlantic Area (NATO)
CINFO - Chief of Information
CIO - Combat Intelligence Officer
- Congress of Industrial Organizations
CIP - Calibration In Process
- Civilian Instruction Program
- Clean and In Place
- Configuration Identification Package
- Contract Implementation Plan
- Conversion In Place
- Cost Improvement Program
CIPASH - Committee for International Program in Atmospheric Sciences and Hydrology
CIR - Central Input Recording Program
- Conformance Inspection Record
- Critical Item Report
- Customer Inspection Record
CIRA - Committee International Reference Atmosphere
CIRAD - Corporation for Information Systems Research and Development
CIRC - Circumference
CIRVIS - Communication Instructions for Reporting Vital Intelligence Sightings

CIS	- Christmas Island Station
	- Command Intelligence and Status
	- Communication and Instrumentation System
C&IS	- Communication and Instrumentation System (CIS preferred)
CISCO	- Communications Integrated System Control Office
	- Compass Integrated System Compiler
CISCS	- Construction Interface Surveillance Control Section
CIT	- California Institute of Technology (CALTEC preferred)
	- Call-In Time
	- Carnegie Institute of Technology
CITCE	- Comite International de Thermodynamique et de Cinetique Electro-Chimiques (International Committe of Electro-Chemical Thermodynamics and Kinetics)
CITE	- Capsule Integrated Test Equipment
CITEL	- Committee for Inter-American Telecommunications
CIVENGRLAB	- Civil Engineering Laboratory
CIW	- Command Intelligence and Weather
CJO	- Communications Jamming Operator
CJS	- Canadian Joint Staff
CJTS	- Commander, Joint Task Force
CKMTA	- Cape Kennedy Missile Test Annex (formerly CCMTA)
CKMTC	- Cape Kennedy Missile Test Center (formerly CCMTC)
CKN	- Cape Kennedy (formerly CNV) (remote site)
CKT	- Circuit
CL	- Carload
	- Centerline
	- Centiliter
	- Check List
	- Closed Loop
	- Contact Lost
	- Cruiser, Light
C/L	- Chaff Loading
CLA	- Clear and Add
CLAC	- Closed Loop Approach Control
CLAM	- Chemical Low-Altitude Missile
CLAMP	- Chemical Low-Altitude Missile, Puny
CLARK	- Combat Launch and Recovery Kit
CLASS	- Computer-Based Laboratory for Automated School Systems
CLC	- Conversion and Check Limit
	- Course Line Computer
CLCS	- Cable Launch Control System.
	- Closed-Loop Control System
CLEAR	- Compiler, Loader, Executive Program, Assembler, Routines
CLEM	- Cargo Lunar Excursion Module
CLFC	- Climb Forward Component
CLFT	- Closed Loop Flight Test Language

■ **CLI**

CLI	- Card and Light Gun Input
CLIP	- Compiler and Language for Information Processing
CLM	- Circumlunar Mission
	- Clear Magnitude
CLO	- Closeout
CLOT	- Closed Loop Operations Test
CLR	- Conference Letter Report
	- Coordination Letter Report
CLS	- Clear and Subtract
	- Command Liaison and Surveillance and Keying
	- Controlled Leakage System
CLSS	- Communication Link Subsystem
CLU	- Central Logic Unit
CLUMP	- COMPOOL Look-Up Memory Print
CLUS	- Continental Limits of United States
CM	- Center Matched
	- Center of Mass
	- Centimeter
	- Command Module
	- Computer Module
	- Construction and Machinery
	- Controlled Mine Field
	- Core Memory
	- Countermeasure
	- Countermortar
	- Coverage Masker
C/M	- Command Module
C&M	- Control and Monitor
CMA	- Canadian Medical Association
	- Civil-Military Affairs
	- Communication Managers Association
CMAS	- Council for Military Aircraft Standards
CMC	- Commandant of the Marine Corps
CMD	- Core Memory Driver
CM&D	- Countermeasures and Deception
CMDAC	- Current Mode Digital-to-Analog Converter
CME	- Commercial Measuring Equipment
CMEC	- Convectron-Microsyn Erection Circuit
CMF	- Coherent Memory Filter
	- Combat Mission Folder
	- Command Message Formulator
	- Compare Full Words
CMH	- Countermeasures Homing
CMHA	- Confidential—Modified Handling Authorized (formerly CONFMOD)
CMI	- Command Maintenance Inspection

CML	- Chemical
	- Compare Left Half Words
	- Current Mode Logic
CMLC	- Chemical Corps
	- Civilian-Military Liaison Committee
CMM	- Communications and Telemetry
	- Compare Mask
CMMR	- Confirmed and Made a Matter of Record
CMO	- Contigency Mode of Operation
CMP	- Compression
	- Computed Maximum Pressure
	- Computer
	- Controlled Materials Plan
CMPC	- Chicago Molded Products Corporation
CMPS	- Centimeter per Second
CMPT	- Component
CMR	- Common Mode Rejection
	- Communications Moon Relay
	- Compare Right Half Words
	- Contract Management Region
CMRCS	- Command Module Reaction Control System
CMS	- Command Module Simulator
	- Corporate Management System
CMT	- Command Module Technician
	- Corrected Mean Temperature
CMTM	- Communications and Telemetry
CMU	- Compatibility Mock-Up
CMVM	- Contact-Making Voltmeter
CN	- Change Notice
	- Compass North
	- Control Number
C/N	- Carrier to Noise
C&N	- Communications and Navigation
CNAS	- Civil Navigation Aids System
CNES	- Centre National d'Etudes Spatiales
CNET	- Centre National d'Etudes des Telecommunications
CNGB	- Chief, National Guard Bureau
CNI	- Communication-Navigation-Identification
CNJC	- Cable Network Joint Committee
CNL	- Control and Liaison
C&N LAB	- Communication and Navigation Laboratory
CNO	- Chief of Naval Operations
CNP	- Chief of Naval Personnel
	- Chopped Nylon Phenolic
CNRC	- Canadian National Research Council
CNRS	- Comite National pour les Recherches Scientifiques

■ CNS

CNS	- Control Network System
CNT	- Canadian National Telegraph
	- Celestial Navigation Trainer
CNV	- Cape Canaveral (remote site) (now CKN)
CO	- Carbon Monoxide
	- Change Order
	- Checkout
	- Close-Open
	- Commanding Officer
	- Company
	- Crystal Oscillator
C/O	- Case Of
	- Checkout
	- Cutoff
C-O	- Crew-Operated
COA	- Cognizant Operating Authority
	- Comptroller of the Army
	- Constant Output Amplifier
COAD	- Coordinate Adder
COBI	- Coded Biphase
COBOL	- Common Business-Oriented Language
COBRA	- Compatible On-Board Ranging
COC	- Code Operations Coordinator
	- Combat Operations Center
	- Command and Control System
	- Complete Operational Capability
COD	- Carrier-Onboard-Delivery
	- Carrier-On-Deck
CODAN	- Carrier-Operated Device, Anti-Noise
	- Coded Analysis
CODAP	- Control Data Assembly Program
CODAR	- Correlation Display Analyzing and Recording
CODASYL	- Conference on Data Systems Language
CODEL	- Computer Developments Limited Automatic Coding System
CODIC	- Color Difference Computer
CODIPHASE	- Coherent Digital Phased Array System
CODIT	- Computer Direct to Telegraph
CODN	- Component Operational Data Notice
CODORAC	- Coded Doppler Radar Command
COE	- Cab Over Engine
	- Corps of Engineers (CE preferred)
COED	- Computer-Operated Electronic Display
COESA	- Committee on Extension to the Standard Atmosphere
COF	- Computer Operations Facility
COFEC	- Cause of Failure, Effect, and Correction

COG	- Combat Operations Group
	- Computer Operations Group
	- Crab-Oriented Gyro
COGO	- Coordinate Geometry Program
COGS	- Continuous Orbital Guidance System
COHO	- Coherent Oscillator
COI	- Communication Operation Instructions
COIN	- Command Information
	- Committee on Information Needs
	- Counter-Insurgency
COL	- Computer Oriented Language
COLA	- Cost of Living Allowance
COLASL	- Compiler—Los Alamos Scientific Laboratory
COLIDAR	- Coherent Light Detection and Ranging
COLT	- Communication Line Terminator
COM	- Complement Magnitude
COMAC	- Continuous Multiple-Access Collator
COMAR	- Computer, Aerial Reconnaissance
COMATS	- Commander, Military Air Transport Service
COMCARIBSEAFRON	- Commander, Caribbean Sea Frontier
COMCM	- Communications Countermeasures and Deception
COMCRUDESLANT	- Commander, Cruiser Destroyer Forces, Atlantic Fleet
COMCRUDESPAC	- Commander, Cruiser Destroyer Forces, Pacific Fleet
COMEASTSEAFRON	- Commander, Eastern Sea Frontier
COMECON	- Council for Mutual Economic Assistance
COMEIGHT	- Commandant, Eighth Naval District
COMELEVEN	- Commandant, Eleventh Naval District
COMENT	- Command Evaluation and Training
	- Computer-Operated Management Evaluation Technique
COMFIRSTFLT	- Commander, First Fleet (Pacific)
COMFIVE	- Commandant, Fifth Naval District
COMFOUR	- Commandant, Fourth Naval District
COMFOURTEEN	- Commandant, Fourteenth Naval District
COMHAWAIISEAFRON	- Commander, Hawaiian Sea Frontier
COMINCH	- Commander in Chief, United States Fleet (formerly CINCUS)
COMINT	- Communications Intelligence
COMIT	- Computing System Massachusetts Institute of Technology
COMJAM	- Communications Jamming
COMLOGNET	- Combat Logistics Network
COMMCEN	- Communications Center
COMMEL	- Communications Electronics
COMINLANT	- Commander, Mine Force, Atlantic Fleet
COMINPAC	- Commander, Mine Force, Pacific Fleet
COMMUNICAT	- Communications Satellite

■ COMMZ

COMMZ - Communications Zone
COMNAVAIRLANT - Commander, Naval Air Force, Atlantic Fleet
COMNAVAIRPAC - Commander, Naval Air Force, Pacific Fleet
COMNAVFORJAP - Commander, Naval Forces, Japan
COMNAVFORPHIL - Commander, Naval Forces, Phillipines
COMNAVFORMARIANAS - Commander, Naval Forces, Marianas
COMNEED - Communications Need
COMNET - Communications Network
COMNINE - Commandant, Ninth Naval District
COMONE - Commandant, First Naval District
COMP - Comparator
       - Compiler
       - Computer
COMPACT - Compact Multi-Purpose Automatically-Controlled Transportable Reactor
COMPHIBLANT - Commander, Amphibious Force, Atlantic Fleet
COMPHIBPAC - Commander, Amphibious Force, Pacific Fleet
COMPLAN - Communications Plan
COMPOOL - Communications Pool
COMPRNC - Commandant, Potomac River Naval Command
COMRAT - Commuted Ration
COMSAT - Communications Satellite
       - Communications Satellite Corporation
COMSEC - Communications Security
COMSECFLT - Commander, Second Fleet (Atlantic)
COMSERVLANT - Commander, Service Force, Atlantic Fleet
COMSERVPAC - Commander, Service Force, Pacific Fleet
COMSEVENTEEN - Commandant, Seventeenth Naval District
COMSEVENTHFLT - Commander, Seventh Fleet (Pacific)
COMSIX - Commandant, Sixth Naval District
COMSIXFLT - Commander, Sixth Fleet (Atlantic)
COMSRNC - Commandant, Severn River Naval Command
COMSS - Compare String with String
COMSTS - Commander, Military Sea Transportation Service
COMSUBLANT - Commander, Submarine Force, Atlantic Fleet
COMSUBPAC - Commander, Submarine Force, Pacific Fleet
COMSW - Compare String with Word
COMTHIRTEEN - Commandant, Thirteenth Naval District
COMTHREE - Commandant, Third Naval District
COMTRAN - Commercial Translator
COMTWELVE - Commandant, Twelfth Naval District
COMWESTSEAFRON - Commander, Western Sea Frontier
COMZ - Communications Zone (COMMZ preferred)
CONAC - Continental Air Command
CONAD - Continental Air Defense Command
CONALOG - Contact Analog

CONALT - USCONARC Alternate Headquarters
CONARC - Continental Army Command (USCONARC preferred)
CONELRAD - Control of Electromagnetic Radiation
CONFMOD - Confidential—Modified Handling Authorized (now CMHA)
CONLUS - Continental Limits of the United States
CONOBJTR - Conscientious Objector
CONPY - Contact Party
CONREP - USCONARC Emergency Relocation Plan
CONSIM - Console Simulator
CONTRAIL - Condensation Trail
CONTRANS - Conceptual Through Random-Net Simulation
CONUS - Continental United States
CONV - Converter
CONVAIR - Consolidated-Vultee Aircraft Corporation (now General Dynamics)
CONVEL - Constant Velocity
COO - Chicago Operations Office
COOP - Continuity of Operations and Plans
COORS - Communications Outage Restoration System
COP - Coefficient of Performance
 - Control for Operational Programs
COPAN - Command Post Alerting Network
COPE - Computer Operator Proficiency Examination
COPERS - Commission Preparatoire Europeen de Recherches Spatiales
COR - Carrier Operated Relay
 - Combat Operations Report
 - Contracting Officer's Representative
CORA - Conditional Response Analog Machine
CORAD - Correlation Radar
CORAL - Command Radio Link
 - Correlation Radio Link
CORC - Chief of Reserve Components
 - Cornell Computing Language
CORDIC - Coordinate Rotation Digital Computer
CORDP - Correlated Radar Data Printout
CORDPO-SORD - Correlated Radar Data Printout—Separation of Radar Data
CORE - Computer Oriented Reporting Efficiency
COREN - Corps of Engineers (CE preferred)
CORENG - Corps of Engineers (CE preferred)
CORG - Combat Operations Research Group
CORREGATE - Correctable Gate
COS - Civilian Occupational Speciality
COSAR - Compression Scanning Array Radar
COSEAL - Compass System Extensively Altered
COSI-KON - Crimp-On, Snap-In Contacts

- **COSMIC**

    COSMIC   - Code Name Given to Identify NATO Documents
                - Command Operations Simulation Model with Interrogation Control
    COSMON - Component Open-Short Monitor
    COSOS   - Conference on Self-Organizing Systems
    COSPAR - Committee on Space Research
    COST   - Congressional Office of Science and Technology
    COTAR   - Correlation Tracking and Ranging System
    COTAR-AME - Correlation Tracking and Ranging Angle Measuring Equipment
    COTAR-DAS - Correlation Tracking and Ranging Data Acquisition System
    COTAR-DME - Correlation Tracking and Ranging Data Measuring Equipment
    COTAT   - Correlation Tracking and Triangulation
              - Cosine Tracking and Triangulation
    COTR   - Contracting Officer's Technical Representative
    COV   - Cutout Valve
    COW   - Cooperative Observational Week
    COZI   - Communications Zone Indicator
    CP   - Candlepower
           - Center of Pressure
           - Check Parity
           - Chemically Pure
           - Circular Pitch
           - Circular Polarization
           - Clock Phase
           - Clock Pulse
           - Command Post
           - Command Pulse
           - Communications Processor
           - Concrete Piercing
           - Conflict Probe
           - Control Panel
           - Control Point
           - Control Programmer
           - Co-Pilot
           - Core Prime
           - Cost and Performance
    C/P   - Cartesian to Polar
           - Constant Power
    C-P   - Cartesian to Polar
    C&P   - Contracts and Pricing
    CPA   - Closest Point of Approach
           - Component Processing Area
           - Contractor Processing Area
           - Cost, Planning, and Appraisal

CPAWS	- Computer Planner and Aircraft Weighing Scales
CPC	- Card Programmed Calculator
	- Cartesian-to-Polar Converter
	- Clock Pulsed Control
	- Computer Process Control
	- Cycle Program Control
	- Cycle Program Counter
CPD	- Calls per Day
	- Cards per Day
	- Central Procurement Division
	- Combat Potential Display
	- Command Processor Distributor
	- Consolidated Programming Document
CPDC	- Command Processor Distributor Control
	- Computer Program Development Center
CPDD	- Command Post Digital Display
CPDS	- Command Processor Distributor Storage
CPE	- Central Processing Element
	- Central Programmer and Evaluator
	- Circular Probable Error
CPFF	- Cost Plus Fixed Fee
CPG	- Command Processing Group
	- Control Programming Group
CPI	- Characters per Inch
CPIF	- Cost Plus Incentive Fee
CPIP	- Computer Program Implementation Process
CPL	- Computer Program Library
CPM	- Cards per Minute
	- Command Processor Module
	- Counts per Minute
	- Critical Path Method
	- Cycles per Minute
CPO	- Central Planning Office
	- Combat Operations Center Processor
	- Command Pulse Output
	- Component Pilot Overhaul
CPP	- Chemical Processing Plant
	- Complete Purchased Part
CPPS	- Critical Path Planning and Scheduling
CPR	- Combat Operations Center Prelude
	- Contract Procurement Request
CPS	- Cathode Potential Stabilized
	- Center Poises
	- Central Processing System
	- Critical Path Schedule
	- Cycles per Second

■ CPSG

CPSG	– Common Power Supply Group
CPSU	– Communist Party of the Soviet Union
CPT	– Charge Parity Time
	– Crew Procedures Trainer
	– Critical Path Technique
CPTA	– Computer Programming and Testing Activity
CPT/CTL	– Crew Procedures Trainer/Combat Training Launch
CPU	– Central Processing Unit
CPX	– Command Post Exchange
	– Command Post Exercise
CQ	– Camera Quality
	– Commercial Quality
CR	– Card Ready
	– Cathode Ray
	– Change Release
	– Change Request
	– Cold Rolled
	– Complete Round
	– Conditional Release
	– Continuous Rod
	– Contractor's Response
	– Controlled Rectifier
	– Control Relay
	– Customer's Request
C/R	– Command Receiver
C&R	– Construction and Repair
	– Convoy and Routing
CRA	– Conditional Release Authorization
	– Controlled Rupture Accuracy
CRAF	– Civil Reserve Air Fleet
CRAFT	– Changing Radio Automatic Frequency Transmission
CRAM	– Card Random Access Memory
	– Contractual Requirements Analysis and Management
CRAS	– Coder and Random Access Switch
CRB	– Central Radio Bureau
	– Combat Reports Branch
CRC	– Cambridge Research Center
	– Collins Radio Company
	– Combat Reporting Center
	– Complete-Round Chart
	– Control and Reporting Center
	– Cost Reimbursement Contract
CRCA	– Central Records Control Area
CRCL	– Contractor Recommended Change List
CRD	– Capacitor-Resistor Diode Network
	– Chief of Research and Development

CRDF	- Cathode-Ray Direction Finder
CRE	- Combat Readiness Evaluation
CRECON	- Counter Reconnaissance
CRES	- Corrosion Resistant Steel
CREWTAF	- Crew Training Air Force
CRICKET	- Cold Rocket Instrument Carrying Kit
CRIG	- Capacitor Rate-Integrating Gyroscope
CRIME	- Censorship Records and Information, Middle East
CRIS	- Command Retrieval Information System
CRISD	- Center for Research in System Development
CRISS	- Center for Research in System Sciences
CRM	- Confusion Reflector Material
	- Counter Radar Missile
CRMR	- Continuous-Reading Meter Relay
CRO	- Cathode-Ray Oscilloscope
	- Cathode-Ray Oscillogram
CRP	- Collimation and Registration Program
	- Combat Report Post
	- Compulsary Report Point
	- Control and Reporting Post
CRPL	- Central Radio Propagation Laboratory
CRQ	- Convert by Replacement from MQ
CRRB	- Central Reference Room Bulletin
CRREL	- Cold Regions Research and Engineering Laboratory
CRS	- Calibration Requirement Summary
	- Contract Requirements Schedule
	- Corrosion Resistant Steel
	- Course
	- Critical Requirements System
CRT	- Cathode-Ray Tube
	- Charactron Tube
	- Combat Readiness Training
CRTS	- Controllable Radar Target Simulator
CRUDESPAC	- Cruiser-Destroyer Forces, Pacific
CRULANT	- Cruiser Forces, Atlantic
CRWPC	- Canadian Radio Wave Propagation Committee
CRYPTA	- Cryptoanalysis
CRYPTO	- Cryptography
CS	- Card Station
	- Center Section
	- Chief of Staff
	- Cirrostratus
	- Civil Service
	- Close Support
	- Coding Specification
	- Command System

- **CS**

CS	– Common Steel
	– Communications Systems
	– Comprehensive System
	– Controlled Stock
	– Control Signal
	– Control Switch
	– Core Shift
	– Creation Sheet
	– Current Series
	– Cycle Shift
C/S	– Call Signal
	– Configuration Summary
C&S	– Costs and Schedules
CSA	– Canadian Standards Association
	– Chief of Staff, Army
	– Computer System Analyst
	– Contractor Supply Area
	– Contractor Support Area
CSAF	– Chief of Staff, Air Force
CSAGI	– Comite Special de l'Annee Geophysique Internationale
CSAR	– Communications Satellite Advanced Research
CSAT	– Combined Systems Acceptance Test
CSB	– Concrete Splash Block
CSC	– Cadmium-Sulfide Cell
	– Civil Service Commission
	– Command and Staff College
	– Communication Simulator Console
	– Communications Systems Center
C-Scope	– Cathode-Ray Screen
CSD	– Computer Systems Director
	– Contract Support Detachment
	– Core Shift Driver
	– Crew Systems Division (formerly LSD)
CSDS	– Command Ship Data System
CSE	– Core Storage Element
CSED	– Coordinated Ship Electronics Design
CSF	– Central Switching Facility
	– Contractor Supply Facility
	– Contractor Support Facility
CSG	– Combat Service Group
	– Communications Study Group
CSGS	– Combat Center Simulation Generation System
CSIG	– Control Systems Integration Group
CSIGO	– Chief Signal Officer
CSIRO	– Commonwealth Scientific and Industrial Research Organization

CSL	- Command Systems Laboratory
	- COMPOOL Sensitive Language
	- Computer Sensitive Language
	- Control Systems Laboratory
	- Critical Situation Log
CSM	- Clock and Simulation Tape Maintenance
	- Command and Service Module
C&SM	- Command and Service Module (CSM preferred)
CSMT	- CSM (Clock and Simulation Tape Maintenance) Message Type
CSO	- Complex Safety Officer
	- Cross Servicing Order
CSP	- Coder Sequential Pulse
CSPM	- Communications Security Publications Memorandum
CSPO	- Control Systems Procurement Office
CSQ	- Coastal Sentry, Quebec
CSR	- Check Status Register
	- Control Shift Register
	- Corporate Sponsored Research
	- Critical Storage Report
CSS	- Clock Subsystem
	- Communications Subsystem
	- Contractor Storage Site
	- Crew Safety System
	- Control Stick Steering
	- Cryogenic Storage System
CSSB	- Compatible Single Sideband
CSSG	- Combat Service Support Group
CST	- Central Standard Time
	- Combined Systems Test
	- Complex Safety Technician
CS/T	- Combined Station and Tower
CSTA	- Combat Surveillance and Target Acquisition
	- Consolidated Station
CSTACTC	- Combat Surveillance and Target Acquisition Training Command
CSTI	- Control Stick Tie-In
CSTU	- Combined Systems Test Unit
CSU	- Clear and Subtract
	- Command Security Unit
CSW	- Command Surveillance and Weather
CSWY	- Causeway
CT	- Checkout Tape
	- Close Triplet
	- Combat Team
	- Command Transmitter
	- Commercial Translator

■ CT

CT	– Communications Technician
	– Communications Terminal
	– Compound Terminal
	– Constant Temperature
	– Contract Transfer
	– Control Transformer
	– Counter
	– Crosstrail
	– Current Transformer
C/T	– Command Transmitter
CTA	– Call Time Adjustor
CTAF	– Crew Training Air Force
CTB	– Commercial Traffic Bulletin
CTC	– Carbon Tetrachloride
	– Central Tracking Center
	– Central Traffic Control
	– Channel Traffic Control
	– Contract Technical Compliance
CTCA	– Channel and Traffic Control Agency
CTCC	– Continental Division Transport Control Center
CTCF	– Channel and Technical Control Facilities
CTCI	– Contract Technical Compliance Inspection
CTCU	– Channel Traffic Control Unit
CTDA	– Critical Turning Distance Add
CT&DDS	– Central Timing and Data Distribution System
CTE	– Cable Termination Equipment
	– Central Timing Equipment
	– Coefficient of Thermal Expansion
	– Commander, Task Element
	– Contractor Training Equipment
CTF	– Chlorine Trifluoride
	– Commander, Task Force
CTFE	– Chlorotrifluoroethylene
CTG	– Commander, Task Group
CTI	– California Technical Industries
CTL	– Combat Training Launch
	– Component Test Laboratory
	– Confidence Training Launch
	– Core Transistor Logic
CTLI	– Combat Training Launch Instrumentation
CTN	– Cable Termination Network
	– Canton Island (remote site)
CTO	– Commercial Transportation Officer
	– Crew Training Officer
CTOC	– Corps Tactical Operations Center
C to C	– Center to Center

CTOF	– Calculated Time Over Fix
CTP	– Command Translator and Programmer
	– Confidence Test Program
CTR	– Controlled Thermonuclear Reaction
CTS	– Central Track Sort
	– Communication and Tracking Subsystem
	– Contract Technical Service
CTSP	– Contract Technical Service Personnel
CTT	– Card-to-Tape Tape
	– Crosstelling Technician
CTU	– Centrigrade Thermal Unit
	– Central Timing Unit
	– Combat Training Unit
	– Commander, Task Unit
	– Compatibility Test Unit
CTV	– Constant Tangential Velocity
	– Control Test Vehicle
CTVW	– Continuous-Tension Viscosity Wrapping
CU	– Close-Up
	– Cubic
CU CM	– Cubic Centimeter
CUE	– Command Uplink Electronics
	– Computer Update Equipment
	– Correlating Users Exchange
CUERL	– Columbia University Electronic Research Laboratory
CU FT	– Cubic Foot
CUG	– Common Users Group
CUH	– Control User Handbooks
CU IN	– Cubic Inch
CU M	– Cubic Meter
CU MM	– Cubic Millimeter
CUP	– Casting for Ultra Purity
CUPS	– Consolidated Unit Personnel Section
CUS	– Continental United States
CU&S	– Control, Utility, and Support Group
CUSRPG	– Canada-United States Regional Planning Group
CU YD	– Cubic Yard
CV	– Credit Voucher
	– Collection Voucher
	– Combat Vehicle
	– Continuously Variable
	– Counter Voltage
	– Credit Voucher
CVA	– Chance-Vought Aircraft
CVE	– Complete Vehicle Erector
CVK	– Centerline Vertical Keel

- **CVN**

CVN	– Casualty Vulnerability Number
CVPDS	– Command Video Prelaunch Distribution System
CVR	– Change Verification Record
	– Controlled Visual Rules
	– Convert by Replacement from AC
CVSG	– Channel Verification Signal Generator
CW	– Chemical Warfare
	– Clockwise
	– Complied With
	– Continuous Wave
C/W	– Complied With
C-W	– Curtiss-Wright Corporation
C&W	– Control and Warning
CWA	– Communications Workers of America
CWAR	– Continuous Wave Acquisition Radar
CWBW	– Chemical Warfare—Biological Warfare
CWE	– Current Working Estimate
CW/FM	– Continuous Wave Frequency Modulated
CWG	– Communications Working Group
	– Constant Wear Garment
CWIF	– Continuous Wave Intermediate Frequency
CWMTU	– Cold Weather Materiel Test Unit
CWO	– Communication Watch Officer
CWOP	– Cold Weather Operations
CX	– Control Transmitter
CX-HLS	– Cargo/Transport Experimental—Heavy Logistics System
CXI	– Crosstell Input
CXO	– Crosstell Out to Automatic DC
CY	– Calendar Year
CYBORGS	– Cybernetic Organisms
CYI	– Canary Islands (remote site)
CZ	– Canal Zone
	– Combat Zone

D

D	- Datum
	- Deci
	- Density
	- Depth
	- Diameter
	- Digit
	- Digital
	- Dimensional
	- Diode
	- Distance
	- Double
	- Drag
	- Drone
	- Dyne
DA	- Data Available
	- Deacon-Arrow
	- Decimal Add
	- Decimal to Analog
	- Density Altitude
	- Department of the Army
	- Detonation Altitude
	- Detroit Arsenal
	- Dip Angle
	- Direct Action
	- Director of Aircraft
	- Double Amplitude
D/A	- Digital to Analog
D-A	- Deacon-Arrow
	- Digital to Analog
D&A	- Detail and Assembly
DAAG	- Damage Assessment Advisory Group
DAATCO	- Department of the Army Air Traffic Coordinating Officer
DAB	- Deacon-Arrow Ballistic

■ DAC

DAC	- Damage Assessment Center
	- Data Analysis Center
	- Design Advisory Group
	- Disassemble COMPOOL
	- Display Analysis Console
	- Douglas Aircraft Company
	- Digital-to-Analog Converter
DACAPS	- Data Collection and Processing System
DACAS	- Damage Assessment and Casualty Report
DACCC	- Defense Area Communications Control Center
DACCC-AL	- Defense Area Communications Control Center, Alaska
DACCC-CON	- Defense Area Communications Control Center, CONUS
DACCC-EUR	- Defense Area Communications Control Center, Europe
DACCC-PAC	- Defense Area Communications Control Center, Pacific
DACL	- Dynamic Analysis and Control Laboratory
DACO	- Douglas Aircraft Company
DACOM	- Datascope Computer Output Microfilmer
DACON	- Data Controller
	- Digital-to-Analog Converter
DACOS	- Deputy Assistant Chief of Staff
DACPO	- Data Count Printout
DAD	- Damage Assessment Department
	- Data Dictionary Assembler
DADEE	- Dynamic Analog Differential Equation Equalizer
DADIC	- Data Dictionary
DADIT	- Daystrom Analog-to-Digital Integrating Translator
DADS	- Detroit Air Defense Sector
DAE	- Data Acquisition Equipment
DAEMON	- Data Adaptive Evaluator and Monitor
DAF	- Data File Program
	- Department of the Air Force
DAFSC	- Duty Air Force Speciality Code
DAFT	- Digital/Analog Function Table
DAG	- Deutsche Astronautische Gasellschaft
DAGC	- Delayed Automatic Gain Control
DAGMAR	- Drift and Ground Speed Measuring Airborne Radar
DAI	- Drift Angle Indicator
DAISY	- Data Acquisition and Interpretation System
	- Data Analysis of the Interpreter System
DAISY 201	- Double-Precision Automatic Interpretive System
DAL	- Design Approval Layout
DALT	- Drop Altitude
DAM	- Data Association Message
	- Descriptor Attribute Matrix
DAMIT	- Data Analysis Computer Program by Massachusetts Institute of Technology

DAMP	- Department of the Army Materiel Program
	- Downrange Anti-Missile Measurement Program
DAMS	- Defensive Anti-Missile System
DAMWO	- Department of the Army Modification Work Order
DAN	- Deacon and Nike
DAO	- Division Ammunition Officer
DA&P	- Data Analysis and Processing
DAPE	- Developed Armament Probable Error
DAR	- Defense Acquisition Radar
DARA	- Deutsche Arbeitgemeinschaft Fur Rechen-Anlagen
DARAC	- Damped Aerodynamic Righting Attitude Control
DARB	- Distressed Airman Recovery Beacon
DARE	- Damage Assessment Reduction and Evaluation
	- Data Automatic Reduction Equipment
	- Data Automation Research and Experimentation
	- DOVAP Automatic Reduction Equipment
DART	- Data Analysis Reduction Tape
	- Development Advanced Rate Technique
	- Diode Automatic Reliability Tester
	- Director and Response Tester
DARTS	- Digital Azimuth Range Tracking System
DAS	- Data Acquisition System
	- Data Automation System
	- Datatron Assembly System
	- Decision Assist System
	- Digital Attenuator System
	- Director of Administration
DASA	- Defense Atomic Support Agency
DASADD	- Defense Atomic Support Agency Data Division
DASAT	- Data Selector and Tagger
DASC	- Defense Automotive Supply Center
	- Direct Air Support Center
DAS/COTAR	- Data Acquisition System—Correlation Tracking and Ranging
DASH	- Drone Anti-Submarine Helicopter
DASL	- Department of Army Strategic Logistics
DASO	- Department of Army Special Order
DASP	- Directorate of Advanced Systems Planning
DAST	- Division for Advanced Systems Technology
DASTARD	- Destroyer Anti-Submarine Transportable Array Detector
DAT	- Drone-Assisted Torpedo
DATA	- Defense Air Transportation Administration
DATAC	- Defense and Tactical Armament Control
	- Digital Automatic Test and Classifier
DATACOM	- Data Communications
DATAR	- Digital Automatic Tracking and Ranging

■ DATA-STOR

DATA-STOR	- Data Storage
DATDC	- Data Analysis and Technique Development Center
DATICO	- Digital Automatic Tape Intelligence Checkout
DATMOBAS	- David Taylor Model Basin
DATOR	- Digital Data, Auxiliary Storage, Track Display, Outputs, and Radar Display
DATS	- Dynamic Accuracy Test System
DAU	- Data Acquisition Unit
DAVC	- Delayed Automatic Volume Control
DAVI	- Dynamic Anti-Resonant Vibration Isolator
DAVID	- Defense of Airborne Vehicles in Depth
DAW	- Directorate of Atomic Warfare
DAWG	- Dynamic Air War Game
DAZD	- Double Anode Zener Diode
DB	- Dead Band
	- Decibel
	- Departmental Bulletin
	- Depth Bomb
	- Double Biased
	- Double Braid
D/B	- Decimal to Binary
D-B	- Decimal to Binary
DBA	- Department Budget Analyst
	- Doing Business As
DBB	- Detector Balance Bias
DBC	- Digital-to-Binary Converter
DBHP	- Drawbar Horsepower
DBI	- Differential Bearing Indicator
DBIL	- Data Base Input Language
DBL	- Data Base Load
DBM	- Decibel referred to one Milliwatt
DBP	- Drawbar Pull
DBRN	- Decibel above Reference Noise
DBUT	- Data Base Update Time
DBV	- Decibel referred to one Volt
DBW	- Decibel referred to one Watt
	- Design Bandwidth
	- Differential Ballistic Wind
DC	- Damage Control
	- Data Central
	- Decade Counter
	- Decimal (Dewey) Classification
	- Deployment Chutes
	- Depth Charge
	- Development Center
	- Digital Computer

DC	- Direct Current
	- Direct Cycle
	- Directional Coupler
	- Direction Center
	- Discrepancy Check
	- Display Clerk
	- Display Console
	- District of Columbia
	- Double Contact
	- Dow Corning Corporation
	- The Dow Chemical Company
D/C	- Drift Correction
D&C	- Displays and Controls
DCA	- Defense Communications Agency
	- Deputy Commander, Army
	- Design Change Authorization
	- Digital Computers Association
	- Direction Center Active
	- Discrepancy Control Area
	- Document Change Analysis
	- Document Change Authorization
	- Doppler Count Accumulator
	- Drift Correction Angle
DCAC	- Design Change Approval Committee
DCAI	- Defense Communications Agency Instruction
DCAS	- Data Collection and Analysis System
	- Deputy Commander for Aerospace Systems
DCC	- Data Communications Channel
	- Data Condition Code
	- Design Change Coordination
	- Design Consistency Committee
	- Development Control Center
	- Direct Current Clamp
	- Discrimination and Control Computer
	- Double-Cotton-Covered Wire
DCCC	- Defense Communications Control Center
	- Design Change Coordination Committee
DCCS	- Defense Communication Control System
DCCU	- Decommutator Conditioning Unit
DCD	- Defense Communications Department
	- Design Coordinations Document
	- Double-Channel Duplex
DCDS	- Dual Command Destruct System
DCE	- Directorate of Communications-Electronics
DCF	- Direction Center Function
DCFEM	- Dynamic Crossed-Field Electron Multiplication

■ DCG

DCG	– Data Coordination Group
	– Diode Capacitor Gate
DCIA	– Direction Center Initial Appearance
DCIB	– Data Communication Input Buffer
DCIP	– Data Correction Indicator Panel
DCIS	– Downrange Computer Input System
DCL	– Digital Computer Laboratory
	– Dual Cycle Left
DCLCS	– Data Conversion and Limit Check Submodule
DCM	– Directorate for Classification Management
DCN	– Design Change Notice
	– Drawing Change Notice
DCNO	– Deputy Chief of Naval Operations
DCNO/A	– Deputy Chief of Naval Operations, Air
DCNO/D	– Deputy Chief of Naval Operations, Development
DCNO/FO&R	– Deputy Chief of Naval Operations, Fleet Operations and Readiness
DCNO/L	– Deputy Chief of Naval Operations, Logistics
DCNO/P&P	– Deputy Chief of Naval Operations, Plans and Policy
DCO	– Data Control Office
	– Detailed Checkout
	– Development Contract Officer
	– District Communications Officer
	– Drawing Change Order
	– Dynamic Checkout
DCOP	– Detailed Checkout Procedure
DCOS	– Data Communication Output Selector
	– Downrange Computer Output System
DCP	– Design Change Proposal
	– Development Control Program
	– Digital Computer Programming
	– Display Control Panel
DCPG	– Direction Center Programming Group
DCPR	– Direction Center Processor for Remote Combat Center
DCR	– Data Conversion Receiver
	– Design Change Request
	– Detail Condition Register
	– Development Council for Research
	– Digital Conversion Receiver
	– Direct Current Restorer
	– Disposition of Contract Request
	– Document Change Request
	– Drawing Change Request
DCRM	– Discrepancy Check Request Memorandum
DCS	– Data Conditioning System
	– Data Control Systems

DCS	- Data Conversion System
	- Defense Communications Service
	- Defense Communications System
	- Deputy Chief of Staff
	- Design Control Specification
	- Destruct Command System
	- Digital Command System
	- Direction Center Standby
	- Double-Channel Simplex
DCSC	- Defense Construction Supply Center
DCS/C	- Deputy Chief of Staff, Comptroller (AFSC)
DCS/FF	- Deputy Chief of Staff, Flight Facilities (AFSC)
DCS/L	- Deputy Chief of Staff, Logistics (AFSC)
DCSLOG	- Deputy Chief of Staff, Logistics (Army)
DCS/O	- Deputy Chief of Staff, Objectives (AFSC)
	- Deputy Chief of Staff, Operations (Air Force; now DCS/P&R)
DCSOPS	- Deputy Chief of Staff, Military Operations (Army)
DCS/P	- Deputy Chief of Staff, Personnel (AFSC and Air Force)
DCSPER	- Deputy Chief of Staff, Personnel (Army)
DCS/P&P	- Deputy Chief of Staff, Plans and Programs (Air Force)
DCS/P&R	- Deputy Chief of Staff, Programs and Requirements (Air Force; formerly DCS/O)
DCS/R&D	- Deputy Chief of Staff, Research and Development (Air Force, formerly DCS/R&T)
DCS/R&T	- Deputy Chief of Staff, Research and Technology (Air Force; formerly DCS/R&D)
DCS/S	- Deputy Chief of Staff, Systems (Air Force)
DCS/S&L	- Deputy Chief of Staff, Systems and Logistics (Air Force)
DCS/T	- Deputy Chief of Staff, Telecommunications (AFSC)
DCT	- Data Conversion Transmitter
	- Divide Check Test
DCTL	- Direct Coupled Transistor Logic
DCU	- Data Control Unit
	- Decade Counting Unit
	- Decimal Counting Unit
	- Digital Counting Unit
	- Display and Control Unit
DCV	- Design Change Verification
	- Direct Current, Volts
DCW	- Digital Display Control and Warning Light
DCX	- Direct Current Experiment
DD	- Data Demand
	- Decimal Divide
	- Deep Drawn
	- Delay Driver
	- Department of Defense (DOD preferred)

■ DD

DD	– Destroyer
	– Digital Data
	– Digital Display
	– Document Distribution
	– Double Drift
	– Due Date
DDA	– Digital Differential Analyzer
	– Digital Display Alarm
	– Drawing Departure Authorization
DDAS	– Digital Data Acquisition System
DDC	– Data Display Central
	– Defense Documentation Center (formerly ASTIA)
	– Development and Display Center
	– Digital Data Converter
	– Direct Data Channel
	– Display Data Controller
	– Dual Diversity Comparator
DDCE	– Digital Data Conversion Equipment
DDD	– Deadline Delivery Date
	– Diesel Direct Drive
	– Digital Display Detection
	– Direct Distance Dialing
DDE	– Destroyer, Escort
DDG	– Digital Data Group
	– Digital Display Generator
DDGE	– Digital Display Generator Element
DDI	– Depth Deviation Indicator
DDIS	– Document Data Indexing Set
DDL	– Digital Data Line
DDM	– Derived Delta Modulation
	– Difference in Depth and Modulation
	– Digital Display Makeup
DDN	– Documentation Development Notification
DDOCE	– Digital Data Output Conversion Element
DDP	– Data Distribution Panel
	– Debriefing Display Program
	– Deputy Director, Programs
	– Digital Data Processor
DDPE	– Digital Data Processing Equipment
DDR	– Digital Data Receiver
	– Drawing Data Requirement
DDRC	– Document Drawing Revision Change
	– Drawing Data Required for Change
DDRE	– Director of Defense Research and Engineering
DDR&E	– Director of Defense Research and Engineering (DDRE preferred)

DEER

DDRR	– Directional Discontinuity Ring Radiator
DDS	– Data Display System
	– Data Distribution System
	– Deputy Director, Systems
	– Digital Data Servo
	– Digital Data System
	– Digital Display Scope
	– Doppler Detection System
DDT	– Design Development Test
	– Digital Data Transmitter
	– Doppler Data Translator
DDTE	– Digital Data Terminal Equipment
DDTESS	– Digital Data Terminal Equipment Service Submodule
DDTS	– Direct Dialing Telephone System
DDU	– Digital Distributing Unit
	– Disc Data Unit
DE	– Damage Evaluation
	– Date of Entry
	– Date of Extension
	– Deflection Error
	– Diesel-Electric
	– Digital Encoder
	– Display Electronics
	– Display Equipment
	– Division Entry
	– Double End
DEA	– Display Electronic Assembly
DEADS	– Detroit Air Defense Sector
DEAL	– Detachment Equipment Authorization List
DEAS	– Directorate of Engineering, Aeronautical Systems
DEBM	– Directorate of Engineering, Ballistic Missiles
DEBS	– Display Exercise for Battle Staff
DEC	– Deceleration
	– Decrease Decimal
DECA	– Display Electronic Control Assembly
DECAP	– Decentralized Air Defense Programming
DECM	– Defense Electronic Countermeasure
DECON	– Decontaminate
DECOR	– Digital Electronic Continuous Ranging
DECR	– Document Error/Clarification Request
DECUF	– Defense Capability Under Fallout
DECUS	– Digital Equipment Computer Users Society
DED	– Defense Electronics Division
DEDD	– Diesel Electric Direct Drive
DEE	– Digital Evaluation Equipment
DEER	– Directional Explosive Echo Ranging

■ DEES

DEES	- Directorate of Engineering, Electronic Systems
DEFA	- Direction des Etudes et Fabrications d'Armement
DEFCON	- Defense Readiness Condition
DEFREPNAMA/USRO	- Defense Representative North Atlantic and Mediterranean Areas/United States Regional Office
DEFT	- Dynamic Error-Free Transmission
DEG	- Degree
DEG&DEP	- Degaussing and Deperming
DEGN	- Diethylene Glycol Dinitrate
DEI	- Design Engineering Inspection
	- Development Engineering Inspection
DEIMOS	- Development and Investigation of a Military Orbital System
DELTA	- Detailed Labor and Time Analysis
DELTIC	- Delay Line Time Compression
DEMATRON	- Distributed Emission Magnetron Amplifier
DEMO	- Demolition
DEMOB	- Demobilization
DEN	- Dow Epoxy-Novalac
DEP	- Deflection Error Probable
	- Departure Point
	- Deposit
DEPI	- Differential Equation Pseudo-Code Interpreter
DEPSECDEF	- Deputy Secretary of Defense
DEPSUM	- Deployment Summary
DEPT	- Department
DER	- Destroyer Escort Radar
	- Document Error Report
DERAX	- Detection and Ranging
DER&TS	- Directorate of Engineering, Ranges and Test Support
DESC	- Defense Electronics Supply Center
DESLANT	- Destroyer Forces, Atlantic
DESOIL	- Diesel Oil
DESS	- Dual Environment Safety Switch
DET	- Damage Evaluation Team
DETA	- Diethylene Triamine
DEUCE	- Digital Electronic Universal Calculating Engine
DEV	- Deviation
DEVR	- Distortion-Eliminating Voltage Regulator
DEW	- Distant Early Warning
DEWIZ	- Distant Early Warning Identification Zone
DEXAN	- Digital Experimental Airborne Navigator
DF	- Decimal Fraction
	- Defensive Fire
	- Defogging
	- Degree of Freedom
	- Direct Flight

DF	- Direction Finder
	- Direction Finding
	- Disposition Form
	- Double Feeder
	- Drawing File
	- Drop Forge
	- Dual Facility
D/F	- Direction Finder
	- Direction Finding
DFA	- Detonation Fragmentation and Air Blast
	- Division Final Appearance
DFAWS	- Direct Fire Anti-Tank Weapon System
DFD	- Digital Flight Display
DFF	- Division Final Fade
DFING	- Direction Finding
DFOD	- Defense Field Operations Department
DFP	- Deviant Flight Plan
DFS	- Dynamic Flight Simulator
DFSR	- Directorate of Flight Safety Research
DG	- Degaussing
	- Directional Gyroscope
	- Disc Grind
	- Double Glass
	- Double Groove
DGBC	- Digital Geoballistic Computer
D-GDGE	- Distributor to Group Display Generator Electronics
DGDP	- Double Groove, Double Petticoat
DGP	- Data Generation Program
	- Directional Gyroscope Position
DGR	- Directorate of Geophysics Research
DGS	- Data Ground Station
DGSC	- Defense General Supply Center
DGZ	- Desired Ground Zero
DGZPRO	- Desired Ground Zero Tape Prepare Program
DH	- Desired Heading
DHE	- Data Handling Equipment
DI	- Dark Ignition
	- De-Icing
	- Demand Indicator
	- District Inspector
	- Director of Intelligence
	- Due In
DIA	- Date of Initial Appointment
	- Defense Intelligence Agency
DIAC	- Defense Industry Advisory Council (formerly Mutual Assistance Program)

■ DIALGOL

DIALGOL	- Dialect of Algorithmic Language
DIAN	- Decca Integrated Airborne Navigator
DIANE	- Digital Integrated Attack and Navigation Equipment
DIAPER	- Division Adaptation Personnel
DIBA	- Digital Internal Ballistic Analyzer
DICBM	- Defense Intercontinental Ballistic Missile
DICE	- Development Interim Control Equipment
	- Dynamic Inputs to CC (Control Center) Equipment
DID	- Daily Intelligence Digest
	- Digital Information Detection
	- Digital Information Display
	- Direct Inward Dialing
DIDAP	- Digital Data Processor
DIE	- Department of Industrial Engineering
	- Document of Industrial Engineering
DIGEST	- Diebold Generator for Statistical Tabulation
DIGIPLOT	- Digital Plotting System
DIGIRALT	- Digital Radar Altimeter
DIGS	- Deputy Inspector General of Safety
DIM	- Depot Industrial Maintenance
	- Design Information Manual
	- Difference Magnitude
DIMUS	- Digital Multi-Beam Steering
DINA	- Diethylene Glycol Nitromethane
	- Direct Noise Amplification
DINFIA	- Direccion Nacional de Fabricaciones e Investigaciones Aeronauticas (Argentina)
DINFOS	- Defense Information School
DIO	- District Intelligence Officer
DIP	- Display Information Processor
DIPEC	- Defense Industrial Plant Equipment Center
DIPS	- Dietary Information Processing System
	- Dual Impact Prediction System
DIR	- Depot Inspection and Repair
DIRAFIED	- Director Armed Forces Information and Education
DIRAM	- Digital Range Machine
DIRCOL	- Direction Cosine Linkage
DIRNRL	- Director Naval Research Laboratory
DIRNSA	- Director National Security Agency
DIRS	- Division Integrated Record System
DIS	- Data Input Supervisor
	- Defense Intelligence School
DISC	- Defense Industrial Supply Center
DISCOM	- Digital Selective Communications
DISS	- Data Input Subsystem
DISTRAM	- Digital Space Trajectory Measurement System

DJ	- Drill Jig
DL	- Data Link
	- Deadline
	- Dead Load
	- Deciliter
	- Delay Line
	- Difference of Latitude
	- Drawing List
D/L	- Data Link
DLA	- Data Link Address
DLCC	- Division Logistics Control Center
DLI	- Defense Language Institute
DLM	- Destination Load Model
DLO	- Delayed Output
	- Difference in Longitude
	- Double Local Oscillator
DLOC	- Division Logistical Operation Center
DLS	- Data Link Set
	- Data Link Simulator
DLSC	- Defense Logistics Service Center
DLSM	- Deputy Logistic Support Manager
DLT	- Data Link Terminal
	- Data Link Translator
DLTM	- Data Link Test Message
DLTO	- Dog-Leg-To-Orbit
DM	- Decimal Multiply
	- Decimeter
	- Demand Meter
	- Design Manual
	- Destroyer, Minelayer
	- Diesel-Mechanical
	- Doppler Missile
D/M	- Demodulate-Modulate
	- Demodulator-Modulator
D&M	- Deployment and Mobility
DMC	- Dead Man Controls
	- Digital Micro-Circuit
DME	- Depot Maintenance Equipment
	- Design Margin Evaluation
	- Direct Mission Equipment
	- Distance Measuring Equipment
DME/COTAR	- Distance Measuring Equipment—Correlation Tracking and Ranging
DMET	- Distance Measuring Equipment, Terminal
DMF	- Depot Maintenance Facility
DMF-R	- Depot Maintenance Facility-Recycle

■ DMG

DMG	– Defense Marketing Group
DMIC	– Defense Metals Information Center
DMJM	– Daniels, Mann, Johnson, and Mendenhall
DMNI	– Device Multiplexing Non-Synchronized Inputs
DMOS	– Duty Military Occupational Speciality
DMP	– Display Makeup
	– Display Masking Parameters
DMPI	– Desired Mean Point of Impact
DMR	– Defective Material Report
	– Design Modification Request
DMS	– Data Multiplexing System
	– Defense Materials System
	– Deviation from Mean Standard
	– Digital Multiplexing Synchronizer
DMSC	– Defense Medical Supply Center
DMSS	– Data Multiplex Subsystem
DMT	– Daily Metabolic Turnover
DMU	– Data Measurement Unit
	– Dynamic Mockup
DMX	– Data Multiplex
DMZ	– Demilitarized Zone
DN	– Department of the Navy
DNA	– Desoxyribonucleic Acid
	– Does not Apply
DNB	– Departure-from-Nucleate Boiling
DNC	– Director of Naval Communications
DNCCC	– Defense National Communications Control Center
DNCCCS	– Defense National Communications Control Center System
DNI	– Director of Naval Intelligence
DNIF	– Duty Not Involving Flying
DNP	– Dummy Nose Plug
DNR	– Downrange (also D/R)
DNS	– Doppler Navigation System
DO	– Defense Order
	– Deputy for Operations
	– Director of Operations
	– Dissolved Oxygen
	– Dispersing Officer
	– Due Out
DOA	– Dead on Arrival
	– Department of the Army
	– Duty Orbital Analyst
DOC	– Data and Operations Center
	– Date of Change
	– Decimal-to-Octal Conversion
	– Demonstration of Operational Capability

DP

DOC	– Department of Commerce
	– Direct Operating Cost
DOD	– Date on Dock
	– Department of Defense
	– Direct Outward Dialing
DODCLPMI	– Department of Defense Consolidated List of Principal Military Items
DODD	– Department of Defense Directive
DODDAC	– Department of Defense Damage Assessment Center
DODIC	– Department of Defense Identification Code
DODRE	– Department of Defense Research and Engineering
DOES	– Demonstration Operations and Evaluation Support
DOETS	– Dual-Object Electronic Tracking System
DOF	– Degree of Freedom
	– Demonstration of Operational Feasibility
	– Direction of Flight
DOFL	– Diamond Ordnance Fuze Laboratory
DOI	– Division Office Instruction
DOL	– Dynamic Octal Load
DOLARS	– Digital Off-Line Automatic Recording System
	– Doppler Location and Ranging System
DON	– Department of the Navy
DOO	– Director, Office of Oceanography
DOP	– Division Operating Procedure
DOPLOC	– Doppler Phase Lock
DOPP	– Doppler
DORAN	– Doppler Range and Navigation
DORIS	– Direct Order Recording and Invoicing System
DORRI	– UDOP Correction Routine
DOS	– Department of State
DOT	– Department of the Treasury
	– Dictionary of Occupational Titles
	– Director of Operations and Training
	– Double Offset Tactic
	– Duplex One-Tape
DOTF	– Director of Operations and Training Fighters
DOTW	– Director of Operations and Training Weapons
DOVAP	– Doppler Velocity and Position
DP	– By Direction of the President
	– Dashpot
	– Data Processing
	– Deck Penetrating
	– Deep Penetration
	– Deflection Plate
	– Design Proof
	– Dew Point

■ DP

DP	– Diametrical Pitch
	– Distribution Point
	– Double Pole
	– Double Propellant
	– Double Purpose
	– Dual Purpose
DPA	– Data Processing Activities
	– Design Proposal Approval
D-PAT	– Drum-Programmed Automatic Tester
DPB	– Data Processing Branch
	– Destruct Package Building
DPBC	– Double-Pole Both Connected
DPC	– Data Processing Center
DPCA	– Displaced Phase Center Antenna
DPD	– Data Processing Department
	– Digital Phase Difference
	– Digit Plane Driver
DPDT	– Double-Pole Double-Throw
DPE	– Data Processing Equipment
DPF	– Data Processing Facility
DPFC	– Double-Pole Front Connected
DPI	– Data Processing Installation
DPIF	– Destruct Package Installation Facility
DPL	– Division Programming Leader
DPM	– Development Program Manager
DPMA	– Data Processing Management Association
DPMOAP	– Society of Data Processing Machine Operators and Programmers
DPO	– Development Part Order
	– Development Planning Objective
	– Double-Pulse Operation
DPP	– Dampproofing
	– Detailed Pass Plan
DPPO	– District Publications and Printing Office
DPS	– Data Processing System
	– Delayed Printer Simulator
	– Drogue Parachute System
DPSC	– Defense Petroleum Supply Center
DPSS	– Data Processing Subsystem
	– Display Presentation Subsystem
DPST	– Double-Pole Single-Throw
DPT	– Departure Control
	– Design Proof Test
DPTS	– Digital Programming Test Set
DPTT	– Double-Pole Triple-Throw

DPU	- Data Processing Unit
	- Document Processing Unit
DPV	- Dry Pipe Valve
DPWO	- District Public Works Officer
DQCM	- Data Quality Control Monitor
DQM	- Data Quality Monitor
DR	- Data Receiver
	- Dead Reckoning
	- Demolition Rocket
	- Detection Radar
	- Discrepancy Report
	- Discrimination Radar
	- Dress Rehearsal
	- Drill Rod
D/R	- Data Receiver
	- Directional Radio
	- Downrange
DRA	- Dead Reckoning Analyzer
	- Drum Read Amplifier
DR&A	- Data Reduction and Analysis
DRADS	- Degradation of Radar Defense Systems
DRAI	- Dead Reckoning Analyzer Indicator
DRANS	- Data Reduction and Analysis Subsystem
DRAT	- Data Reduction Analysis Tape
DRB	- Defence Research Board (Canadian)
	- Design Review Board
DRC	- Defense Research Corporation
DRCCC	- Defense Regional Communications Control Center
DR&CWG	- Data Reduction and Computing Working Group
DRD	- Directorate of Research and Development
	- Division of Reactor Development
	- Drum Read Driver
DRE	- Data Reduction Equipment
DRED	- Data Router and Error Detector
DRET	- Direct Re-Entry Telemetry
DRETS	- Direct Re-Entry Telemetry System
DREWS	- Direct Readout Equatorial Weather Satellite
DRI	- Denver Research Institute
DRIFT	- Diversity Receiving Instrumentation for Telemetry
DRIP	- Data Reduction for Interception Program
	- Data Reduction Input Program
DRL/UT	- Defense Research Laboratory, University of Texas
DRM	- Drafting Room Manual
DRN	- Daily Report Notice
	- Data Reference Number

■ DRO

DRO	- Defense Research Office
	- Destructive Readout
DROMDI	- Direct Readout Miss-Distance Indicator
DRON	- Data Reduction
DROS	- Direct Readout Satellite
DRP	- Data Reduction Program
	- Dead Reckoning Plotter
	- Digital Recording Process
	- Discoverer Research Program
	- Division of Reactor Development
DRR	- Development Revision Record
	- Digital Radar Relay
DRRL	- Digital Radar Relay Link
DRS	- Data Recording System
	- Discrepancy Report Squawk
	- Downrange Ship
DRSA	- Data Recording System Analyst
DRSC	- Direct Radar Scope Camera
DRT	- Dead Reckoning Tracer
	- Drawing Release Ticket
DRTE	- Defense Research Telecommunications Establishment
DRTR	- Dead Reckoning Trainer
DRV	- Developmental Re-Entry Vehicle
DRWG	- Data Reduction Working Group
	- Drawing
DS	- Decanning Scuttle
	- Decimal Subtract
	- Decoder Simulator
	- Departed Station
	- Depth Sounding
	- Detached Service
	- Developmental Silo
	- Diode Switch
	- Direct Support
	- Discarding Sabot
	- DYNA-SOAR
D/S	- Data Set
	- Direct Ship
	- Dispatch Section
D&S	- Display and Storage
DSA	- Defense Supply Agency
DSAS	- Direct Support Aviation Section
DSB	- Double Sideband
DSB/SC	- Double Sideband, Suppressed-Carrier
DSC	- Data Sub-Central
	- Data Synchronizer Channel

DSC	- Discone Antenna
	- Document Service Center
DSCD	- Design Specification Control Drawing
DSD	- Defense Systems Department
DSE	- Data Storage Equipment
	- Development Support Equipment
	- Directorate of System Engineering
DSF	- Data Scanning and Formatting
	- Design Safety Factor
DSI	- Directorate of Scientific Intelligence
DSIF	- Deep Space Instrumentation Facility
DSIR	- Department of Scientific and Industrial Research
DSL	- Deep Scattering Layer
	- Dual Shift Left
DSM	- Defense Suppression Missile
	- Directorate of System Management
DSMG	- Designated Systems Management Group
DSN	- Deep Space Network
DSOTS	- Demonstration Site Operational Test Series
DSP	- Disassemble Sequence Parameters
	- DODDAC System Program
	- Dynamic Sequence Parameters
DSPE	- DODDAC System Program Evaluation
DSPL	- Douglas Space Physics Laboratory
DSPR	- Defense Supply Procurement Regulation
DSR	- Data Scanning and Routing
	- Data Specification Request
	- Document Status Report
	- Dual Shift Right
DSRG	- Data System Review Group
DS/RPIE	- Direct Support—Real Property Installed Equipment
DSS	- Deep Space Station
	- Digital Subsystem
DSSC	- Defense Subsistence Supply Center
DSSO	- Defense Surplus Sales Office
DSSTP	- Development Site System Training Program
DST	- Daylight Savings Time
DSTL	- Division System Training Leader
DSU	- Data Synchronizer Unit
	- Direct Support Unit
DSV	- Douglas Space Vehicle
DT	- Dead Time
	- Dive Time
	- Double Throw
	- Down Time
D/T	- Deputy for Technology

■ D&T

D&T	- Development and Test
DTA	- Differential Thermal Analysis
	- Division Technical Assistant
DTC	- Division Traffic Center
	- Dynamic Test Chamber
DTCS	- Digital Test Command System
DTE	- Data Terminal Equipment
	- Data Transmitting Equipment
	- Depot Tooling Equipment
	- Display Tester Element
DTG	- Date-Time Group
DTIE	- Division of Technical Information Extension
DTL	- Diode-Transistor Logic
DTMB	- David Taylor Model Basin
DTMS	- Defense Traffic Management Service
	- Digital Test Monitor System
DTO	- Detailed Test Objective
	- Director of Tactical Operations
	- Document Test Objectives
DTOC	- Division Tactical Operations Center
DTOS	- Detailed Test Objectives
DTP	- Detailed Test Plan
	- Drum Timing Pulse
DTPB	- Divider Time Pulse Distributor
DTR	- Data Telemetry Register
	- Definite Time Relay
	- Demand Totalizing Relay
	- Digital Telemetering Register
DTRE	- Defense Telecommunications Research Establishment
DTRM	- Dual-Thrust Rocket Motor
DT/RSS	- Data Transmission/Recording Subsystem
DTS	- Data Transmission System
	- Date Time Stamp
	- Dix Tracking Station
	- Double-Throw Switch
D/T&S	- Deputy for Test and Support
DTT	- Design Tool Test
DTU	- Digital Telemetry Unit
DTV	- Drop Test Vehicle
DTVC	- Digital Transmission and Verification Converter
DTVM	- Differential Thermocouple Voltmeter
DUADS	- Duluth Air Defense Sector
DUALGRAN	- Dual Granulation
DUNC	- Deep Underwater Nuclear Counting Device
DUO	- Datatron Users Organization
DUSA	- Deputy Undersecretary of the Army

DUSAA	- Davidson United States Army Airfield
DUSC	- Deep Underwater Support Center
DV	- Debit Voucher
DVD	- Delta Velocity Display
	- Design Verification Demonstration
	- Differential Velocity Display
DVFR	- Defense Visual Flight Rules
DVH	- Divide or Halt
DVM	- Digital Voltmeter
DVO	- Delta Velocity On/Off
DVP	- Divide or Proceed
DVT	- Design Verification Test
DVTP	- Divide Time Pulse
DVU	- Delta Velocity Ullage
DW	- Double Weight
	- Drum Writer
D/W	- Direct Writing
	- Double Weight
DWD	- Drum Write Driver
DWF	- Duty Weather Forecaster
DWG	- Drawing
DWI	- Descriptor Work Index
DWT	- Dead Weight
DX	- Distant Reception
DXTZ	- Display Crosstell Zones
DYANA	- Dynamics Analyzer-Programmer
DYNAMO	- Dynamic Action Management Operation
DYNA-MOWS	- Dynamic Manned Oribtal Weapon System
DYNA-SOAR	- Dynamic-Soaring
DYNM	- Dynamotor
DYNO	- Dynamometer
DYSTAC	- Dynamic Storage Analog Computer
DZ	- Drop Zone
DZA	- Drop Zone Area

E

E	- East
	- Enamel
	- Energy
EA	- Easterline Angus
	- Enemy Aircraft
E/A	- Enemy Area
EAA	- Experimental Aircraft Association
EAC	- Eastern Air Command
	- Educational Advisory Committee
	- Expected Approach Clearance
EAD	- Effective Air Distance
	- Extended Active Duty
EADC	- Eastern Air Defense Command
EADCC	- Eastern Air Defense Control Center
EADF	- Eastern Air Defense Force
EAES	- European Atomic Energy Society
EAF	- Earth's Armed Forces
EAFB	- Edwards Air Force Base
	- Eglin Air Force Base
	- Ellington Air Force Base
	- Ellsworth Air Force Base
EAGER	- Electronic Audit Gager
EAGLE	- Elevation Angle Guidance Landing Equipment
EAI	- Electronic Associates, Incorporated
EAL	- Eastern Air Lines
	- Expected Average Life
EAM	- Electric Accounting Machine
	- Electronic-Automatic Machine
EAME	- Europe-Africa-Middle East
EAMR	- Engineering Advance Material Release
EANF	- Establishment of Air Navigation Facilities
EAO	- Experimental Aerodynamics Section (USSR)
EAP	- Effective Air Path

EAPD	- Eastern Air Procurement District
EAR	- Electronic Audio Recognition
EAS	- Equivalent Air Speed
EASCOMINT	- Extended Air Surveillance Communications Intercept
EASE	- Engineering Automatic System for Solving Equations
EAST	- Evaluation and Subsystem Training
EASTAF	- Eastern Transport Air Force
EAT	- Earliest Arrival Time
EAX	- Electronic Automatic Exchange
EBA	- Engineering Budget Authorization
EBIC	- Electron Bombardment Induced Conductivity
EBICON	- Electron Bombardment Induced Conductivity
EBMA	- Engine Booster Maintenance Area
EBPA	- Electron Beam Parametric Amplifier
EBR	- Efficacite Biologique Relative (French for RBE)
	- Experimental Breeder Reactor
EBS	- Energy Band Structure
EBU	- European Broadcasting Union
EBW	- Electron Beam Welding
	- Exploding Bridge Wire
EBWR	- Experimental Boiling Water Reactor
EC	- Edge Connector
	- Emergency Capability
	- Engine Contractor
	- Engineering Change
	- Error Correcting
	- Evaluation Center
E&C	- Electronics and Control
ECA	- Electronic Control Assembly
	- Engineering Change Analysis
	- Etched Card Assembly
	- Extended Coverage Altitude
ECAC	- Electromagnetic Compatibility Analysis Center
	- Engineering College Administrative Council
ECAG	- Equipment Change Analysis Group
ECAP	- Electronic Control Assembly—Pitch
ECAPS	- Emergency Capability System
ECAR	- Electronic Control Assembly—Roll
ECAY	- Electronic Control Assembly—Yaw
ECB	- Equipment Coordination Brief
ECC	- Emergency Combat Capability
	- Emergency Combat Condition
	- European Coordinating Committee
ECCM	- Electronic Counter-Countermeasure
ECCO	- Ethyl Cellulose and Caster Oil
ECCT	- Error Correction Console Technician

■ **ECD**

ECD	- Engineering Control Drawing
	- Entry Corridor Display
	- Estimated Completion Date
ECDL	- Emergency Carbon Dioxide Limit
ECDT	- Electro-Chemical Diffused Transistor
ECE	- Engineering Cost Estimate
ECET	- Electronic Control Assembly—Engine Thrust
ECG	- Electrocardiograph
	- Electrocardiography
	- Electrocardiogram
	- Electronic Components Group
ECI	- Electronic Communications Incorporated
ECIS	- Earth-Centered Inertial System
ECK	- Emergency Communications Key
ECL	- Eddy Current Loss
	- Electronic Components Laboratory
	- Equipment Component List
ECLO	- Emitter-Coupled Logic Operator
ECM	- Electric Cipher Machine
	- Electric Coding Machine
	- Electrochemical Machining
	- Electronic Countermeasure
	- Engineering Change Memorandum
	- European Common Market
EC&M	- Environmental Control and Mechanism
ECMA	- European Computer Manufacturers Association
ECME	- Electronic Checkout Maintenance Equipment
	- Electronics Countermeasure Environment
ECMP	- Electronic Countermeasures Program
ECMR	- Eastern Contract Management Region
ECN	- Engineering Change Notice
ECO	- Electron Coupled Oscillator
	- Electronic Checkout
	- Engine Cutoff
	- Engineering Change Order
ECOMS	- Early Capability Orbital Manned Station
ECP	- Electronic Circuit Protector
	- Electronic Control Products
	- Engineering Change Proposal
ECPD	- Engineers' Council for Professional Development
ECPY	- Electronic Control Assembly—Pitch and Yaw
ECR	- Engineering Change Request
	- Equipment Certification Requirement
	- Executive Control Routine
	- Extended Coverage Range

ECS	- Engine Control System
	- Engineering Costs and Schedules
	- Environmental Control System
	- Error Connection Servo
	- Executive Control System
	- Exospheric Composition Studies
ECSA	- European Communications Security Agency
ECSC	- European Coal and Steel Community
ECSS	- Energy Conversion and Storage System
ECT	- Estimated Completion Time
ECTL	- Emitter-Coupled Transistor Logic
ECU	- Electrical Conversion Unit
	- Environmental Control Unit
	- Extreme Close-Up
ED	- Engineering Design
	- Error Detecting
	- Expanded Display
E-D	- Expansion-Deflection
E&D	- Evaluation and Development
EDA	- Electronic Differential Analyzer
	- Electronic Digital Analyzer
	- Execution Damage Assessment
EDAC	- Error Detection and Correction
EDC	- Electronic Digital Computer
	- Engagement Direction Center
	- Engineering Design Change
	- Engineering Design Coordination
	- Estimated Date of Completion
	- European Defense Community
EDCP	- Engineering Design Change Proposal
EDCSA	- Effective Date of Change of Strength Accountability
EDD	- Estimated Delivery Date
EDDF	- Error Detection and Decision Feedback
EDDIE	- Environmental Distribution of Dynamic Item Entries
EDDS	- Early Docking Demonstration System
EDE	- Electronic Design Engineering
EDFR	- Effective Date of Federal Recognition
EDGE	- Electronic Data Gathering Equipment
	- Experimental Display Generation
EDHE	- Experimental Data Handling Equipment
EDI	- Electronic Devices, Incorporated
EDICT	- Engineering Department Interface Control Task
EDITAR	- Electronic Digital Tracking and Ranging
EDITP	- Engineering Development Integration Test Program
EDL	- Electronics Defense Laboratory
	- Engineering Development Laboratory

■ EDLCC

EDLCC	- Electronic Data Local Communications Complex
	- Electronic Data Local Control Center
EDLO	- Engineering Division Liaison Office
EDLP	- Engineering Development Laboratory Program
EDM	- Electrical Discharge Machine
	- Electrical Discharge Machining
	- Engineering Design Modification
ED&M	- Electronic Design and Manufacture
EDN	- Electrical Design News
EDO	- Effective Diameter of Objective
	- Engineering Duties Only
EDP	- Effective Directives and Plans
	- Electronic Data Processing
	- Engineering Development Program
	- Environment Determination Program
	- Equipment Deadlined for Parts
	- Experimental Dynamic Processor
EDPC	- Electronic Data Processing Center
EDPE	- Electronic Data Processing Equipment
EDPEO	- Electronic Data Processing Equipment Office
EDPM	- Electronic Data Processing Machine
EDPR	- Engineering Development Part Release
EDPS	- Electronic Data Processing System
EDR	- Electronic Decoy Rocket
	- Engineering Design Review
	- Equivalent Direct Radiation
EDRB	- Engineering Design Review Board
EDRCC	- Electronic Data Remote Communications Complex
EDS	- Electronic Design Section
	- Emergency Detection System
	- Engagement Direction Station
EDSAC	- Electronic Data Storage Automatic Computer
EDT	- Eastern Daylight Time
EDTCC	- Electronic Data Traffic Control Center
	- Electronic Data Transmission Communications Center
EDTR	- Experimental, Developmental, Test and Research Awards
EDVAC	- Electronic Digital-Vernier Analog Computer
	- Electronic Discrete Variable Automatic Computer
EDVAP	- Electronic Digital-Vernier Analog Plotter
EE	- External Environment
E&E	- Evasion and Escape
EEC	- European Economic Community
EECT	- End Evening Civil Twilight
EED	- Electromagnetic Explosive Device
EEG	- Electroencephalograph
	- Electroencephalography
	- Electroencephalogram

EEI	- Edison Electric Institute
	- Electrical-Electronic
	- Essential Elements of Information
EEM	- Earth Entry Module
EENT	- End Evening Nautical Twilight
EER	- Explosive Echo Ranging
EET	- Education Equivalency Test
	- End of Evening Twight
	- Equivalent Exposure Time
EETF	- Electronic Environmental Test Facility
EF	- Each Face
	- Elevation Finder
	- Entered From
	- Extra Fine
EFAL	- Electronic Flash Approach Lighting
EFC	- Electronic Frequency Control
	- Equivalent Full Charge
	- European Federation of Corrosion
EFICON	- Estimating Finance Control
EFL	- Effective Focal Length
EFRC	- Edwards Flight Research Center
EFSE	- Engineering Factory Support Equipment
EFSSS	- Engine Failure Sensing and Shutdown System
EFT	- Engineering Flight Test
EFTA	- European Free Trade Association
EFTO	- Encrypt for Transmission Only
EFTWA	- Engineering Flight Test Work Authorization
EGAD	- Electronegative Gas Detector
EGADS	- Electronic Ground Automatic Destruct Sequencer
EGAL	- Elevation Guidance for Approach and Landing
EGC	- Exposure Growth Curves
EGECON	- Electronic Geographic Coordinate Navigation
EGL	- Eglin Air Force Base, Florida (remote site)
EGO	- Eccentric Geophysical Observatory
	- Eccentric Orbiting Geophysical Observatory (preferred)
EGRESS	- Evaluation of Glide Re-Entry Structural System
EGSE	- Electronic Ground Support Equipment
EGT	- Elaspsed Ground Time
	- Estimated Ground Time
	- Exhaust Gas Temperature
EHF	- Extremely High Frequency (band 11)
EHP	- Effective Horsepower
	- Electrical Horsepower
EHT	- Extra High Tension
EHV	- Extra High Voltage
EHY	- Engage High Yield

■ EI

EI	- Electromagnetic Interference
	- Electronic Interface
	- Electronic Interference
	- Engineering Instruction
E-I	- Electromagnetic Interference
EIA	- Electronic Industries Association (formerly RETMA)
EIAC	- Electronic Industries Association of Canada
EIC	- Engineering Information Center
	- Engineering Institute of Canada
	- Equipment Installation and Checkout
EICM	- Employer's Inventory of Critical Manpower
EIB	- Export-Import Bank
EIME	- Electronic Instrument Manufacturers Exhibit
EIN	- Engineer Intelligence Note
EI/NI	- Electron Irradiation and Neutron Irradiation
EIRC	- Economic and Industrial Research Corporation
EIS	- Eyes in the Sky
EIT	- Electrical Information Test
EJC	- Engineers Joint Council
EJCC	- Eastern Joint Computer Conference
EJS	- Engineering Job Sheet
EKC	- Eastman Kodak Company
EKG	- Electrocardiograph
	- Electrocardiography
	- Electrocardiogram
EKS	- Electrocardiogram Simulator
EL	- Education Level
	- Elastic Limit
	- Electro-Luminescence
	- Elevation
ELC	- Emergency Launch Capability
ELCA	- Earth Landing Control Area
ELCS	- Experimental Labor Control System
ELD	- East Longitude Date
ELDO	- European Launcher Development Organization
ELE	- Electronic Launching Equipment
ELECOM	- Electronic Computing System
ELF	- Extremely Low Frequency
ELFC	- Electro-Luminescent Ferroelectric Cell
ELGMT	- Erector-Launcher, Guided Missile, Transportable
ELI	- Equitable Life Interpreter
	- Extra Low Interstitials
ELINT	- Electronic Intelligence
ELIT	- Electronics Information Test
ELLA	- European Long Lines Agency
ELM	- Element Load Module

ELMINT	- Electromagnetic Intelligence
ELOI	- Emergency Letter of Instructions
ELP	- English Language Program
ELR	- Engineering Liaison Request
	- Experimental Launching Round
ELRAC	- Electronic Reconnaissance Accessory
ELS	- Earth Landing System
	- Element Summary Program
ELSEC	- Electronic Security
ELSIE	- Electronic Letter Sorting and Indicating Equipment (British)
ELSSE	- Electronic Sky Screen Equipment
ELVIS	- Electro-Luminescent Vertical Indication System
	- Electro-Visual System
EM	- Educational Manual
	- Electro-Mechanical
	- Emergency Maintenance
	- Engineering Model
	- External Memorandum
E/M	- ECM (Electronic Countermeasure) Malfunction
	- Escape Motor
EMA	- Electronic Missile Acquisition
E-MAD	- Engine Maintenance, Assembly, and Disassembly Building
EMAR	- Experimental Memory Address Register
EMC	- Engineered Military Circuit
	- Equivalent Mission Cycle
	- Essential Minimum Control
EMCCC	- European Military Communications Coordinating Committee
EMCON	- Emission Control
EMD	- Electric Motor Driven
	- Entry Monitor Display
EMDS	- Electronic Material Data Service
EMF	- Electromotive Force
	- Electronic Manufacturing Facility
	- Equipment Maintenance Facility
EMFE	- Engine Make Factory Equipment
EMG	- Electromyograph
	- Electromyography
	- Electromyogram
EMGE	- Electronic Maintenance Ground Equipment
EMI	- Electric Music Instruments
	- Electromagnetic Interference
EMIP	- Experimental Manned Interceptor Program
EMIT	- Engineering Management Information Technique
EML	- Equipment Modification List
EMLR	- Engineering Manufacturing Liaison Release
EMM	- Electrical and Mechanical Maintenance

■ EMO

EMO	- Emergency Off
EMOS	- Earth Mean Orbital Speed
EMP	- Electromagnetic Pulse
EMPIRE	- Early Manned Planetary Interruptionless Round-Trip Expedition
EMQ	- Electromagnetic Quiet
EMR	- Electromagnetic Radiation
	- Electro-Mechanical Research, Incorporated
EMRSC	- Experimental Medical Research Support Center
EMS	- Electromagnetic Surveillance
EMSS	- Emergency Mission Support System
EMT	- Early Missile Test
	- Exact Manning Table
EMTECH	- Electromagnetic Technology
EMTED	- Electromagnetic Test and Evaluation Data System
EMU	- Electromagnetic Unit
EMWA	- Engineering Minor Work Authorization
ENC	- Electron-Nuclear Coupling
ENCA	- European Naval Communications Agency
END	- End of Data
ENDOR	- Electron-Nuclear Double Resonance
ENEA	- European Nuclear Energy Agency
ENIAC	- Electronic Numerical Integrator and Calculator
ENK	- Enter Keys
ENSI	- Equivalent-Noise-Sideband Input
ENT	- Equivalent Noise Temperature
ENTAC	- Engin Teleguide Anti-Char
EO	- Earth Orbit
	- Engineering Order
	- European Office
	- Executive Order
	- Explosive Ordnance
E&O	- Engineering and Operations
EOA	- Effective On or About
	- End of Address
EOAR	- European Office of Aerospace Research
EOB	- End of Block
EOC	- Electronics Operations Center
	- Engine Out Capability
EOCP	- Engine Out of Commission for Parts
EOCR	- Enginnering Order Control Record
EOD	- End of Data
	- Explosive Ordnance Disposal
EODP	- Engineering Order Delayed for Parts
EOF	- End of File

EOGO	- Eccentric Orbiting Geophysical Observatory (EGO preferred)
EOIEC	- Effects of Initial Entry Conditions
EOL	- Earth Orbit Launch
	- End of Life
EOLM	- Electro-Optical Light Modulator
EOLT	- End of Logical Tape
EOM	- Earth Orbital Mission
	- End of Message
	- End of Month
EOMS	- Earth-Orbital Military Satellite
EOP	- Experimental Operating Procedure
EOPF	- End of Powered Flight
EOPR	- Engineering Order Purchase Request
EOQ	- Economic Order Quantity
EOQC	- European Organization for Quality Control
EOR	- Earth Orbit Rendezvous
	- End of Record
	- End of Reel
	- Explosive Ordnance Reconnaissance
E&OR	- Estimate and Order Review
EORA	- Explosive Ordnance Reconnaissance Agent
EORBS	- Earth Orbiting Recoverable Biological Satellite
EORQ	- Engineering Order Request for Quotation
EOS	- Electro-Optical Systems, Incorporated
EOSA	- Explosive Ordnance Safety Approval
EOT	- End of Tape
EOTS	- Electron Optic Tracking System
EOVM	- End of Valid Message
EP	- Electrically Polarized
	- Electro-Pneumatic
	- Entry Point
	- Etched Plate
	- Explosion Proof
EPA	- Electron Probe Analyzer
	- Engineering Personnel Authorization
	- Engineering Project Authorization
	- Equipment Problem Area
EPAM	- Elementary Perceiver and Memorizer
EPC	- Electronic Program Control
EPCCS	- Emergency Positive Control Communications System
EPCO	- Emergency Power Cutoff
EPD	- Earliest Possible Date
EP&D	- Electrical Power and Distribution
EPDS	- Electrical Power Distribution System
EPEC	- Emerson Programmer-Evaluator-Controller

■ EPG

EPG	- Eniwetok Proving Ground
EPI	- Earth Path Indicator
	- Electronic Position Indicator
	- Expanded Position Indicator
EPIC	- Electric Properties Information Center
EPIL	- Experimental Preflight Inspection Letter
EPIREPT	- Epidemiological Report
EPL	- Electronic Products Laboratories
EPOE	- End Piece of Equipment
EPR	- Electron Paramagnetic Resonance
	- Emergency Parts Release Requisition
	- Engineering Parts Release
	- Engineering Purchase Request
EPRA	- Electronic Production Resources Agency
EPRS	- Engineering Parts Release System
EPS	- Electrical Power System
	- Emergency Power Supply
	- Energetic Particles Satellite
	- Experimental Prototype Silo
EP&S	- Educational Programs and Services
EPSTF	- Electrical Power System Test Facility
EPT	- Environmental Proof Test
EPU	- Electrical Power Unit
	- Emergency Power Unit
EPUT	- Events per Unit Time
EQA	- Equipment Quality Analysis
EQRD	- Equipment Ready Date
ER	- Echo Ranging
	- Electronic Reconnaissance
	- Emergency Rescue
	- Enroute
	- Event Record
E/R	- Emergency Recovery
	- Emergency Rescue
ERA	- Engineering Research Associates
	- Exclusive OR to Accumulator
ERAR	- Experience Retention Action Request
ERAS	- Electronic Reconnaissance Access Set
ERB	- Edwards Rocket Base
	- Equipment Review Board
ERBM	- Extended Range Ballistic Missile
ERC	- Electronics Research Center
	- Expendability and Repair Code
ERCS	- Emergency Rocket Communications System
ERD	- Electronic Research Directorate
	- Engineering Release Date
	- Estimated Receival Date

ERDL	- Electronic Research and Development Laboratory
	- Experimental Reactor Development Laboratory
ERETS	- Edwards Rocket Engine Test Station
	- Experimental Rocket Engine Test Station
ERFA	- European Radio Frequency Agency
ERFPI	- Extended Range Floating Point Interpretive System
ERG	- Electroretinograph
	- Electroretinography
	- Electroretinogram
ERI	- Engineering Receiving Inspection
	- Engineering Research Institute
ERIT	- Engineering Receiving, Inspection Test
ERL	- Event Record Log
ERM	- Engineering Report Manual
ERMA	- Electronic Recording Method of Accounting
ERN	- Experimental Reference Number
ERNIE	- Electronic Random Numbering and Indicating Equipment
ERO	- European Research Office
EROS	- Experimental Reflector Orbital Shot
ERP	- Educational Reimbursement Program
	- Effective Radiated Power
	- Eye Reference Point
ERS	- Emergency Relocation Site
	- Engineering Release Schedule
	- Entry and Recovery Simulation
ERSCP	- End Refueling and Start Climb Point
ERU	- Earth Rate Unit
	- Engineering Release Unit
ES	- Echo Sounding
	- Electrochemical Society
	- Electronic Standard
	- Electrostatic Storage
	- Experimental Sector
E/S	- Earth to Space
E-S	- Earth to Space
ESA	- Electrical Surge Arrester
	- Engineering Study Authorization
	- Explosive Safety Approval
ESAR	- Electronically Steerable Array Radar
ESB	- Electrical Systems Branch
	- Engine Storage Building
ESC	- Electronic Systems Center
	- External System Complex
ESCAPE	- Expansion Symbolic Compiling Assembly Program for Engineering
E-SCAT	- Emergency Security Control of Air Traffic

■ ESCO

ESCO	- Engineer Supply Control Office
ESCS	- Electronic Spacecraft Simulator
ESD	- Electronic Systems Division (of AFSC)
	- Electrostatic Storage Deflection
	- Estimated Shipping Date
ESE	- Engineering Support Equipment
ESEG	- Electronic Systems Engineering Group
ESG	- Electrostatically Suspended Gyroscope
	- Empiric Studies Group
	- Expanded Sweep Generator
ESGA	- Electronally Suspended Gyroscope Accelerometer
ESGSSFDB	- Empiric Studies Group Simulated SAC Force Data Base
ESHP	- Equivalent Shaft Horsepower
ESL	- Engineering Societies Library
	- Evans Signal Laboratory
ESLO	- European Space Launcher Organization
ESM	- Environmental Systems Monitor
ESMU	- Electronic Systems Mock-Up
ESN	- Event Schedule Network
ESNA	- Elastic Stop Nut Corporation of America
ESO	- Electronic Supply Office
	- Electronic Supply Officer
ESP	- Elastic Strength Pressure
E&SP	- Equipment and Spare Parts
ESPAR	- Electronically Steerable Phased Array Radar
ESPT	- Executive Sequence Parameter Table
ESR	- Electron Spin Resonance
	- Event Storage Record
ESRO	- European Space Research Organization
ESS	- Electrical Supervisory Subassembly
	- Electronic Science Section
	- Electronic Sequence Switching
	- Electron Spin Spectra
	- Emergency Survival System
	- Entry Survival System
	- Experimental SAGE Sector
ESSA	- Electronic Scanning and Stabilizing Antenna
ESSFL	- Electron Steady-State Fermi Level
ESSPO	- Electronic Supporting Systems Project Office
ESSU	- Electronic Selective Switching Unit
EST	- Eastern Standard Time
	- Electrostatic Storage Tube
	- Employment and Suitability Testing
E&ST	- Employment and Suitability Testing
ESTE	- Engineering Special Test Equipment
ESTEC	- European Space Technical Center

ESTF	- Electronic System Test Facility
ESU	- Electrostatic Unit
	- Engineering Service Unit
ESV	- Earth Satellite Vehicle
	- Emergency Shutoff Valve
ESWA	- Engineering Shop Work Authorization
ET	- Elapsed Time
	- Energy Transfer
	- Engineering Test
	- Ephemeris Time
	- Erection Torquer
	- Escape Tower
	- Evaluation Test
E/T	- Engineering Test
	- Escape Tower
ETA	- Estimated Time of Arrival
	- Estimated Time to Acquisition
ETAS	- Effective True Air Speed
ETAWG	- Engineering Test Area Working Group
ETC	- Eggleston Tape Compare
	- Estimated Time of Completion
ETCG	- Elapsed-Time Code Generator
ETD	- Estimated Time of Departure
ETDP	- Emergency Traffic Disposition Plan
ETE	- Estimated Time Enroute
	- External Test Equipment
ETF	- Eglin Test Facility
	- Engine Test Facility
ETFS	- ECM (Electronic Countermeasure) Transmitter Frequency Setup
ETI	- Elapsed Time Indicator
	- Estimated Time of Interception
ETIC	- Estimated Time in Commission
ETL	- Electronic Technology Laboratory
	- Engineering Test Laboratory
ETM	- Electronic Test and Maintenance
	- Engineering Test Model
	- Enter Trapping Mode
ETMG	- Electron Tube Management Group
ETMWG	- Electronic Trajectory Measurements Working Group
ETN	- Equipment Table Nomenclature
ETO	- Engineering Test Order
	- Estimated Time Over
	- Ethylene Oxide
	- European Theater of Operations
ETOC	- Estimated Time of Correction

■ E to E

E to E	- End to End
ETP	- Electrical Tough Pitch
	- Engineering Test Program
	- Estimated Time of Penetration
	- Experimental Test Procedure
ETPD	- Estimated Time of Parachute Deployment
ETPR	- Engineering Test Part Release
ETPS	- Engineering Test Program Spares
ETPY	- ECA (Electronic Control Assembly) Thrust Vector—Pitch and Yaw
ETR	- Eastern Test Range (AFETR preferred)
	- Engineering Test Reactor
	- Engineering Test Record
	- Estimated Time for Refueling
	- Estimated Time of Return
	- Export Traffic Release
ETRO	- Estimated Time of Return to Operation
ETS	- Edwards Test Station
	- Electronic Timing Set
	- Engagement Tracking Station
	- Engine Test Stand
	- Environment Table Simulation
	- Equivalent Target Size
ETSAL	- Electronic Terms for Space Age Language
ETSE	- Engineering Test Support Equipment
ETT	- End of Tape Test
	- Estimated Time to Track
ETTI	- End Translation Time Indicator
ETV	- Educational Television
ETVC	- Environmental Test Vacuum Center
ETVM	- Electrostatic Transistorized Voltmeter
EUCLID	- Experimental Use Computer, London Integrated Display
EUCOM	- European Command
EUR	- Emergency Unsatisfactory Report
EURATOM	- European Atomic Energy Community (Communaute Europeene de l'Energie Atomique)
EUSA	- Eighth United States Army
EUSEC	- Conference of Engineering Societies of Western Europe and the United States
EUT	- Equipment Under Test
EUV	- Extreme Ultraviolet
EV	- Electron Volt
EVA	- Electronic Velocity Analyzer
	- Elevation Versus Amplitude
	- Extra-Vehicular Activity
	- Extra-Vehicular Astronaut

EVAL	- Evaluation
EVATA	- Electronic Visual-Aural Training Aid
EVIL	- Elevation Versus Integrated Log
EVM	- Evacuation Mission
EVRATOM	- European Atomic Energy Community (USSR)
EVS	- Extra-Vehicular Suit
EVSC	- Extra-Vehicular Suit Communications
EVSS	- Extra-Vehicular Space Suit
EVSTC	- Extra-Vehicular Suit Telecommunications
EW	- Early Warning
	- Electronic Warfare
EWA	- Engineering Work Authorization
EWASER	- Electromagnetic Wave Amplification by Stimulated Emission of Radiation
EW/CRP	- Early Warning/Control and Reporting Post
EWCS	- Experimental Weapons Control System
EWG	- Equipment Working Group
EWO	- Emergency War Order
	- Engineering Work Order
EWP	- Emergency War Plan
	- Expecting Wire Phenomenon
EWR	- Early Warning Radar
	- Engineering Work Request
	- Entwicklungsring
EWS	- Extended Work Sheet
EWT	- Effects of Weapons on Targets
EWTAD	- Early Warning Threat Analysis Display
EWTO	- Engineering Work Transfer Order
EWTR	- Engineering Work Transfer Request
EWW	- Extended Work Week
EXO	- Executive Officer
EXOBIOLOGY	- Exoterrestrial Biology
EXOLIFE	- Exoterrestrial Life
EXOS	- Executive Office of the Secretary
EXP	- Expansion
EXPERT	- Expanded Program Evaluation and Review Technique
EXTAL	- Extra Time Allowance
EXTRADOP	- Extended Range Doppler

F	- Fahrenheit
	- Farad
	- Fighter
	- Fire
	- Flat
	- Fluorine
	- Force
	- Frequency
	- Fuel
FA	- Field Artillery
	- Flight Acceptance
	- Free Aperature
	- Frequency Agility
	- Friendly Aircraft
	- Fully Automatic
F/A	- Field Activities
	- Final Assembly
	- Fuel to Air
F&A	- Flight and Aerodynamics
	- Fore and Aft
FAA	- Federal Aviation Agency (formerly CAA)
	- Fleet Air Arm
FAAWTC	- Fleet Anti-Air Warfare Training Center
FAB	- Fleet Air Base
FABMDS	- Field Army Ballistic Missile Defense System
FABU	- Fuel Additive Blender Unit
FAC	- Federal Aviation Commission
	- Field Accelerator
	- First Alarm Code
	- Foward Air Controller
FA&C	- Field Assembly and Checkout
FACI	- First Article Configuration Inspection
FACS	- Floating Decimal Abstract Coding System

134

FAR

FACT	- Factual Compiler
	- Fighter Aircraft Code Type
	- Flexible Automatic Circuit Tester
	- Flight Acceptance Composite Test
	- Forecast and Control Technique
	- Fully Automatic Compiler-Translator
	- Fully Automatic Compiling Technique
FACTOR	- Fourteen-O-One Automatically Controlled Test Optimizing Routine
FACTS	- Field Army Calibration Team Support
FAD	- Fairchild Aircraft Division
	- Floating Add
FADAC	- Field Artillery Data Computer
FADTC	- Fleet Air Defense Training Center
FAE	- Final Approach Equipment
FAETUPAC	- Fleet Airborne Electronic Training Unit, Pacific
FAGMS-S	- Field Army Guided Missile System—Sergeant
FAGS	- Federation of Astronomical and Geophysical Services
FAGUPAC	- Fleet Air Gunnery Unit, Pacific
FAI	- Federation Aeronautique Internationale
	- First Article Inspection
FAIN	- First Article Inspection Notice
FAIN/SR	- First Article Inspection Notice—Status Report
FAIR	- Fabrication, Assembly, and Inspection Records
	- First Article Inspection Report
	- Fleet Aircraft
FAIS	- Force Air, Intelligence Study
FAIT	- First Article Inspection Tag
FAJ	- Final Assembly Jig
FAK	- Flyaway Kit
FAM	- Field Artillery Missile
	- Flight Acceptance Meeting
	- Floating Add Magnitude
	- Frequency Assignment Model
FAMU	- Fuel Additive Mixture Unit
FAO	- Finish All Over
	- Food and Agriculture Organization of the United Nations
FAP	- Floating Point Arithmetic Package
	- FORTRAN Assembly Program
	- Frequency Allocation Panel
FAPUSMCEB	- Frequency Allocation Panel United States Military Communication Electronics Board
FAR	- Failure Analysis Report
	- False Alarm Rate
	- First Alarm Register
	- Forward Acquisition Radar

■ FARET

FARET	- Fast Reactor Test Facility
FARGO	- Fourteen-O-One Automatic Report Generating Operation
FAS	- Flight Advisory Service
FASO	- Forward Airfield Supply Organization
FASRON	- Fleet Aircraft Service Squadron
FAST	- Facility for Automatic Sorting and Testing
	- Fast Automatic Shuttle Transfer
	- Field Data Applications, Systems, and Techniques
	- Flight Advisory Service Testing
	- Four Address to SOAP (Symbolic Optimum Assembly Program) Translator
	- Fourteen-O-One Automated System of Testing
FASTAR	- Frequency Angle Scanning, Tracking, and Ranging
FASTI	- Fast Access to Systems Technical Information
FASTP	- Foreign Area Specialist Training Program
FASTROM	- Falling Sphere Trajectory Measurement
FASTRON	- Fleet Aircraft Service Squadron
FAT	- Factory Acceptance Test
	- Fast Automatic Transfer
	- Final Assembly Test
	- Flight Acceptance Test
FATDL	- Frequency and Time-Division Data Link
FATE	- Force Application Tactics Evaluation
FATOC	- Field Army Tactical Operation Center
FAV	- Fixed Angle Variable
FAW	- Forward Area Weapons
FAWS	- Flight Advisory Weather Service
FAWTC	- Fleet Anti-Warfare Training Center
FAX	- Facsimile Transmission
	- Fuel-Air Explosive
FB	- Fighter Bomber
	- Film Bulletin
	- Flat Bar
	- Fuse Block
F&B	- Fill and Bleed
	- Fire and Bilge
FBM	- Feet Board Measure
	- Fleet Ballistic Missile
FBMS	- Fleet Ballistic Missile System
FBN	- Flightborne
FBOC	- Fighter Base Operations Center
FBOE	- Frequency Band of Emission
FBP	- Field Barometric Pressure
FBRL	- Final Bomb Release Line
FBS	- Fighter Bomber Squadron
FBW	- Fly-By-Wire

FC	- Ferrite Core
	- Fighter Center
	- Filter Center
	- Fire Code
	- Fire Control
	- Fixed Camera
	- Flight Center
	- Flight Control
	- Foot-Candle
	- Fractocumulus
	- Front Connected
	- Functional Code
F/C	- Ferrite Core
	- Flight Certificate
	- Flight Control
	- Fuel Cell
	- Functional Checkout
F-C	- Foot-Candle
FCA	- Fire Control Area
FCBR	- Factory Change Board Report
FCC	- Federal Communications Commission
	- Fighter Control Center
	- Flight Communications Center
	- Flight Control Center
FCCO	- Flight Change Control Order
FCD	- Flight Control Division (formerly FCOB)
	- Formal Change Draft
	- Frequency Control Division
	- Fuel Cells Display
FCDA	- Federal Civil Defense Administration
FCDR	- Failure and Consumption Data Report
FCE	- Fire Control Equipment
FCG	- Facility Change Group
FCH	- Flight Controllers Handbook
FCIR	- Facility Change Initiation Request
FCL	- Feedback Control Loop
	- Feeder Control Logic
	- Flight Control Laboratory
	- Full Cycle Left
	- Fuze Cavity Lined
FCO	- Field Change Order
	- Flight Crew Operations (formerly FCOD)
	- Frequency Control Officer
	- Fuel Cutoff
	- Functional Checkout
FCOB	- Flight Crew Operations Branch (now Flight Control Division)

■ FCOD

FCOD	- Flight Crew Operations Division (now Flight Crew Operations)
FCOV	- Facility Checkout Vehicle
FCP	- Facility Change Proposal
	- Facility Control Program
FCPC	- Fleet Computer Programming Center
FCPCP	- Fleet Computer Programming Center, Pacific
FCR	- Facility Change Request
	- Fire Control Radar
	- Fuse Current Rating
FCRL SYS	- Flight Control Ready Light System
FCS	- Federal Catalog System
	- Feedback Control System
	- Fire Control System
	- Flight Control System
	- Functional Checkout Set
FCST	- Federal Council for Science and Technology
FCT	- Filament Center Tap
	- Flight Certification Test
	- Flight Crew Trainer
	- Fraction Thereof
FCTS	- Flight Crew Trainer Simulator
FCVN	- Fatal Casualties Vulnerability Number
FCWG	- Frequency Coordination Working Group
FD	- Flight Directive
	- Flight Director
	- Free Drop
	- Frequency Diversity
	- Frequency Division
	- Frequency Doubler
	- Full Duplex
F/D	- Face of Drawing
	- Field of Drawing
	- Full Duplex
FDA	- Food and Drug Administration
FDAI	- Flight Director Attitude Indicator
FDB	- Field Dynamic Braking
FDC	- Fire Department Connection
	- Fire Direction Center
FDCA	- Federal Defense Communications Authority
FDDL	- Frequency Division Data Link
FDE	- Field Decelerator
	- Flight Data Entry System
FDG	- Fractional Doppler Gate
FDH	- Floating Divide or Halt

FDI	- Field Discharge
	- Flight Direction Indicator
	- Flight Direction Instrument
FDL	- Flight Datum Line
FDM	- Frequency Division Multiplex
FDO	- Flight Dynamics Officer
FDP	- Floating Divide or Proceed
FDR	- Fast Death Response
	- Field Discussion Report
	- Fix Dump Reducer
	- Frequency Diversity Radar
FDRI	- Flight Director Rate Indicator
FDS	- Fighter Data Storage
FDT	- Full Duplex Teletype
FDU	- Frequency Determining Unit
FDX	- Full Duplex
FE	- Far East
	- Fighter Escort
F/E	- Flight Engineer
FEA	- Failure Effects Analysis
FEAF	- Far East Air Force
FEAMCOM	- Far Eastern Air Materiel Command
FEAT	- Frequency of Every Allowable Term
FEB	- Flying Evaluation Board
	- Forward Equipment Bay
	- Functional Electronic Block
FEBS	- Functional Electronic Blocks
FEC	- Far East Command
	- Federal Electric Company
FECM	- Firm Engineering Change Memo
FED	- Federal
	- Fuel Element Department
FED SPEC	- Federal Specification
FED STD	- Federal Standard
FEE	- Failure Effects Evaluation
FEI	- Firing Error Indicator
	- Flight Error Instrumentation
FEIA	- Flight Engineers International Association
FEO	- Field Engineering Order
FEP	- Fluorinated Ethylene Propylene
FEPC	- Fair Employment Practices Code
FER	- Forward Engine Room
FERO	- Far East Research Office
FES	- Fast Erection System
FESE	- Field-Enhanced Secondary Emission
FET	- Field-Effect Transistor

■ FETS

FETS	- Field-Effect Transistors
FEX	- Fleet Exercise
FEXT	- Frame Time for Extrapolation
FF	- Fixed Fee
	- Flip-Flop
	- Florida Facility
	- Folding Fin
	- Ford Foundation
	- Forward Firing
	- Free Fall
	- Free Flight
	- Free Flyaround
	- Front Focal Length
	- Fuel Flow
F-F	- Flip-Flop
F&F	- Fire and Flushing
FFAR	- Folding Fin Aircraft Rocket
	- Forward Firing Aircraft Rocket
FFB	- Free-Fall Bomb
FFC	- Fault and Facilities Control
	- Federal Facilities Corporation
FFD	- Field Forcing, Decreasing
FFEC	- Field-Free Emission Current
FFI	- Field Forcing, Increasing
FFMI	- Fighter Fuel Manually Inserted
FFNC	- First Fix Not Converted
FFOB	- Fighter Fuel Onboard
FFP	- Field Protective
FFPC	- Firm Fixed Price Contract
FFR	- Field Reversing
	- Flash Format Program
	- Fuze Frequency Rating
FFRR	- Full Frequency Range Recording
FFS	- Flight Following Service
FFSA	- Field Functional System Assembly
FFSAC	- Field Functional Systems Assembly and Checkout
FFT	- For Further Transfer
	- Free-Fall Test
FFTV	- Free-Flight Test Vehicle
FFW	- Field Weakening
FG	- Flow Gauge
	- Foreground
	- Free Gyroscope
	- Fuel Gage
FGD	- Fine Grain Data
FGMD	- Fairchild Guided Missile Division

FGP	- Facility Grounding Point
	- Fallout Grid Program
FGS	- Fort Greely Station
FH	- Flat Head
FHP	- Fractional Horsepower
	- Friction Horsepower
FHS	- Forward Heat Shield
FHTE	- Flight Hardware Test Equipment
FI	- Fighter Interceptor
F/I	- Field Intensity
FIA	- Financial Inventory Accounting
FIAD	- Flame Ionization Analyzer and Detector
FIAN	- P. N. Lebedev Institute of Physics (USSR)
FIARE	- Flight Investigation of Apollo Re-Entry Environment
FIAT	- Field Information Agency, Technical
FIB	- Field Installation Branch
FIBAS	- Field Installation Branch Adaption Section
FIBCS	- Field Installation Branch Control Section
FIC	- Fleet Intelligence Center
	- Flight Information Center
	- Frequency Interference Control
FICA	- Federal Insurance Contributions Act
FICC	- Frequency Interference Control Center
FID	- Federation Internationale de Documentation
	- Field Instrumentation Division
	- Flame Ionization Detector
FIDO	- Flight Dynamics Officer
	- Fog Investigation and Dispersal Operation
	- Functions Input Diagnostic Output
FIER	- Foundation for Instrumentation, Education, and Research, Incorporated
FIFO	- First In, First Out
	- Floating Input, Floating Output
FII	- Federal Item Identification
FIIG	- Federal Item Identification Guide
FIIN	- Federal Item Identification Number
FIL	- Franklin Institute Laboratories
FILCEN	- Filter Center
FILS	- Flarescan Instrument Landing System
FIM	- Field Instructions Memorandum
FIMG	- Facilities Installation Monitoring Group
FINA	- Following Items Not Available
FINE	- Fighter Inertial Navigation Equipment
FIP	- Fleet Introduction Program
FIR	- Flight Information Region
	- Fuel Indicator Reading

■ FIR

FIR	- Full Indicator Reading
	- Functional Items Replacement
	- Further Information Request
FIRE	- Flight Investigation Re-Entry Environment
FIRETRAC	- Firing Error Trajectory Recorder and Computer
FIRST	- Fabrication of Inflatable Re-Entry Structures for Test
FIS	- Fighter Interceptor Squadron
FITS	- Fourteen-O-One Input-Output Tape System
FIU	- Forward Interpretation Unit
FIW	- Fighter Interceptor Wing
FK	- Faker
FL	- Filter
	- Flight Line
	- Flow Line
	- Focal Length
	- Foot-Lambert
F-L	- Foot-Lambert
FLAC	- Florida Automatic Computer
FLAG	- Fixed Link Aerospace to Ground
FLAIR	- Fundamental Land-Air Integrated Research
FLAK	- Fliegerabwehrkanone (German antiaircraft cannon)
FLAME	- Facility Laboratory for Ablative Materials Evaluation
FLAT	- Flight Plan Aided Tracking
FL-BE	- Filter-Band Eliminator
FL-BP	- Filter-Bandpass
FLCDG	- Flow Control Data Generation
FLD	- Field
FLDO	- Field Officer
	- Final Limit Down
FLEA	- Flux Logic Element Array
FLEACT	- Fleet Activity
FLEASWSCOL	- Fleet Anti-Submarine Warfare School
FLENUMWEAFAC	- Fleet Numerical Weather Facility
FLF	- Final Limit Forward
FLFT	- Full Load Frame Time
FLH	- Final Limit Hoist
FL-HP	- Filter-High Pass
FLI	- Flight Leader Identification
FLICON	- Flight Control
FLID	- Find or List the Identifications
FLIDAP	- Flight Data Position
FLIDEN	- Flight Data Entry
FLINT	- Floating Interpretative Language
FLIOP	- Flight Operations Planner
FLIP	- Film Library Instantaneous Program
	- Flight Information Publication

FLIP	- Flight Launched Infrared Probe
	- Flight Plan
	- Floating Indexed Point Arithmetic
	- Floating Instrument Platform
	- Floating Point Interpretative Program
FLL	- Final Limit, Lower
FL-LP	- Filter-Low Pass
FLO	- Functional Line Organization
FLOGAIR	- Fleet Logistics Air Wing
FLOOD	- Fleet Observation of Oceanographic Data
FLOP	- Floating Octal Point
FLOT	- Flotilla
FLOX	- Fluorine Plus Liquid Oxygen
FLPA	- Flight Level Pressure Altitude
FLR	- Final Limit, Reverse
	- Flight Line Recorder
FLRNG	- Flash Ranging
FLSA	- Fair Labor Standards Act
FLSC	- Flexible Linear Shaped Charge
FLT	- Flight
FLTREADREP	- Fleet Readiness Representative
FLU	- Final Limit, Up
FLYTAF	- Flying Training Air Force
FM	- Fan Marker
	- Field Manual
	- Frequency Modulation
	- Frequency Multiplex
FMA	- Failure Mode Analysis
	- Frequency Modulator Altimeter
FMACC	- Foreign Military Assistance Coordinating Committee
FMASC	- Foreign Military Assistance Steering Committee
FMB	- Federal Maritime Board
FMC	- Food Machinery and Chemical Corporation
FMCR	- Fleet Marine Corps Reserve
FMCS	- Federal Mediation and Conciliation Service
FM/CW	- Frequency Modulated Continuous Wave
FMD&C	- Flight Mechanics, Dynamics, and Control
FME	- Field Maintenance Equipment
FMEP	- Friction Mean Effective Pressure
FMF	- Fleet Marine Force
FMFB	- Frequency Modulation Feedback
FMFLANT	- Fleet Marine Force, Atlantic
FM-FM	- Frequency Modulation-Frequency Modulation
FMFPAC	- Fleet Marine Force, Pacific
FMIC	- Frequency Monitoring and Interference Control
FML	- Feedback, Multiple Loop

FM-PM

FM-PM	- Frequency Modulation-Phase Modulation
FMR	- Frequency Modulated Radar
	- Frequency Modulated Receiver
FMS	- Floating Machine Shop
	- Free Machining Steel
FMT	- Frequency Modulated Transmitter
FMTR	- Florida Missile Test Range
FMTS	- Field Maintenance Test Station
FMX	- Frequency Modulated Transmitter
FN	- Field Note
	- Flat Nose
FNH	- Flashless Nonhygroscopic
FNP	- Fusion Point
FNS	- Far North Station
FNWF	- Fleet Numerical Weather Facility
FO	- Fallout
	- Fast Operating
	- Field Order
	- Flat Oval
	- Flight Order
	- Forward Observer
	- Fuel Oil
F/O	- Follow-On
	- Fuel to Oxidizer
FOA	- Foreign Operations Administration
FOAL	- Force Allocation
FOB	- Field Operations Branch
	- Free Onboard
	- Freight Onboard
FOBTSU	- Forward Observer Target Survey Unit
FOC	- Flight Operations Center
FOCC	- Fleet Operations Control Center
FOCI	- First Operational Computer Installation
FOD	- Factory on Dock
	- Field Operations Department
	- Flight Operations Division
FOF	- Flight Operations Facility
FofF	- Field of Fire
FOG	- Field Operations Group
FOI	- Field Operations Intelligence
	- Fighter Officer for Interceptors
FOIR	- Field-of-Interest Register
FOLNOAVAL	- Following Items Not Available
FOM	- Field Operations Memorandum
	- Fighter Officer for Missiles
FONCON	- Telephone Conversation

FOOSP	- Fourteen-O-One Statistical Program
FOPREP	- Fallout Preparation Program
FORAST	- Formula Assembler Translator
FORC 2	- Formula Coder and Automatic Coding System
FORDS	- Floating Oceanographic Research and Development Station
FORGEN	- Force Generation
FORTOCOM	- FORTRAN I Compiler
FORTRAN	- Formula Translation System
FORTRANSIT	- FORTRAN and IT System
FORTRUNCIBLE	- FORTRAN Style RUNCIBLE
FOS	- Fighter Operations Simulator
	- Flight Operations Support
FOSDIC	- Film Optical Sensing Device for Input to Computers
FOT	- Fuel Oil Transfer
FOUO	- For Official Use Only
FOV	- Field of View
FP	- Feedback Positive
	- Feedback Potentiometer
	- Fixed Price
	- Flame Proof
	- Flashless Propellant
	- Flash Point
	- Flat Point
	- Forward Perpendicular
	- Freezing Point
	- Freight and Passenger Vehicle
	- Fuel Pressure
	- Full Period
F/P	- Flat Pattern
	- Full Period
FPA	- Foreign Policy Association
FPC	- Federal Power Commission
	- Flight Plan Code
	- Forty Pound Charge
FPD	- Flight Propulsion Division
FPI	- Faded Prior to Intercept
	- Fixed Price Incentive
	- Fuel Pressure Indicator
FPID	- Fixed Price Incentive with Delay Firm Target
FPIF	- Fixed Price Incentive with Initially Firm Target
FPIS	- Forward Propagation by Ionospheric Scatter
FPL	- Final Protective Line
	- Fire Plug
	- Frequency Phase Lock
FPM	- Federal Personnel Manual
	- Feet per Minute

■ FPO

FPO	- Fixed Path of Operation
	- Fleet Post Office
	- Future Projects Office
FPPC	- Flight Plan Processing Center
FPR	- Film Production Request
	- Fixed Price Redetermination
FPRE	- Flat-Plate Reference Enthalpy
FPR-E	- Fixed Price with Retroactive Estimate
FPS	- Feet per Second
	- Fixed Radar Tracking
	- Flight Plan Station
	- Flight Preparation Sheet
	- Foot-Pound-Second
	- Frames per Second
FPT	- Fitted Parts Tag
	- Flight Plan Talker
	- Full Power Trial
FPTS	- Forward Propagation by Tropospheric Scatter
FPTYP	- Flight Plan Type
FPU	- First Production Unit
	- Flight Plan Unassociated
FQ	- Flight Qualification
FQT	- Flight Qualification Test
FR	- Failure Report
	- Fast Release
	- Fighter Reconnaissance
	- Fighter Return
	- Fuel Ratio
	- Full Rate
FRA	- Fleet Reserve Association
FRAM	- Fleet Rehabilitation and Modernization Program
FRAT	- First Recorded Appearance Time
FRC	- Failure Recurrence Control
	- Federal Radiation Council
	- Flight Research Center
FRD	- Formerly Restricted Data
FRDI	- Flight Research and Development Instrumentation
FRE	- Field Representative, Europe
FRED	- Fast Relocatable Editing Dump
FREL	- Feltman Research and Engineering Laboratories
FREQ	- Frequency
FRESCAN	- Frequency Scanning
FRESCANNER	- Frequency Scanning Radar
FRESH	- Foil Research Supercavitating Hydrofoil
FRF	- Flight Readiness Firing
FRFE	- Field Representative, Far East

FRFT	- Flight Readiness Firing Test
FRINGE	- File and Report Information Processing Generator
FRMC	- Frame Counter
FRN	- Floating Round
FROB	- Flash Radar Order of Battle
FROG	- Free-Rocket-Over-Ground
FROST	- Food Reserves on Space Trip
FRS	- Failure Reporting System
	- Frequency Response Survey
FRSBY	- Friendly, Round Robin, Special, Bee and Yoke Tracks
FRT	- Fighter Recovery and Telling
	- Flight Rating Test
FRU	- Fleet Radio Unit
FRW	- Fixed Radio Control
FS	- Far Side
	- Feasibility Study
	- Feedback, Stabilized
	- Federal Specification
	- Federal Standard
	- Field Service
	- Film Strip
	- Fin Stabilized
	- Firing Station
	- First Step
	- Flying Status
	- Full Scale
FSA	- Federal Security Agency
	- Field Services Administrator
	- Fine Structure Analysis
	- Fire Support Area
	- Foreign Service Availability
FSB	- Federal Specifications Board
	- Floating Subtract
FSC	- Federal Supply Classification
FSCC	- Fire Support Coordination Center
FSCI	- Federal Supply Code Identification
	- Frequency Space Characteristic Impedance
FSCL	- Fire Support Coordination Line
FSCT	- Floyd Satellite Communications Terminal
FSD	- Federal Systems Division
	- Flying Spot Digitizer
	- Fuel Supply Depot
FSE	- Factory Support Equipment
	- Field Support Equipment
FSF	- Flight Safety Foundation
FSG	- Flight Strip Generator

■ **FSI**

FSI	- Federal Stock Item
FSK	- Frequency Shift Keying
FSL	- Flight Simulation Laboratory
FSM	- Field Strength Meter
	- Floating Subtract Magnitude
FSMWO	- Field Service Modification Work Order
FSN	- Federal Stock Number
	- Fiscal Station Number
FSO	- Field Service Operation
FSP	- Frequency Standard, Primary
FSPC	- Field-Site Production Capability
FSPO	- Force Structure Planning Objective
FSPRS	- Field-Site Production and Reduction System
FSPS	- Field-Site Production System
FSPSC	- Field-Site Production Study Committee
FSR	- Field Service Representative
	- Field Strength Radio
	- Fin-Stabilized Rocket
	- Frequency Scan Radar
FSS	- Federal Supply Schedule
	- Flight Service Station
	- Full Scale Station
FST	- Frequency Shift Transmission
	- Full Store
FSTC	- Foreign Science and Technology Center
FSTE	- Fixed Systems Test Equipment
FSTR	- Field Service Technical Report
FSU	- Field Storage Unit
	- Full Scale Unit
FSV	- Final Stage Vehicle
FT	- Firing Table
	- Flight Test
	- Flush Threshold
	- Foot
	- Full Time
	- Fume Tight
	- Functional Test
F/T	- Functional Test
FTB	- Functional Test Bulletin
FTC	- Fast Time Constant
	- Federal Trade Commission
	- Flight Test Center
	- Flight Test Control
	- Flight Time Capability
FT-C	- Foot-Candle
FTCC	- Flight Test Coordinating Committee

FTCP	- Flight Test Change Proposal
FTD	- Field Training Detachment
	- Flight Test Directive
	- Foreign Technology Division (of AFSC)
	- Functional Test Document
FTE	- Factory Test Equipment
	- Functional Test Equipment
FTF	- Forward Transfer Function
FTG	- Field Training Group
FTL	- Facility Tape Loading
	- Federal Telecommunications Laboratory
FT-L	- Foot-Lambert
FT-LB	- Foot-Pound
FTLP	- Final Turn Lead Pursuit
FTM	- Flight Test Missile
	- Flight Training Missile
FTMC	- Frequency and Time Measurement Counter
FTO	- Flight Test Operations
F to F	- Face to Face
FTP	- Field Test Program
	- Flight Test Procedure
	- Flight Test Program
	- Functional Test Procedure
FTR	- Fixed Transom
FTRW	- Flight Test Reports Writer
FTS	- Federal Telecommunications System
	- Functional Test Specification
FTU	- Flight Test Umpire
	- Flight Test Unit
	- Functional Test Unit
FTV	- Flight Test Vehicle
FTW	- Federation of Telephone Workers
FTWG	- Flight Test Work Group
FTWS	- Flight Test Work Sheet
FTX	- Field Training Exercise
FU	- Firing Unit
FUA	- Fire Unit Analyzer
FUBAR	- Fouled Up Beyond All Recognition
FUD	- Fire Up Decoder
FUIF	- Fire Unit Integration Facility
FURAN	- Furfuryl Alcohol and Aniline
FV	- Flush Valve
	- Flux Valve
	- Front View
FVR	- Fuse Voltage Rating

■ FW

FW	- Fire Wall
	- Frame Synchronization Word
	- Full Wave
FWC	- Factory Work Code
FWDBL	- Forward Bomb Line
FWG	- Facility Working Group
	- Factory Work Group
FWHM	- Full Width at Half Maximum
FWR	- Full-Wave Rectification
	- Full-Wave Rectifier
FWTT	- Fixed Wing Tactical Transport
FY	- Fiscal Year
FYI	- For Your Information
FYIG	- For Your Information and Guidance
FYMP	- Five Year Materiel Program

G

G	- Gallon
	- Gas
	- Giga
	- Glass
	- Gram
	- Gravity
	- Grid
GA	- Gas Amplification
	- Glide Angle
	- Ground Alert
	- Ground Area
	- Ground to Air
G/A	- Glide Angle
	- Ground to Air
	- Guidance Amplifier
G&A	- General and Administrative
	- Geophysics and Astronomy
GAC	- Geological Association of Canada
	- Goodyear Aerospace Corporation
GADL	- Ground-to-Air Data Link
GADR	- Guided Air Defense Rocket
GADS	- Goose Air Defense Sector
GAEC	- Grumman Aircraft Engineering Corporation
GAFI	- State Astrophysical Institute (USSR)
GAFPG	- General Aviation Facilities Planning Group
G/A/G	- Ground to Air to Ground
GAI	- Gate Alarm Indicator
	- Main Astronomical Institute (USSR)
GAINS	- Gimballess Analytic Inertial Navigation System
	- Global Airborne Integrated Navigation System
GAISh	- P. K. Shternberg State Astronomical Institute (USSR)
GAIT	- Government and Industry Team
GAL	- Gallon

- **GALCIT**

GALCIT	- Guggenheim Aeronautical Laboratory of California Institute of Technology
GAM	- Guided Aircraft Missile
	- Ground-to-Air Missile (now SAM)
GAMAD	- Gas Monitor and Adjuster
GAMM	- Gesellschaft für Angewandte Mathematik und Mechanik (German Association for Applied Mathematics and Mechanics)
GAMMA	- Generalized Automatic Method of Matrix Assembly
GAO	- General Accounting Office
	- Main Astronomical Observatory (USSR)
GAP	- Government Aircraft Plant
GAPA	- Ground-to-Air Pilotless Aircraft
GAPL	- Group Assembly Parts List
GAR	- Growth Analysis and Review
	- Guided Aircraft Rocket
GARD	- Gamma Atomic Radiation Detector
	- Grumman-Alderson Research Dummies
GARD-TRAK	- Gamma Absorption and Radiation Detection Tracking
GARIOA	- Government and Relief in Occupied Areas
GAS	- General Automotive Service
	- Government of America Samoa
GASL	- General Applied Science Laboratory
GASP	- GRATIS and Simulator Production System
	- Gravity and Sun Pointing
GASR	- Guided Air-to-Surface Rocket
GAT	- General Aviation Transponder
	- Generalized Algebraic Translator
	- Georgetown Automatic Translator
	- Greenwich Apparent Time
	- Ground-to-Air Transmitter
GAT-2	- Generalized Algebraic Translator, 2nd Version
GATE	- Generalized Algebraic Translator Extended
GATE-20	- Generalized Algebraic Translator Extended for Bendix G-20
GATR	- Ground-Air Transmitter-Receiver
GATT	- General Agreement on Tariffs and Trade
	- Ground-to-Air Transmitter Terminal
GATU	- Geophysical Automatic Tracker Unit
GAW	- Guaranteed Annual Wage
GB	- Glide Bomb
GBAAFB	- Grand Bahama Auxiliary Air Force Base
GBI	- Grand Bahama Island
	- Ground Backup Instrument
GBL	- Government Bill of Lading
GBR	- Gun, Bomb, and Rocket

GC	- Gigacycles
	- General Counsel
	- Ground Control
G-C	- Guanine-Cytosine
G&C	- Guidance and Control
GCA	- Ground Controlled Approach
	- Group Capacity Analysis
	- Guidance Coupler Assembly
GC&A	- Guidance, Control, and Airframe
GCAL	- Gram Calorie
GCC	- Ground Control Center
G&CC	- Guidance and Control Coupler
GCD	- Gain Control Driver
	- Greatest Common Divided
GCE	- Ground Control Equipment
G&CEP	- Guidance and Control Equipment Performance
GCG	- Guidance Control Group
GCI	- Ground Controlled Intercept
GCIS	- Ground Controlled Intercept Squadron
GCL	- Ground Controlled Landing
GCM	- Government Communications Manager
GCO	- Guidance Control Officer
GCOE	- Ground Control Operational Equipment
GCPS	- Gigacycles per Second
GCR	- Gaseous Core Reactor
	- Ground Controlled Radar
GCRB	- Gas-Cooled Reactor Branch
GCRE	- Gas-Cooled Reactor Experiment
G/CS	- Guidance and Control System
GCT	- Greenwich Civil Time
GCTS	- Ground Communications Tracking System
GCU	- Gyroscope Coupling Unit
GD	- General Dynamics
GDA	- Gun Defended Area
GD-A	- General Dynamics—Astronautics
GD-A	- General Dynamics-Astronautics
GDC	- General Dynamics Corporation
	- Ground Digit Control
GD/C	- General Dynamics/Convair
GD/D	- General Dynamics/Daingerfield
GD/E	- General Dynamics/Electronics
GDF	- Gas Dynamic Facility
GD/FW	- General Dynamics/Forth Worth
GDG	- Group Display Generator
GDMT	- Gemini Detailed Maneuver Table
GDOP	- Geometric Dilution of Precision

■ GDP

GDP	- General Development Plan
GD/P	- General Dynamics/Pomona
GE	- Gas Ejection
	- General Electric Company
GE-ANPD	- General Electric Aircraft Nuclear Propulsion Department
GE/B	- General Electric/Burroughs
GEC	- General Electrodynamics Corporation
GECOM	- General Compiler
GECS	- Ground Environmental Control System
GED	- Gasoline Engine Driven
GEDL	- General Electric Data Link
GEEIA	- Ground Electronics Engineering Installation Agency
GEEP	- General Electric Electronic Processor
GEERS	- Groupe Europeen d'Etudes pour les Recherches Spatiales
GEESE	- General Electric Electronic System Evaluator
GEF	- Ground Equipment Failure
GEFS	- General Electric Flame Site
GEM	- Ground Effects Machine
	- Guidance Evaluation Missile
GEN	- General
	- Generator
GENDARME	- Generalized Data Reduction, Manipulation, Evaluation
GENDEP	- General Depot
GENESIS	- General NORAD Environment Simulation and Subsystem
GENEX	- General NORAD Environment Simulation Subsystem Executive Program
GEOD	- Geodetic
GEOL	- Geological
GEON	- Gryo-Erected Optical Navigation
GEOREF	- Geographical Reference (World Geographic Reference System)
GEOREFS	- Geographical Reference System
GEOS	- Geodetic Satellite
GEOSCAN	- Ground-Based Electronic Omnidirectional Satellite Communications Antenna
GEP	- Goddard Experimental Package
	- Ground Environment Program
GEPAC	- General Electric Programmable Automatic Computer
GEPURS	- General Electric General Purpose
GERM	- Ground Effects Research Machine
GERSIS	- General Electric Range Safety Instrumentation System
GERTS	- General Electric Radio Tracking System
GERV	- General Electric Re-Entry Vehicle
GESOC	- General Electric Satellite Orbit Control
GET	- Ground Elapsed Time
GETIS	- Ground Environment Team of the International Staff (NATO)

GETOL	- Ground Effect Take-Off and Landing
GETR	- General Electric Test Reactor
GETS	- Generalized Electronic Trouble Shooting
	- General Track Simulation Program
	- Ground Equipment Test Set
GETWS	- Get Word from String
GEV	- Giga-Electron-Volt
	- Ground Effects Vehicle
GEVIC	- General Electric Variable Increment Computer
GF	- Generator Field
GFA	- Government Furnished Article
	- Gunfire Area
GFADS	- Grand Forks Air Defense Sector
GFAE	- Government Furnished Aircraft Equipment
GFC	- Gas Filled Counter
	- Government Furnished Container
GFD	- Gap-Filler Data
GFE	- Government Furnished Equipment
GFF	- Government Furnished Facilities
GFI	- Gap-Filler Input
GFIP	- Gross Fault Indicator Panel
GFM	- Government Furnished Material
	- Gravitational Field Measurements
GFO	- Gap-Filler Output
GFP	- Government Furnished Property
GFP&S	- Government Furnished Property and Service
GFR	- Gap-Filler Radar
	- Government Facility Request
GFRC	- Gas-Flow Radiation Counter
GFSS	- Gunfire Support Ship
GFV	- Guided Flight Vehicle
GG	- Gas Generator
	- Ground to Ground
GGE	- Ground Guidance Equipment
GGM	- Ground-to-Ground Missile (now SSM)
GGO	- Main Geophysical Observatory (USSR)
GGR	- Ground Gunnery Range
GH	- Grid Heading
GH_2	- Gaseous Hydrogen
GHA	- Greenwich Hour Angle
	- Ground Hazard Area
GHE	- Gaseous Helium
	- Ground Handling Equipment
GHOST	- Global Horizontal Sounding Technique
GHP	- Geographical Position (also GP)
GHQ	- General Headquarters

■ GHT

GHT	- Golden Hour Tango
GI	- Galvanized Iron
	- General Input
	- General Instrument
	- Goddard, Incorporated
	- Government Issue
	- Grid Interval
GIANT	- General Item and Table
GICS	- Global Instrumentation Control System
	- Ground Instrumentation and Communication System
GID	- General Installation Dolly
GIE	- Ground Instrumentation Equipment
GIER	- General Industrial Equipment Reserve
GIF	- Gulf IT to FORTRAN Translator
GIFS	- Guggenheim Institute of Flight Structures
GIGI	- Gamma Inspection of Grain Integrity
GIM	- State Institute of Meteorology (USSR)
GIMRADA	- Geodesy, Intelligence, and Mapping Research and Development Agency
GIPS	- Ground Information Processing System
GIPSE	- Gravity-Independent Photo-Synthetic Gas Exchanger
GIRD	- Gruppa Isutcheniya Reaktivnovo Dvisheniya (Russian Group for Investigation of Reactive Motion)
GIRLS	- Generalized Information Retrieval and Listing System
GIS	- Ground Instrumentation System
GIT	- General Information Test
GIUK	- Greenland-Iceland-United Kingdom
GKI	- General Kinetics, Incorporated
GL	- Gun Laying
GLA	- General Laboratory Associates, Incorporated
GLADS	- Great Falls Air Defense Sector
GLAKES	- Great Lakes
GLANCE	- Global Lightweight Air Navigation Computing Equipment
GLC	- Gas-Liquid Chromatography
GLDS	- Gemini Launch Data System
GLE	- Government Loaned Equipment
GLEEP	- Graphite Low Energy Experimental Pile
GLIPAR	- Guide Line Identification Program for Anti-Missile Research
GLM	- Government Loaned Material
GLO	- Ground Liaison Officer
GLOBECOM	- Global Communications
GLOCOM	- Global Communications
GLOMB	- Glide Bomb
GLOPAC	- Gyroscopic Low Power Attitude Control
GLOTRAC	- Global Tracking Network

GLV	- Gemini Launch Vehicle
	- Glove Valve
GM	- General Motors
	- Gravity Meter
	- Grid Modulation
	- Guided Missile
G/M	- Groups per Message
GMA	- Government Modification Authorization
GMAB	- Guided Missile Assembly Building
GMAT	- Greenwich Mean Astronomical Time
GMAV	- Grumman Avenger
GMC	- General Motors Corporation
	- Gun Motor Carriage
GMCM	- Guided Missile Countermeasure
GM-CM	- Gram-Centimeter
GMD	- Ground Meteorological Detection
GMDEP	- Guided Missile Data Exchange Program
GM/DRL	- General Motors Defense Research Laboratories
GME	- Gimbal Module Electronics
GMEVALU	- Guided Missile Evaluation Unit
GMFLT	- Generic Master Flight Library Tape
GMGRU	- Guided Missile Group
GM/LOF	- Guided Missile Line of Flight
GMMO	- Guided Missile Maintenance Officer
GMOO	- Guided Missile Operations Officer
GMR	- Ground Mapping Radar
GMRD	- Guided Missile Research and Development
	- Guided Missiles Range Division
GMS	- Gemini Mission Simulator
	- General Military Science
GMSQUAD	- Guided Missile Squadron
GMSRON	- Guided Missile Service Squadron
GMT	- Greenwich Mean Time
GMU	- Guided Missile Unit
GN	- Gaseous Nitrogen
	- Grid Navigation
	- Grid Neutralization
	- Grid North
	- Guidance and Navigation
G/N	- General Note
G&N	- Guidance and Navigation
GN_2	- Gaseous Nitrogen
GNAL	- Georgia Nuclear Aircraft Laboratory
GNC	- Guidance and Navigation Computer
GNE	- Guidance and Navigation Electronics
GNEC	- General Nuclear Engineering Corporation

■ GNP

GNP	- Gross National Product
GNS	- Goose NORAD Sector
G&NS	- Guidance and Navigation System
GO	- General Order
	- General Output
GO_2	- Gaseous Oxygen
GOA	- General Operating Agency
	- Government of Australia
GOC	- Ground Observer Corps
GOCI	- General Operator-Computer Interaction
GOCO	- Government-Owned, Contractor-Operated
GOCR	- Gated-Off Controlled Rectifier
GOE	- Ground Operating Equipment
GOGO	- Government-Owned, Government-Operated
GOIE	- Government Owned Industrial Equipment
GOM	- Government Owned Material
GOP	- Ground Observer Post
	- Group of Paths
GOPL	- General Outpost Line
GOR	- General Operational Requirement
	- Ground Operational Requirement
GOREPS	- Ground Observer Report-Severe Weather
GOS	- General Operating Specification
GOSNIIGVF	- State Scientific-Research Institute of the Civil Air Fleet (USSR)
GOSS	- Ground Operational Support System
GOST	- All-Union State Standard (USSR)
GOTRAN	- Load and Go FORTRAN
GOVT	- Government
GOX	- Gaseous Oxygen
GP	- Gang Punch
	- Gaseous Propellant
	- Gas Pressure
	- Generalized Programming
	- General Purpose
	- Geographical Position
	- Glide Path
	- Group
	- Gun Pointer
GPA	- General Purpose Amplifier
	- Guidance Prearm
GPATS	- General Purpose Automatic Test System
GPC	- General Purpose Capsule
	- Government Publications Center
GPD	- Gallons per Day
	- Gimbal Position Display
	- GOSS Program Directorate

GPDC	- General Purpose Digital Computer
GPERF	- Ground Passive Electronic Reconnaissance Facility
GPES	- Ground Proximity Extraction System
GPG	- General Planning Group
	- Grains per Gallon
GPH	- Gallons per Hour
GPI	- General Precision, Incorporated
	- Gimbal Position Indicator
	- Ground Position Indicator
GPL	- General Precision Laboratory
GPLD	- Government Property Lost or Damaged
GPM	- Gallons per Minute
	- Groups per Message
	- Groups per Minute
GPO	- Gemini Project Office
	- Government Printing Office
GPS	- Gallons per Second
	- General Problem Solver
GPSF	- General Purpose Simulation Facility
GPSS	- General Process Simulation Studies
GPSSM	- General Purpose Surface-to-Surface Missile
GPSU	- Ground Power Supply Unit
GPU	- Gas Power Unit
GPV	- General Purpose Vehicle
GPX	- Generalized Programming Extended
GR	- General Reconnaissance
	- Germanium Rectifiers
	- Grid Resistor
	- Grid Return
	- Ground Run
	- Gunnery Range
GRADS	- Grand Forks Air Defense Sector
GRAPHDEN	- Graphical Data Entry
GRASER	- Gamma Ray Amplification by Simulated Emission of Radiation
GRASS	- Gamma-Ray Ablation Sensing System
GRATIS	- Generation, Reduction, and Training Input System
GRD	- Geophysics Research Directorate
	- Ground Detector
GRDPRO	- Grid Procedure
GREB	- Galactic Radiation Experiment Background Satellite
GRF	- Group Repetition Frequency
GRFO	- Gun Range Finder Operator
GRNC	- Group Not Counted
GRS	- Gamma Ray Spectrometry
GRT	- Gamma Ray Telescope
	- Generation of Recording Parameter

■ GRU

GRU	- Central Intelligence Administration (USSR)
GRWT	- Gross Weight
GS	- General Service
	- General Staff
	- General Support
	- Ground Speed
	- Ground Stabilized
	- Guide Slope
GSA	- General Service Administration
	- Ground Safety Approval
	- Guidance System Analyst
GSAP	- Gun Sight Aiming Point
GSC	- General Staff Corps
GSD	- General Situation Display
	- General Supply Depot
GSDFJ	- Ground Self Defense Force Japan
GSDS	- Goldstone Duplicate Standard
GSE	- General Systems Engineering
	- Ground Support Equipment
GSE-BI	- Ground Support Equipment—Base Installation
GSE-M	- Ground Support Equipment—Mechanical
GSE-ME	- Ground Support Equipment—Maintenance Equipment
GSEMF	- Ground Support Equipment—Maintenance Facility
GSE-S	- Ground Support Equipment—Structure
GSE-SE	- Ground Support Equipment—Support Equipment
GSE-SS	- Ground Support Equipment—Strategic System
	- Ground Support Equipment—System and Service
GSE-T&H	- Ground Support Equipment—Transportation and Handling
GSE-TS	- Ground Support Equipment—Test Stand
GSE-WSR	- Ground Support Equipment—Weapon System Requirement
GSFC	- Goddard Space Flight Center
GSHE	- Government Support Handling Equipment
GSI	- Government Serial Identification
	- Government Source Inspection
	- Ground Speed Indicator
GSME	- Ground Support Maintenance Equipment
GSO	- General Staff Officer
	- Ground Safety Officer
GSOR	- General Staff Operational Requirements
GSP	- Guidance Signal Processor
	- Ground Safety Plan
GSPO	- Ground Systems Project Officer
GSP-R	- Guidance Signal Processor-Repeater
GSR	- Galvonic Skin Response
	- Glide Slope Receiver
	- Group Selection Register

GSS	- Geodetic Stationary Satellite
	- Global Surveillance System
	- Ground Support System
GSSC	- General Systems Simulation Center
	- Ground Support Simulation Computer
GSSF	- General Service Stock Fund
	- Ground Special Security Forces
GSSO	- General Store Supply Office
GST	- General Survey Test
	- Greenwich Sidereal Time
	- Ground Special Tools
GSTE	- Ground Support Test Equipment
GSTP	- Ground System Test Procedure
GSV	- Globe Stop Valve
	- Greenwich Mean Time (USSR)
	- Guided Space Vehicle
GT	- Gas Tight
	- Gate Tube
	- Glide Torpedo
	- Grease Trap
	- Grid Track
	- Gross Tons
G-T	- Gemini-Titan
GTA	- Graphic Training Aid
GTC	- Gain Time Control
GTD	- Gas Turbine Engine (USSR)
GTE	- Ground Test Equipment
GT&E	- General Telephone and Electronics
GTG	- Gas Turbine Generator
	- Ground Timing Generator
GTI	- Grand Turk Island
GTK	- Grand Turk (GTI preferred)
GTM	- Ground Test Missile
GTP	- General Test Plan
	- Ground Test Plan
	- Ground Test Program
GTR	- Ground Test Reactor
GT&R	- General Tire and Rubber Company
GTTC	- Gulf Transportation Terminal Command
GTTF	- Gas-Turbine Test Facility
GTV	- Gate Valve
	- Ground Transport Vehicle
	- Guidance Test Vehicle
	- Guided Test Vehicle
GTW	- Gross Take-Off Weight
GU	- Guidance Unit

■ GUISE

GUISE	- Guidance System Evaluation
GUSTO	- Guidance Using Stable Tuning Oscillations
GV	- Group Velocity
GVS	- Ground Vibration Survey
GvsT	- Deceleration units of Gravity versus Time
GvsV	- Deceleration units of Gravity versus Velocity
GW	- Gravity Wave
	- Guerrilla Warfare
	- Guided Weapon
GWC	- Global Weather Central
GWR	- General Work Release
GWT	- Gross Weight
GYI	- Grand Canary Island (remote site)
GYM	- Guaymas, Mexico (remote site)
GYRO	- Gyroscope
GVS	- Gas Vortex System
	- Ground Vibration Survey
GZ	- Ground Zero
GZT	- Greenwich Zone Time
GZTPRD	- Ground Zero Tape Read Program

H

H	- Hard
	- Hardware
	- Hatch
	- Heavy
	- Hecto
	- Height
	- Helicopter
	- Henry
	- High
	- Hostile
	- Hydrogen
	- Hyperbolic
H+	- Minutes Past The Hour
HA	- Hand Actuated
	- Heavy Artillery
	- High Altitude
	- Hour Angle
H/A	- Hazardous Area
HAA	- Heavy Antiaircraft Artillery
	- High Altitude Abort
HAATC	- High Altitude Air Traffic Control
HAAW	- Heavy Anti-Tank Assault Weapon
HAB	- High Altitude Bombing
HAC	- Hughes Aircraft Company
HAD	- High Accuracy Data System
	- High Altitude Density Rocket (Australia)
	- Hypersonic Aerothermal Dynamics
HADC	- Holloman Air Development Center
HADD	- Hawaiian Air Defense Division
HADES	- Hypersonic Air Data Entry System
HADTS	- High Accuracy Data Transmission System
HAF	- High Altitude Facility

■ HAFB

HAFB	- Hill Air Force Base
	- Holloman Air Force Base
HAIRS	- High Altitude Test and Evaluation of Infrared Sources
HAM	- Heavy Automotive Maintenance
	- High Activity Mode
	- High-Speed Automatic Monitor
HAMD	- Helicopter Ambulance Medical Detachment
HAMOS	- High Altitude Synoptic Meteorological Observation
HAP	- High Altitude Probe
HAPO	- Hanford Atomic Products Operation
HAR	- Harbor
HARAO	- Hartford Aircraft Reactors Area Office
HARD	- High Altitude Relay Point
HARDEFDIV	- Harbor Defense Division
HARDTS	- High Accuracy Radar Data Transmission System
HARE	- High Altitude Recombination-Energy Propulsion
HARK	- Hardened Re-Entry Kill
HARM	- High Altitude Recovery Mission
HARP	- High Altitude Relay Point
	- High Altitude Research Project
HAS	- High Altitude Sampler
HASCO	- Helicopter and Airplane Services Corporation
HASP	- High Altitude Sampling Program (Air Force)
	- High Altitude Sounding Projectile (Navy)
	- High Altitude Space Probe
HAT	- Handover Transmitter
	- High Altitude Test
	- High Altitude Testing Rocket (Australia)
HATS	- High Altitude Terrain Contour Data Sensor
HAW	- Heavy Anti-Tank Weapon
	- Kauai Island, Hawaii (remote site)
HAWK	- Homing All The Way Killer
HAYSTAQ	- Have You Stored Answers to Questions
HAZ	- Heat-Affected Zone
HB	- Handbook
	- Heavy Bombardment
	- Height of Burst
	- Homing Beacon
HBS	- Harbor Boat Service
HBW	- Hot Bridge Wire
HC	- Hand Control
	- Hard Copy
	- Heat Content
	- Height Console
	- High Capacity
	- High Carbon
	- Holding Coil

164

H/C	- Hand Carry
HCC	- Hawaii Control Center
HCD	- Hughes Communications Division
HCF	- Highest Common Factor
HCG	- Horizontal Location of Center of Gravity
HCL	- Horizontal Center Line
HCN	- Handbook Change Notice
HCR	- Hard Copy Response
HCU	- Homing Comparator Unit
HD	- Half Duplex
	- Harbor Defense
H/D	- Half Duplex
H&D	- Hardened and Dispersed
HDATZ	- High Density Air Traffic Zone
HDC	- Harbor Defense Command
	- Helicopter Direction Center
HDD	- Human Disorientation Device
HDDS	- High Density Data System
HDEP	- High Density Electronic Packaging
HDI	- Human Development Institute
HDOC	- Handy Dandy Orbital Computer
HDP	- High Density Plasma
	- High Density Polyethylene
	- Horizontal Data Processing
HDT	- Heat Distortion Temperature
HDW	- Hardware
HDX	- Half Duplex
HE	- Handling Equipment
	- Heat Exchanger
	- Heavy Enamel
	- Helium
	- High Energy
	- High Explosive
	- Human Engineering
	- Hydraulic Equipment
	- Hydromagnetic Emissions
H/E	- Heat Exchanger
HEAA	- High Explosive Antiaircraft
HEAP	- High Explosive Armor-Piercing
HEAPS	- Hawaiian Environmental Analysis and Prediction System
HEAT	- High Explosive Anti-Tank
HEAT-T	- High Explosive Anti-Tank with Tracer
HEDCOM	- Headquarters Command
HEDS	- High Explosive, Discarding Sabot
HEEP	- Highway Engineering Exchange Program
HEF	- High Energy Fuel

HEI

HEI	- High Explosive Incendiary
HEISD	- High Explosive Incendiary Self-Destroying
HEIT	- High Explosive Incendiary with Tracer
HEITDISD	- High Explosive Incendiary with Dark Ignition Tracer Self-Destroying
HEL	- Human Engineering Laboratory
HELIOS	- Heteropowered Earth-Launched Inter-Orbital Spacecraft
HELITEAM	- Helicopter Team
HELP	- Helicopter Electronic Landing Path
HELPR	- Handbook of Electronic Parts Reliability
HEM	- Hybrid Electromagnetic Wave
HEMP	- Heuristic Economic Military and Political Model
HEP	- High Explosive Plastic
	- High Explosive with Plugged Tracer Cavity
HEPAT	- High Explosive Plastic Anti-Tank
HEPP	- Hoffman Evaluation Program and Procedure
HEPT	- High Explosive Plastic with Tracer
HERA	- High Explosive Rocket Assisted
HERALDS	- Harbor Echo Ranging and Listening Devices
HERF	- High Energy Rate Forming
HERJ	- High Explosive Ramjet
HERO	- Hazards of Electromagnetic Radiation to Ordnance
HES	- High Explosive with Spotting Charge
HESD	- High Explosive Self-Destroying
HET	- High Explosive with Tracer
HETDI	- High Explosive with Tracer Dark Ignition
HETP	- Height Equivalent to a Theoretical Plate
HETS	- Height Equivalent to a Theoretical Stage
	- Hyper-Environmental Test System
HETSD	- High Explosive with Tracer, Self-Destroying
HEW	- Health, Education, and Welfare
HEX	- High Explosive
HF	- Height Finder
	- Height Finding
	- High Frequency (band 7)
	- Houston Fearless Corporation
	- Human Factors
	- Hydrofluoric Acid
H/F	- Held For
	- Human Factors
H-F	- Hydrogen and Fluorine
HFA	- High Frequency Recovery Antenna
HFC	- Height Finder Central
HFD	- Human Factor Design
	- Human Factors Division
HFDF	- High Frequency Direction Finder

HI-STEP

HFIC	- Human Factors Information Center
HFO	- Height Finder Operator
	- High Frequency Oscillator
HFORL	- Human Factors Operation Research Laboratory
HFS	- Human Factors Society
HFT	- Height Finder Technician
HFX	- High Frequency Transceiver
HG	- Horizontal Gyro
HGA	- High Gain Antenna
HGB	- Hemoglobin
HGE	- Handling Ground Equipment
HGRF	- Hot Gas Radiating Facility
HI	- Hydraulic Institute
HIAC	- High Accuracy Radar
HIAD	- Handbook of Instructions for Aircraft Designers
HIAGSED	- Handbook of Instructions for Aircraft Ground Support Equipment Designers
HIAPSD	- Handbook of Instructions for Aerospace Personnel Subsystem Designers
HIBAL	- High Altitude Balloon
HIBEX	- High Impulse Booster Experiment
HICAPCOM	- High Capacity Communication System
HICOG	- United States High Commissioner for Germany
HICOM	- Allied High Commissioner for Germany
HICOMRY	- High Commissioner of the Ryukyu Islands
HICOMTERPACIS	- High Commissioner, Territory Pacific Islands
HIEX	- High Explosive
HI-FI	- High-Fidelity
HIFOR	- High-Level Weather Forecast
HIG	- Hermetically-Sealed Integrating Gyroscope
HIGED	- Handbook of Instructions for Ground Equipment Designers
HIHOE	- Hydrogen, Ions, Helium, and Oxygen in the Exosphere
HIP	- Hazard Input Program
HIPAR	- High Power Acquisition Radar
HIPERNAS	- High Performance Navigation System
HIPO	- Hospital Indicator for Physicians' Orders
HIPOT	- High Potential Test
HIR	- Hydrostatic Arming Impact Firing Rocket
HIRAC	- High Random Access
HIRAN	- High Precision SHORAN
HIRES	- Hypersonic In-Flight Refueling Subsystem
HIS	- Hardware Interrupt System
HISCS	- Helmet Integrated Sighting and Control System
HISS	- High Intensity Sound Simulator
HI-STEP	- High-Speed Integrated Space Transportation Evaluation Program

■ **HIT**

HIT	- Hypersonic Interference Technology
HITCO	- H.I. Thompson Company
HIVAP	- High Velocity Armor Piercing
HIVOS	- High Vacuum Orbital Simulator
HJ	- Honest John
	- Hose Jacket
HK	- Hunter-Killer
H-K	- Hunter-Killer
HKMP	- Hypervelocity Kill Mechanism
HLAS	- Hot Line Alert System
HLCC	- Hardened Launch Control Center
HLCF	- Hardened Launch Control Facility
HLL	- Hard Lunar Landing
HLTL	- High-Level Transistor Logic
HMC	- Heading Marker Correction
	- Howitzer Motor Carriage
HMED	- Heavy Military Electronics Department
HMG	- Heavy Machine Gun
HMI	- Handbook Maintenance Instructions
HMU	- Hydraulic Mockup
HNI	- Holmes and Naver, Incorporated
HO	- Hydrographic Office
H_2O	- Water
HOA	- Heavy Observation Aircraft
HOB	- Height of Burst
	- Homing on Offset Beacon
HOBS	- High Orbital Bombardment System
HOC	- Handover Coordinator
HOFCO	- Horizontal Functional Checkout
HOI	- Headquarters Office Instruction
HOJ	- Home On Jamming
HOR	- Hydrogen-Oxygen Reaction
HOT	- Hand Over Transmitter
HOTR	- Hydraulic Oil Test Report
HOU	- Houston OIS (Operational Instrumentation System) Facility
HOVI	- Handbook of Overall Instructions
HP	- Hewlett-Packard
	- High Pass
	- High Pressure
	- Horizontal Polarization
	- Horsepower
H/P	- Horizontal Polarization
H-P	- High Pressure
HPA	- High Power Amplifier
HPAG	- High Performance Air to Ground

H/S

HPC	- Hercules Powder Company
	- Houston Petroleum Center
HPCC	- High Performance Control Center
HPD	- Hard Point Defense
	- High Performance Drone
HPF	- Highest Possible Frequency
HP-HR	- Horsepower-Hour
HPM	- Honeycomb Propellant Matrix
HPPS	- Hewlett-Packard Printer Submodule
HPR	- Halt and Proceed
HPRR	- Health Physics Research Reactor
HPT	- High Point
	- High Pressure Test
	- Horizontal Plot Table
HPU	- Hydraulic Power Unit
	- Hydraulic Pumping Unit
HQ	- Headquarters
HQC	- Headquarters Command
HQCOMDUSAF	- Headquarters Command, United States Air Force
HQMC	- Headquarters, Marine Corps
HR	- Hand Reset
	- Historical Record
	- Hit Rate
	- Hot Rolled
	- Hour
H&R	- Hold and Reconsignment
HRAC	- Hypersonic Research Aircraft
HRE	- Homogeneous Reactor Experiment
HRF	- Height-Range Finder
HRI	- Height-Range Indicator
HRIO	- Height-Range Indicator Operator
HRL	- Historical Record Log
	- Horizontal Reference Line
HRP	- Holding and Reconsignment Point
HRRC	- Human Resources Research Center
HRS	- Hovering Rocket System
HRT	- High Resolution Tracker
HRV	- Hypersonic Research Vehicle
HS	- Head Suppression
	- Heat Shield
	- Height Supervisor
	- High Speed
	- Hot Short
	- Hydrogen Sulfide
	- Hydrostatic
H/S	- Heat Shield

■ H-S

H-S	- Hamilton Standards
	- Horizon Scanner
	- Horizon Sensor
H&S	- Headquarters and Service
HSA	- Heat Shield Abort
HSB	- Heat Shield Boost
HSBR	- High-Speed Bombing Radar
HS/C	- House Spacecraft
HSD	- High Speed Data
HSE	- Heat Shield Entry
HSI	- High Speed Impact
	- Horizontal Situation Indicator
HSJ	- Heat Shield Jettison
HSP	- High Speed Printer
HSR	- Heat Shield Recovery
	- High Speed Reader
HSS	- High Speed Steel
HST	- Hawaiian Standard Time
HSU	- Hydraulic Supply Unit
HSZD	- Hermetically-Sealed Zener Diode
HT	- Handling Time
	- Heat Treat
	- Height
	- Height of Target
	- Height Technician
	- Height Telling
	- High Tension
H/T	- Heat Treat
H&T	- Handling and Transportation
HTA	- Heavier Than Air
	- High Temperature Air
HTC	- Half Ton Charge
	- High Temperature Chamber
	- Hughes Tool Company
HTC-AD	- Hughes Tool Company-Aircraft Division
HTCI	- High Tensile Cast Iron
HTD	- Hand Target Designator
HTGR	- High Temperature Gas-Cooled Reactor
HTL	- Helicopter Transportable Launcher
HTM	- High Temperature Materials
HTR	- Halt and Transfer
HTRAC	- Half Track
HTRAP	- Height Reply Analysis Processor
HTRE	- Heat Transfer Reactor Experiment

HTS	- Hawaii Tracking Station
	- Height Telling Surveillance
	- High Tensile Steel
HTSIM	- Height Simulator
HTSUP	- Height Supervisor
HTT	- Heavy Tactical Transport
HTU	- Height of Transfer Unit
HTV	- High-Altitude Test Vehicle
	- Homing Test Vehicle
	- Hypersonic Test Vehicle
HUCR	- Highest Useful Compression Ratio
HUFF-DUFF	- High Frequency Direction Finder
HUGO	- Highly Usable Geophysical Observation
HUK	- Hostile, Unknown, Known
	- Hunter-Killer
HUKB	- Hostile, Unknown, Faker, and Big Photo
HUKFORLANT	- Hunter-Killer Forces, Atlantic
HUKP	- Hostile, Unknown, Faker, Pending Track Identities
HUKS	- Hostile, Unknown, Faker, Special Track Identities
	- Hunter-Killer Submarine
HUMRRO	- Human Resources Research Office
HUS	- Helicopter Utility Squadron
HUTRON	- Helicopter Utility Squadron
HV	- High Velocity
	- High Voltage
	- Hypervelocity
H&V	- Hazard and Vulnerability
HVAP	- High-Velocity Armor-Piercing
	- Hypervelocity Armor-Piercing
HVAPDS	- Hypervelocity, Armor-Piercing, Discarding Sabot
HVAPDSFS	- Hypervelocity, Armor-Piercing, Discarding Sabot, Fin Stabilized
HVAPDS-T	- Hypervelocity, Armor-Piercing, Discarding Sabot with Tracer
HVAR	- High-Velocity Aircraft Rocket
HVAT	- High-Velocity Anti-Tank
HVC	- Hardened Voice Channel
	- Hardened Voice Circuit
HVDC	- High Voltage Direct Current
HVDF	- High and VHF Direction Finding
HVPS	- High Voltage Power Supply
HVR	- High Vacuum Rectifier
HVSCR	- High Voltage Selenium Cartridge Rectifier
HVTP	- High-Velocity Target Practice
	- Hypervelocity Target Practice
HVTPDS	- Hypervelocity Target Practice, Discarding Sabot

■ HVTP-T

HVTP-T	-	Hypervelocity, Target Practice with Tracer
HW	-	Head Wind
	-	Hot Wire
HWT	-	Hypersonic Wind Tunnel
H/X	-	Heat Exchanger
HY	-	Hydraulic
HYCOL	-	Hybrid Computer Link
HYDAC	-	Hybrid Digital-Analog Computer
HYDAPT	-	Hybrid Digital-Analog Pulse Time
HYPERDOP	-	Hyperbolic Doppler
HZ	-	Hazard

I

I	- Current
	- Input
	- Interceptor
	- Inverter
	- Iodine
IA	- Immediately Available
	- Indicated Altitude
	- Initial Appearance
	- Input Axis
	- Intercept Arm
I&A	- Information and Action
IAB	- Instrumentation Analysis Branch
IAC	- Instrument Approach Chart
	- Integration Assembly Checkout
	- International Algebraic Compiler
IACOMS	- International Advisory Committee on Marine Sciences
IACS	- International Annealed Copper Standard
IAD	- Initiation Area Discriminator
	- Interface Analysis Document
	- International Astrophysical Decade
IADC	- Inter-American Defense College
IADB	- Inter-American Defense Board
	- Inter-American Development Bank
IADF	- Icelandic Air Defense Force
IADPC	- Inter-Agency Data Processing Committee
IAE	- Integral of Absolute Error
	- Integrated Absolute Error
IAEA	- International Atomic Energy Agency
IAEC	- International Atomic Energy Commission
IAF	- Information and Forwarding
	- International Astronautical Federation
IAGA	- International Association of Geomagnetism and Aeronomy
IAGC	- Instantaneous Automatic Gain Control

173

■ IAGS

IAGS	- Inter-American Geodetic Society
IAI	- Informational Acquisition and Interpretation
IAL	- International Algebraic Language (now ALGOL)
IAM	- International Association of Machinists
IAMS	- Instantaneous Audience Measurement System
IANEC	- Inter-American Nuclear Energy Commission
IAO	- Internal Automation Operation
IAOD	- In Addition to Other Duties
IAP	- Initial Approach
IAPC	- Inter-American Peace Committee
	- International Administration of Physical Oceanography
IAPO	- Interchangeable at Attach Points Only
	- International Association of Physical Oceanography
IAR	- Inspection Acceptance Record
IAS	- Indicated Air Speed
	- Institute for Advanced Study
	- Institute of Aeronautical Sciences (now AIAA)
	- Instrument Approach System
IASOR	- Ice and Snow on Runway
IASPEI	- International Association of Seismology and Physics of the Earth's Interior
IASR	- Intermediate Altitude Sounding Rocket
IAT	- Indexible Address Tag
	- Indicated Air Temperature
	- Individual Acceptance Test
IATA	- International Air Transport Association
	- Is Amended to Add
IATC	- International Air Traffic Communications
IATCS	- International Air Traffic Communications System
IATD	- Is Amended to Delete
IATR	- Is Amended to Read
IAU	- International Astronomical Union
IAVC	- Instantaneous Automatic Volume Control
IAW	- In Accordance With
IAWR	- Institute for Air Weapons Research
IAZ	- Inner Artillery Zone
IB	- Incendiary Bomb
	- Instruction Book
IBDA	- Indirect Bomb Damage Assessment
IBF	- Institute of Biological Physics (USSR)
IBL	- Intermediate Behavioral Language
IBM	- International Business Machines Corporation
IBMCE	- International Business Machines Customer Engineer
IBRD	- International Bank for Reconstruction and Development
IBRL	- Initial Bomb Release Line
IBS	- Impact Bag System

IC	- ILLIAC Chamber
	- Image Chamber
	- Image Check
	- Incentive Contract
	- Inertial Component
	- Information Center
	- Initial Complement
	- Initial Condition
	- Instruction Counter
	- Integrated Circuits
	- Interceptor Computer
	- Intercommunications (intercom)
	- Interface Control
	- Internal Combustion
I/C	- Intercom
I&C	- Installation and Calibration
	- Installation and Checkout
	- Instrument and Controls
	- Integration and Checkout
ICA	- Industrial Communications Association
	- Item Change Analysis
ICAF	- Industrial College of the Armed Forces
ICAO	- International Civil Aviation Organization
ICAR	- Integrated Command Accounting and Reporting
ICARUS	- Intercontinental Aerospacecraft Range Unlimited System
ICAS	- Intermittent Commerical and Amateur Service
	- International Council of the Aeronautical Sciences
ICBM	- Intercontinental Ballistic Missile
ICBMS	- Intercontinental Ballistic Missile System
ICBS	- Interconnected Business System
ICC	- Instrumentation Control Center
	- International Computation Center
	- Interstate Commerce Commission
IC&C	- Instrumentation Calibration and Checkout
ICCM	- Intercontinental Cruise Missile
ICD	- Interface Control Document
	- Interface Control Drawing
ICDCP	- Interface Control Drawing Change Proposal
ICE	- Input Checking Equipment
	- Internal Combustion Engine
ICED	- Interface Control Envelope Drawing
	- Interface Control Environment Drawing
ICEF	- International Cooperative Emulsion Flight
ICER	- Infrared Cell, Electronically Refrigerated
ICES	- International Committee for Earth Sciences
ICET	- International Center of European Training

■ ICFATCMUTAL

ICFATCMUTAL	- Individual is Cleared for Access to Classified Material Up to and Including
ICGM	- Intercontinental Glide Missile
ICIC	- Interdepartmental Commission on Interplanetary Communications
ICIE	- International Council of Industrial Editors
ICIP	- International Conference on Information Processing
ICL	- Instrumentation Configuration Log
ICM	- Instrumentation and Communications Monitor
	- Intercontinental Missile
ICMP	- Inter-Channel Master Pulse
ICNAF	- International Committee of North American Federation
ICO	- Interagency Committee on Oceanography
	- International Commission for Optics
I&CO	- Installation and Checkout
ICON	- Integrated Control System
ICOSS	- Inertial Command Off-Set System
ICP	- Inventory Control Point
ICPAC	- Instantaneous Compressor Performance Analysis Computer
ICPC	- Inter-Range Communication Planning Committee
ICR	- Instantaneous Center of Rotation
	- Interrupt Control Register
ICS	- Interagency Communications System
	- Intercommunication System
	- Interim Change of Station
	- Interphone Control Station
ICSAL	- Integrated Communications System, Alaska
ICSR	- Inter-American Committee for Space Research
ICSRD	- Interdepartmental Committee on Scientific Research and Development
ICSU	- International Council of Scientific Unions
ICT	- Igniter Circuit Tester
	- Inspection Control Test
	- Insulating Core Transformer
	- Interface Control Tool
	- Interference Compliance Test
	- Internal COMPOOL Table
	- International Critical Tables
	- International Telecommunication Union Radio Conference
ICTE	- Inertial Component Test Equipment
ICTP	- Intensified Combat Training Program
ICU	- Instruction Control Unit
ICUS	- Inside Continental United States
ICV	- Internal Correction Voltage
ICW	- In Compliance With
	- Interrupted Continuous Wave

ICWG	- Interface Control Working Group
ID	- Identification
	- Identity
	- Induced Draft
	- Inside Diameter
	- Intercept Direction
	- Interdivisional
	- Interferometer and Doppler
IDA	- Immediate Damage Assessment
	- Indirect Damage Assessment
	- Input Data Assembler
	- Institute for Defense Analyses
	- Integration and Delivery Area
	- Intercept Distance Aid
	- International Development Association
IDABEE	- Institute of Defense Analysis Compiler
IDACON	- Iterative Differential Analyzer Control
IDAP	- Iterative Differential Analyzer Pinboard
IDAS	- Iterative Differential Analyzer Slave
IDC	- Indoctrination Direction Center
	- Information and Direction Center
	- Interceptor Distance Computer
	- Interdepartmental Communication
IDCN	- Interchangeability Document Change Notice
IDDS	- Instrumentation Data Distribution System
IDE	- Industrial Developed Equipment
IDEP	- Interservice Data Exchange Program
IDES	- Information and Data Exchange System
IDF	- Intermediate Distributing Frame
	- International Distress Frequency
IDFR	- Identified Friendly
IDGIT	- Integrated Data Generation Implementation Technique
IDI	- Improved Data Interchange
	- Industrial Designers Institute
IDIA	- Internal Defense Identification Area
IDIOT	- Instrumentation Digital On-Line Transcriber
IDL	- Identured Drawing List
	- Instrument Development Laboratories
	- Interdiscrepancy Laboratory
IDO	- Identification Officer
	- Internal Distribution Only
IDP	- Input Data Processor
	- Integrated Data Presentation
	- Integrated Data Processing
	- Intermodulation Distortion Percentage
IDPC	- Integrated Data Processing Center

■ IDPG

IDPG	- Impact Data Pulse Generator
IDQA	- Individual Documented Quality Assurance
IDR	- Industrial Data Reduction
IDS	- Identification Section
	- International Data Systems, Incorporated
IDSTO	- Idle Used for Storage
IDT	- Identification Technician
IDTS	- Instrumentation Data Transmission System
IDTSC	- Instrumentation Data Transmission System Controller
IDU	- Indicator Drive Unit
IDWA	- Interdivision Work Authorization
IDWO	- Interdepartmental Work Order
IE	- Infrared Emission
	- Initial Equipment
I&E	- Information and Education
IEA	- Instruments Electronics Automation
IEC	- International Electric Company
	- International Electronics Corporation
	- International Electrotechnical Commission
IECL	- Instrumentation Equipment Configuration Log
IED	- Initial Effective Data
IEE	- Institute of Electrical Engineers (now IEEE)
IEEE	- Institute of Electrical and Electronic Engineers (formerly AIEE, IEE and IRE)
IEI	- Industrial Education Institute
IEL	- Information Exchange List
IEO	- Installation Engineers Office
IER	- Industrial Equipment Reserve
IERC	- Industrial Equipment Reserve Committee
IES	- Illuminating Engineers Society
IESD	- Instrumentation and Electronic Systems Division
IET	- Initial Engine Test
IETF	- Initial Engine Test Firing
IF	- Ice Fog
	- Information Collector
	- Infrared
	- Intermediate Frequency
I/F	- Interface
	- Image-to-Frame Ratio
IFA	- In-Flight Analysis
	- Intermediate Frequency Amplifier
IFB	- Invitation For Bid
IFC	- Indicated Final Cost
	- Instantaneous Frequency Correlation
	- Integrated Fire Control
	- International Finance Corporation

IGY

IFCN	- Interfacility Communication Network
IFCS	- International Federation of Computer Sciences
IFDC	- Integrated Facilities Design Criteria
IFF	- Identification, Friend or Foe
	- Ionized Flow Field
IFIP	- International Federation of Information Processing
IFIPC	- International Federation of Information Processing Congress
IFIPS	- International Federation of Information Processing Societies
IFIS	- Integrated Flight Instrument System
IFLTT	- Intermediate Focal Length Tracking Telescope
IFM	- In-Flight Maintenance
IFP	- S. I. Vavilov Institute of Physical Problems (USSR)
IFPM	- In-Flight Performance Monitoring
IFR	- In-Flight Refueling
	- Instrument Flight Rules
	- Intermediate Frequency Range
IFRB	- International Frequency Registration Board
IFRU	- Interference Rejection Unit
IFS	- Identification Friend or Foe Switching Circuit
	- Integrated Flight System
	- Intermediate Frequency Strip
IFT	- In-Flight Test
	- Intermediate Frequency Transformer
IFTA	- In-Flight Thrust Augmentation
IFTM	- In-Flight Test and Maintenance
IFTS	- In-Flight Test System
IFTSSS	- In-Flight Test System Scan Select
IG	- Inertial Guidance
	- Inner Gimbal
	- Inspector General
	- Instantaneous Grid
IGA	- Inner Gimbal Angle
	- Inner Gimbal Axis
IGAR	- Intermediate Gimbal Axis Rotar
IGAUP	- Interceptor Generation and Umpiring Program
IGC	- Indiana General Corporation
IGE	- Instrumentation Ground Equipment
IGIA	- Interagency Group on International Aviation
IGM	- Inside Gage Marks
IGOR	- Intercept Ground Optical Recorder
IGORTT	- Intercept Ground Optical Recorder Tracking Telescope
IGS	- Inertial Guidance System
	- Inter-Gimbal Subassembly
IGTM	- Inertial-Guided Tactical Missile
IGU	- International Geophysical Union
IGY	- International Geophysical Year

■ IH

IH	- Initial Heading
IHAS	- Integrated Helicopter Avionics System
IHCB	- Industrial Hazards Control Bulletin
IHD	- International Hydrologic Decade (1965-75)
IHER	- International Harvester Engineering Research
IHF	- Inhibit Halt Flip-Flop
IHMR	- International Harvester Manufacturing Research
IHP	- Indicated Horsepower
IHPH	- Indicated Horsepower-Hour
IHSBR	- Improved High-Speed Bombing Radar
IHSEM	- In-House Seminar
IHTU	- Interservices Hovercraft Trials Unit
I/I	- Inventory and Inspection
IIA	- Internal Identification Area
	- Invert Indicators from Accumulator
IID	- Item Identification Document
IIL	- Invert Indicators of the Left Half
IIOE	- International Indian Ocean Expedition
IIP	- Immediate Impact Point
	- Instantaneous Impact Prediction
IIR	- Integrated Instrumentation Radar
	- Inventory and Inspection Report
	- Invert Indicators of the Right Half
IIS	- Invert Indicators from Storage
IISL	- International Institute of Space Law
IIT	- Illinois Institute of Technology
IJAJ	- Intentional Jitter Anti-Jam
IJJU	- Intentional Jitter Jamming Unit
IKOR	- Immediate Knowledge of Results
IL	- Inertial Laboratory
	- Instrumentation Laboratory
	- Intermediate Language
	- Internal Letter
I&L	- Installation and Logistics
ILA	- Instrument Landing Approach
	- International Language for Aviation
ILAAS	- Integrated Light Attack Avionics System
ILABC	- Inter-Laboratory Committee
ILAS	- Instrument Landing Approach System
	- Interrelated Logic Accumulating Scanner
ILC	- International Latex Company
ILCC	- Integrated Launch Control and Checkout
ILCCS	- Integrated Launch Control and Checkout System
ILCF	- Inter-Laboratory Committee on Facilities
ILL	- Impact Limit Lines

ILLCS	- Intra-Launch Facility and Launch Control Facility Cabling Subsystem
ILLIAC	- Illinois (University of Illinois) Integrator and Automatic Computer
ILLIAD	- Illinois (Southern Illinois University) Algorithmic Decoder
ILO	- In Lieu Of
ILOUE	- In Lieu of Until Exhausted
ILPF	- Ideal Low Pass Filter
ILS	- Instrument Landing System
ILSAP	- Instrument Landing System Approach
ILSO	- Incremental Life Support Operations
ILSTAC	- Instrument Landing System and TACAN
ILT	- In Lieu Thereof
IM	- Impulse Modulation
	- Inspection Manual
	- Instrument Man
	- Instrument Measurement
	- Interceptor Missile
	- Interdepartmental Memorandum
	- Intermediate Modulation
	- Item Manger
I&M	- Installation and Maintenance
IMATA	- Independent Military Air Transport Association
IMC	- Image Motion Compensation
	- Instrument Meteorological Conditions
IMCC	- Integrated Mission Control Center
IMCD	- Input Marginal Checking and Distribution
IMCO	- Intergovernmental Maritime Consultative Organization
IMD	- Intercept Monitoring Display
IMDC	- Interceptor Missile Direction Center
IME	- Institute of Mechanical Engineers
IMEO	- Interim Maintenance Engineering Order
IMEP	- Indicated Mean Effective Pressure
IMFSS	- Integrated Missile Flight Safety System
IMGCN	- Interceptor Missile Ground Control Network
IMHEP	- Ideal Man-Helicopter Engineering Project
IMI	- Improved Manned Interceptor
	- Intermediate Manned Interceptor
IMIR	- Interceptor Missile Interrogation Radar
IMIS	- Integrated Management Information System
	- Interim Maneuver Identification System
IML	- Inside Mold Line
IMO	- International Meteorological Organization
IMP	- Impact
	- Impact Prediction
	- Inflatable Micrometeoroid Paraglider

■ IMP

IMP	- Interplanetary Monitoring Platform (Project Explorer)
	- Interplanetary Monitoring Probe
IMPACT	- Integrated Management Planning and Control Technique
	- Integrated Military Pay-Accrued, Controlled, Timely
IMPS	- Interplanetary Monitoring Probes
IMRA	- Infrared Monochromatic Radiation
IMRAN	- International Marine Radio Aids to Navigation
IMS	- Image Motion Simulator
	- Improved Manual System
	- Institute of Management Sciences
	- Institute of Mathematical Statistics
	- Interceptor Mission Sheet
	- Interplanetary Monitor Satellite
IMSOC	- Interceptor Missile Squadron Operations Center
IMSSS	- Interceptor Missile Squadron Supervisors Station
IMT	- Individual Military Training
IMTP	- Industrial Mobilization Training Program
IMU	- Inertial Measurement Unit
	- Interference Mockup
IMUGSE	- Inertial Measurement Unit Ground Support Equipment
IN	- Inch
	- Input
I/N	- Item Number
INACTLANT	- Inactive Fleet, Atlantic
INACTPAC	- Inactive Fleet, Pacific
IND	- Indicator
	- Interceptor Director
INDMAN	- Industrial Manager
INDT	- Interceptor Director Technician
INF	- Infantry
INFO	- Information
INFOCEN	- Information Center
INFRAL	- Information Retrieval Automatic Language
IN-LB	- Inch-Pound
INLR	- Item No Longer Required
INM	- Inspector of Naval Material (INSMAT preferred)
	- Interception Mission
INO	- Issue Necessary Orders
INOP	- Inoperative
INPFO	- Idaho Nuclear Power Field Office
INRA	- Individual Non-Recurrence Action
INS	- Inertial Navigation System
	- Institute of Naval Studies
	- Interceptor Simulator
	- Interstation Noise Suppression
INSACS	- Interstate Airways Communications Station

INSCAIRS	- Instrumentation Calibration Incident Repair Service
INSEC	- Internal Security
INSH	- Inspection Shell
INSMAT	- Inspector of Naval Material
INSSCC	- Interim National Space Surveillance Control Center
INST	- Instruction
INSTAR	- Inertialess Scanning, Tracking, and Ranging
INST REF	- Instrument Reference
INTC	- Intelligence Corps
INTEG	- Integration
INTERDICT	- Interference Detection and Interdiction Countermeasures Team
INTERMAG	- International Nonlinear Magnetics
INTL	- International
INTRAN	- Input Translator
INTREP	- Intelligence Report
INTSUM	- Intelligence Summary
IO	- Information Officer
	- Input-Output
I/O	- Input-Output
IOB	- Input-Output Buffer
IOC	- Initial Operational Capability
	- Intergovernmental Oceanographic Commission
IOCD	- Input-Output under Count Control and Disconnect
IOCP	- Input-Output under Count Control and Proceed
IOCS	- Input-Output Control System
IOD	- Industrial Operations Division
IODC	- Input-Output Delay Counter
IOH	- Item On Hand
IOI	- Interim Operating Instructions
IOL	- Inter-Office Letter
IOM	- Inter-Office Memorandum
IOMS	- Interim Operational Meteorological System
ION	- Institute of Navigation
IOP	- Intelligence Operations
IOPKG	- Input-Output Package
IOPS	- Input-Output Programming System
IOR	- Input-Output Register
IORP	- Input-Output of a Record and Proceed
IORT	- Input-Output of a Record and Transfer
IOS	- Indian Ocean Ship
	- Interceptor Operator Simulator
IOSP	- Input-Output under Signal and Proceed
IOST	- Input-Output under Signal and Transfer
IOT	- Initial Orbit Time
	- Input-Output Check Test

■ IOU

IOU	- Input-Output Utilizer Routine
IP	- Identification of Position
	- Impact Point
	- Impact Predictor
	- Incentive Pay
	- Information Pool
	- Initial Point
	- Instruction Pulses
	- Instructor Pilot
	- Instrumentation Plan
	- Intermediate Pressure
I/P	- Impact Predictor
IPA	- Indicated Pressure Altitude
	- Intermediate Power Amplifier
	- International Phonetic Alphabet
IPADAE	- Integrated Passive Action Detection Acquisition Equipment
IPB	- Illustrated Parts Breakdown
	- Interconnection and Programming Bay
IPBM	- Interplanetary Ballistic Missile
IPC	- Industrial Process Control
IPCCS	- Information Processing in Command and Control Systems
IPD	- Insertion Phase Delay
IPDP	- Intervals of Pulsations of Diminishing Period
IPE	- Industrial Production Equipment
IPET	- Inplant Engineering Test
IPI	- Identified Prior to Intercept
IPL	- Information Processing Language
	- Identured Parts List
	- Information Processing Language
IPM	- Industrial Preparedness Measures
	- Interruptions per Minute
IPN	- Instrumentation Plan Number
IPO	- Installation Planning Order
IPPB	- Incremental Provisioning Parts Breakdown
IPR	- Inspection Planning and Reliability
IPRL	- Interceptor Pilot Research Laboratory
IPS	- Impact Predictor System
	- Inches per Second
	- Initial Program Specification
	- Instrumentation Power Supply
	- Instrumentation Power System
	- Interceptor Pilot Simulator
	- International Pipe Standard
	- Interpretative Programming System
	- Interruptions per Second

IPT	- Inplant Transporter
	- Internal Pipe Thread
IPTP	- Inplant Test Program
IPY	- International Polar Year
IQSY	- International Quiet Sun Year
IR	- Information Retrieval
	- Infrared
	- Inside Radius
	- Inspection Record
	- Inspection Request
	- Inspector's Report
	- Instantaneous Relay
	- Instrumentation Requirement
	- Instrument Reading
	- Instrument Register
	- Interchangeability and Replaceability
	- Interrogator-Responder
IRAH	- Infrared Alternate Head
IRAN	- Inspect and Repair As Necessary
IRAR	- Impulse Response Area Ratio
IRASER	- Infrared Amplification by Stimulated Emission of Radiation
IRASI	- Internal Review and Systems Improvement
IRB	- Inspection Requirements Branch
IRBM	- Intermediate Range Ballistic Missile
IRC	- Infrared Countermeasure
	- Item Responsibility Code
I-R	- Interrogator-Responder
I&R	- Intelligence and Reconnaissance
	- Interchangeability and Replaceability
IRAA&A	- Increase and Replacement of Armor, Armament and Ammunition
IRAC	- Interdepartmental Radio Advisory Council
IRACQ	- Infrared Acquisition Aid
	- Instrumentation Radar and Acquisition Panel
IRCCM	- Infrared Counter-Countermeasure
IRCM	- Infrared Countermeasure
IRCR	- Inspection Records Change Request
IR&D	- Independent Research and Development
IRE	- Institute of Radio Engineers (now IEEE)
IRFNA	- Inhibited Red Fuming Nitric Acid
IRG	- Inertial Rate Gyro
IRGAR	- Infrared Gas Radiation
IRGPG	- Inter-Range Global Planning Group
IRI	- Inspection Records Index
IRIA	- Infrared Information and Analysis Center

■ IRIG

IRIG	- Inertial Rate Integrating Gyroscope (Massachusetts Institute of Technology)
	- Inertial Reference Integrating Gyroscope (NASA)
	- Inter-Range Instrumentation Group
IRIS	- Infrared Information Symposium
IRL	- Information Retrieval Language
IRM	- Infrared Measurements
	- Integrated Range Mission
IRMA	- Infrared Miss-Distance Approximator
IRMP	- Infrared Measurement Program
	- Interservice Radiation Measurement Program
IRN	- Interface Revision Notice (Reference to EO's)
	- Interim Revision Notice (Reference to publications)
IRO	- Industrial Relations Office
	- Interim Range Operations
	- International Refugee Organization
IROAN	- Inspect and Repair Only As Needed
IRON	- Infrared Optical Nose
IRP	- Inertial Reference Package
	- Initial Receiving Point
IRPA	- State All-Union Scientific-Research Institute of Radio Reception and Acoustics (USSR)
IRPM	- Infrared Physical Measurement
IRPP	- Industrial Readiness Planning Program
IRR	- Instrumentation Revision Record
	- Intelligence Radar Reporting
IRRAD	- Infrared Range and Detection
IRRMP	- Infrared Radar Measurement Program
IRS	- Input Read Submodule
	- Integrated Records System
IRSS	- Instrumentation and Range Safety System
	- Integrated Range Safety System
IRT	- Infrared Tracker
	- Initialize Reset Tape
	- Interrogator-Responder-Transducer
IRTWG	- Inter-Range Telemetry Working Group
IRUGRQR	- Item Urgently Required
IS	- Impact Switch
	- Infrared Spectroscopy
	- Initiation Supervisor
	- Instrumentation Summary
	- Instrumentation Squadron
	- Instrumentation System
I/S	- Installation of Systems
ISA	- Ignition and Separation Assembly
	- Instrument Society of America

ISA	- International Security Affairs
	- International Standardization Association
ISAC	- International Security Affairs Committee
ISAP	- Information Sort and Predict
ISAW	- International Society of Aviation Writers
ISB	- Independent Sideband
	- Intelligence Systems Branch
ISC	- Industrial Support Contractor
	- Intrasite Cabling
ISCAN	- Inertialess Steerable Communications Antenna
ISI	- Indian Standards Institute
ISIS	- International Satellites for Ionosphere Studies
ISL	- Artificial Moon Satellite (USSR)
	- Item Study Listings
ISM	- Industrial Security Manual
ISMIS	- Improved SAGE Manned Intercept System
ISO	- International Standardization Organization
ISP	- Specific Impulse
ISPO	- Irradiation Special Purchase Order
ISR	- Instrumentation Status Report
	- International Scientific Report
ISRC	- Independent Safety Review Committee
ISRU	- International Scientific Radio Union (also URSI)
ISS	- Inertial Subsystem
	- Input Subsystem
	- Interface Signal Simulator
	- Institute of Strategic Studies
	- Ion Silicon System
ISSS	- Installation Service Supply Support
IST	- Integrated Systems Test
ISTAR	- Image Storage Translation and Reproduction
ISTMC	- Instrumentation Section Test and Monitor Console
ISU	- Interface Surveillance Unit
	- International Scientific Union
ISZ	- Artificial Earth Satellite (USSR)
IT	- Initiation Technician
	- Insulating Transformer
	- Internal Translator
	- Interrogator-Transponder
I&T	- Identification and Traceability
ITAE	- Integrated Product of Time and Absolute Error
ITAS	- Indicated True Air Speed
ITB	- Invitation to Bid
ITD	- Initial Temperature Difference
ITDE	- Interchannel Time Displacement Error

■ **ITE**

ITE	- Institute of Traffic Engineers
	- Inverse Time Element
ITEF	- Institute of Theoretical and Experimental Physics (USSR)
ITF	- Integrated Test Facility
ITI	- Inspection and Test Instruction
ITL	- Integrate-Transfer-Launch
ITLC	- Integrated Transfer Launch Complex
ITM	- Interceptor Tactical Missile
ITO	- Integrated Task Order
ITOR	- Intercept Target Optical Recorder
ITP	- Individual Training Program
ITR	- Initial Trouble Reports
	- Inverse Time Relay
ITS	- Idaho Test Station
	- Inertial Timing Switch
	- Integrated Trajectory System
	- Interim Table Simulation
	- International Temperature Scale
ITT	- International Telephone and Telegraph Corporation
	- Interpretative Trace and Trap Program
ITTCS	- International Telephone and Telegraph Communications Systems
ITTF	- International Telephone and Telegraph, Federal Division
ITTL	- International Telephone and Telegraph Laboratories
ITT/PMD	- Interpretative Trace and Trap Program Plus Modifications
ITU	- International Telecommunications Union
ITUSAF	- Institute of Technology, United States Air Force
ITZ	- Intertropical Convergence Zone
IU	- Instrumentation Unit
	- Instrument Unit (Saturn unit located between launch vehicle and spacecraft)
	- Interference Unit
I/U	- Instrumentation Unit (IU preferred)
IUA	- Inertial Unit Assembly
IUBS	- International Union of Biological Sciences
IUGG	- International Union of Geodesy and Geophysics
IUPAC	- International Union of Pure and Applied Chemistry, Physics (also IUPAP)
IUPAP	- International Union of Pure and Applied Chemistry, Physics (also IUPAC)
IV	- Initial Velocity
	- Interval
IVD	- Image Velocity Detector
IVDP	- Initial Vector Display Point
IVP	- Interface Verification Procedure

IW	- Indirect Waste
	- In Work
IWDS	- International World Day Service
IWFA	- Inhibited White Fuming Nitric Acid
IWG	- Implementation Work Group
IWISTK	- Issue While in Stock
IWLS	- Iterative Weighted Least Squares
IWS	- Interception Weapons School
IWST	- Integrated Weapon System Training
IWX	- Intelligence-Weather
IX	- Index

J

J	- Jack
	- Joint
	- Joule
	- Junction
JA	- Judge Advocate
	- Jump Address
JAAF	- Joint Action Armed Forces
	- Joint Army-Air Force
JAAFCTB	- Joint Army-Air Force Commerical Traffic Bulletin
JAAFPC	- Joint Army-Air Force Procurement Circular
JABCZ	- JOVIAL Assemble Baby COMPOOL
JACC	- Joint Automatic Control Conference
JACE	- Joint Alternate Command Element
JADB	- Joint Air Defense Board
JADD	- Joint Air Defense Division
JADE	- Japanese Air Defense Environment System
JADF	- Joint Air Defense Force
JADFCOC	- Joint Air Defense Force Combat Operations Center
JADOC	- Joint Air Defense Operation Center
JADW	- Joint Air Defense Wing
JAEIC	- Joint Atomic Energy Intelligence Committee
JAG	- Judge Advocate General
JAIEA	- Joint Atomic Information Exchange Agency
JAL	- Japan Air Lines
JAMAG	- Joint American Military Advisory Group
JAMATO	- Joint Airlines Military Traffic Office
JAMCZ	- JOVIAL Assemble Master COMPOOL
JAMMAT	- Joint Military Mission for Aid to Turkey
JAN	- Joint Army-Navy
JANA	- Joint Army-Navy-Air Force
JANAF	- Joint Army-Navy-Air Force
JANAP	- Joint Army-Navy-Air Force Publication
JANAST	- Joint Army-Navy-Air Force Sea Transportation Message

JANE	- Joint Air Force-Navy Experiment
JANTAB	- Joint Army and Navy Technical Aeronautical Board
JAOC	- Joint Air Operations Center
JASDA	- Julie Automatic Sonic Data Analyzer
JASDF	- Japan Air Self Defense Force
JASG	- Joint Advanced Study Group
JASTOP	- Jet-Assisted Stop
JATO	- Jet-Assisted Take-Off
JATS	- Joint Air Transportation Service
JATSPLAN	- Joint Air Transportation Service Plan
JAWPM	- Joint Atomic Weapons Planning Manual
JB	- Jet Bomb
	- Junction Box
JBTO	- Joint BOMARC Test Organization
JCA	- Joint Communications Agency
JCAE	- Joint Committee on Atomic Energy
JCC	- Joint Communications Center
	- Joint Computer Conference
	- Joint Coordination Center
JCCOMNET	- Joint Coordination Center Communications Network
JCCRG	- Joint Command and Control Requirements Group
JCEADF	- Joint Central Air Defense Force
JCEC	- Joint Communications-Electronics Committee
JCEG	- Joint Communications-Electronics Group
JCI	- Joint Communications Instruction
JCLOT	- Joint Closed Loop Operations Test
J-CLOT	- Joint Closed Loop Operations Test
JCS	- Joint Chiefs of Staff
JCT	- Junction
JDC	- Jet Deflection Control
JDOP	- Joint Development Objectives Plan
JEADF	- Joint Eastern Air Defense Force
JECC	- Joint Economic Committee of Congress
JEDEC	- Joint Electronic Device Engineering Council
JEFM	- Jet Engine Field Maintenance
JEIA	- Joint Electronics Information Agency
JEIPAC	- Japan Electronic Information Processing Automatic Computer
JEOCN	- Joint European Operations Communications Network
JETAM	- Jet Engine Thrust Augmentation Mix
JETAV	- Jet Aviation
JETEC	- Joint Electron Tube Engineering Council
JETP	- Jet Propelled
JFACT	- Joint Flight Acceptance Composite Test
J-FACT	- Joint Flight Acceptance Composite Test
J/FAP	- Joint Frequency Allocation Panel

■ JGN

JGN	- Junction Gate Number
JHU	- Johns Hopkins University
JIC	- Joint Implementation Committee
	- Joint Industry Conference
	- Joint Intelligence Center
JICA	- Joint Intelligence Collecting Agency
JICST	- Japan Information Center of Science and Technology
JIOA	- Joint Intelligence Objectives Agency
JIP	- Joint Input Processing
JIR	- Job Improvement Request
JJY	- Standard Frequency Station, Tokyo, Japan
JLC	- Joint Logistics Committee
JLPG	- Joint Logistics Plans Group
JLRSE	- Joint Long-Range Strategic Estimate
JM	- Johns Manville
J/M	- Jettison Motor
JMB	- Joint Meteorological Board
JMC	- Joint Meteorological Committee
JMD	- Joint Monitor Display
JMRP	- Joint Meteorological Radio Propagation Committee
JMSAC	- Joint Meteorological Satellite Advisory Committee
JMTC	- Joint Military Transportation Committee
JMTG	- Joint Missile Task Group
JMUSDC	- Joint Mexico-United States Defense Commission
JNACC	- Joint Nuclear Accident Coordinating Center
JNW	- Joint Committee on New Weapons and Equipment
JNWP	- Joint Numerical Weather Prediction
JNWPU	- Joint Numerical Weather Prediction Unit
JO	- Job Order
JOC	- Joint Operations Center
JOD	- Joint Occupancy Date
JOHNNIAC	- John (von Neumann) Integrator and Automatic Computer
JOP	- Joint Operating Procedures
JOR	- Joint Operations Requirement
JOSPRO	- Joint Ocean Shipping Procedures
JOSS	- JOHNNIAC Open Shop System
JOVIAL	- Jules Own Version of International Algebraic Language
JP	- Jet Pilot
	- Jet Propellant
	- Jet Propulsion
JPA	- Japan Procurement Agency
JPL	- Jet Propulsion Laboratory
JPO	- Joint Project Office
JPR	- Joint Procurement Regulations
JPTO	- Jet-Propelled Take-Off
JRDB	- Joint Research and Development Board

J/S	- Jamming to Signal
JSAC	- Joint Services Advisory Committee
JSCO	- Journal Status Central Operations Table
JSCP	- Joint Strategic Capabilities Plan
JSGOMRAM	- Joint Study Group on Military Resources Allocation Methodology
JSI	- Journal of Social Issues
JSOP	- Joint Strategic Objectives Plan
JSPD	- Joint Subsidiary Plans Division
JSPG	- Joint Strategic Plans Group
JSS	- Job Schedule Status
JSSC	- Joint Strategic Survey Committee
JSTPA	- Joint Strategic Target Planning Agency
JSTPS	- Joint Strategic Target Planning Staff
JSWPB	- Joint Special Weapons Publications Board
JTA	- Joint Table of Allowance
JTCP	- JOVIAL Test Control Program
JTDS	- Joint Track Data Storage
JTF	- Joint Task Force
JTG	- Joint Task Group
JTR	- Joint Travel Regulation
JTS	- Job Training Standard
JUG	- Joint Users Group
JUSMAG	- Joint United States Military Advisory Group
JUSMAP	- Joint United States Military Advisory and Planning Group
JUSMG	- Joint United States Military Group
JUWTF	- Joint Unconventional Warfare Task Force
JWADF	- Joint Western Air Defense Force
JWG	- Joint Working Group
JWGM	- Joint Working Group Meeting
JWR	- Joint War Room
JWRA	- Joint War Room Annex
JWTC	- Jungle Warfare Training Center

K

K	- Keel
	- Kelvin
	- Key
	- Kilo
	- Kip
	- Klystron
	- Knot
	- Thousand (in reference to kilo)
KAC	- Kennedy Administrative Complex (formerly CAC)
KAFB	- Kirtland Air Force Base
KAI	- Kuybyshev Aviation Institute (USSR)
KAO	- Crimean Astrophysical Observatory (USSR)
KAP	- Commission on Astronomical Equipment (USSR)
KAPL	- Knolls Atomic Power Laboratory
KBS	- Kilobits per Second
KC	- Kilocycle
KCADS	- Kansas City Air Defense Sector
KCAL	- Kilocalorie
KCC	- Keyboard Common Contact
KCS	- Kilocycles per Second
KD	- Knocked Down
KDP	- Known Datum Point
	- Potassium Dihydrogen Phosphate
KE	- Kinetic Energy
K&E	- Keuffel and Esser Company
KEV	- Kiloelectron Volts
KG	- Kilogram
KGCAL	- Kilogram Calorie
KG/CUM	- Kilograms per Cubic Meter
KG/M	- Kilogram-Meter
KGPS	- Kilograms per Second
KGS	- Knots Ground Speed
KH	- Kelsey Hayes Company

KhAI   - Khar'kov Aviation Institute (USSR)
KhGU   - Khar'kov State University (USSR)
KhMII   - Khar'kov Mechanics and Machinery Institute (USSR)
KIAS   - Knots Indicated Air Speed
KIC   - Kollsman Instrument Corporation
KIFIS   - Kollsman Integrated Flight Instrument System
KIGVF   - Kiev Institute of the Civil Air Fleet (USSR)
KILLS   - Ka-Inertial Launch and Leave System
KIM   - Keyboard Input Matrix
KIP   - Kilopound
KIPO   - Keyboard Input Printout
KIS   - Keyboard Input Simulation
KISNOPI - Keyboard Input Simulation—Noise-Problem Input
KISS   - Keep It Simple, Sir
KL   - Kiloliter
KLA   - Klystron Amplifier
KLO   - Klystron Oscillator
KM   - Kilometer
KMC   - Kilomegacycle
KMCS   - Kilomegacycles per Second
KMPD   - Kingston Military Products Division
KMPS   - Kilometers per Second
KN   - Knot
KNK   - Kompates Natrium Kernkraftwerk
KNO   - Kano, Nigeria (remote site)
KO   - Knockout
KOH   - Potassium Hydroxide
KP   - Key Pulsing
         - Key Punch
         - Kill Probability
         - Pocket Dosimeter (USSR)
KPI   - Kips per Square Inch
KPSI   - Kilopounds per Square Inch
KRF   - Knowledge of Results Feedback
KRFT   - Knowledge of Results Feedback Task
KS   - Keystone
KSC   - John F. Kennedy Space Center (formerly Launch Operations Center)
KSI   - Kilopounds per Square Inch
KSL   - Keyboard Simulated Lateraltelling
KSPM   - Klystron Power Supply Modulator
KSR   - Keyboard Send and Receive
KSS   - Kellog Switchboard and Supply
KT   - Kiloton
KTAS   - Knots True Air Speed
KTS   - Kodiak Tracking Station

■ KUCOG

KUCOG	- Kunia Coordinating Group
KUTD	- Keep Up to Date
KV	- Kilovolt
	- Short Wave (USSR)
KVA	- Kilovolt-Ampere
KVAH	- Kilovolt-Ampere Hour
KVAR	- Kilovolt-Ampere Reactive
KVP	- Kilovolt Peak
KW	- Kilowatt
KWAJ	- Kwajalein Island
KWE	- Kilowatts of Electrical Energy
KWH	- Kilowatt-Hour
KW-HR	- Kilowatt-Hour
KWIC	- Key Word in Context
KWIT	- Key Word in Title
KWR	- Kilowatt Reactive
KY	- Keying Device
KYD	- Kiloyard
KYTOON	- Kite Balloon

L

L	- Inductance
	- Lambert
	- Lamp
	- Launcher
	- Left
	- Length
	- Liaison
	- Line
	- Liter
	- Lumen
L-	- Days Before Launch
LA	- Launch Area
	- Lighting Arrester
	- Load Adjuster
	- Low Altitude
LAA	- Light Antiaircraft
	- Low Altitude Attack
LAAC	- Latin American Aviation Conference
LAADS	- Los Angeles Air Defense Sector
LAAFS	- Los Angeles Air Force Station
LAAMBN	- Light Antiaircraft Missile Battalion
LAAR	- Liquid Air Accumulator Rocket
LAB	- Laboratory
	- Low Altitude Bombing
LABIL	- Light Aircraft Binary Information Link
LABPIE	- Low Altitude Bombing Position-Indicating Equipment
LABROC	- Laboratory Rocket Evaluator
LABS	- Low Altitude Bombing System
LAC	- Lockheed Aircraft Corporation
	- Low Complement of Address
	- Lunar Atlas Chart
LACE	- Liquid Air Cycle Engine
LACR	- Low Altitude Coverage Radar

■ LAD

LAD	- Limited Air Defense
	- Los Angeles Division
	- Low Accuracy Data
	- Low Accuracy Designation
LADA	- Light Air Defense Artillery
LADAR	- LASER Detection and Ranging
LADD	- Low-Angle Drogued Delivery
LAE	- Left Arithmetic Element
LAET	- Limiting Actual Exposure Time
LAFB	- Lowry Air Force Base
LAFOKI	- Laboratory of Scientific and Applied Photography and Cinematography (USSR)
LAFTA	- Latin American Free Trade Association
LAGS	- LASER Activated Geodetic Satellite
LAIRS	- Land-Air Integrated Reduction System
LAL	- Low Acceptance Level
LAMCIS	- Los Angeles Multiple Corridor Identification System
LAMMP	- Lower Acceptable Mean Maximum Pressure
LAMP	- Low Altitude Manned Penetrator (now AMP)
	- Lunar Analysis and Mapping Program
LANAC	- Laminar Navigation and Anticollision
LANCRA	- Landing Craft
LANCRAB	- Landing Craft and Bases
LANT	- Atlantic
LANTCOM	- Atlantic Command
LANTFLT	- Atlantic Fleet
LAOD	- Los Angeles Ordnance District
LAP	- Launch Analysis Panel
	- Lesson Assembly Program
LAPDOG	- Low Altitude Pursuit Dive on Ground
LAPFO	- Los Angeles Air Force Procurement Field Office
LAR	- Liquid Aircraft Rocket
	- Local Acquisition Radar
	- Lockheed Assembly Team
LARC	- Large Automatic Research Calculator
	- Light, Amphibious, Resupply Cargo
	- Livermore Atomic Research Computer
	- Libyan-American Reconstruction Commission
LARD	- Load Adjuster Reference Datum
LARDS	- Low Accuracy Radar Data Transmission System
LARF	- Low Altitude Radar Fuzing
LAROO	- Lackland Aircraft Reactors Operations Office
LARR	- Large Area Record Reader
LARVA	- Low-Altitude Research Vehicle
LAS	- Laboratory of Applied Sciences
	- Launch Auxiliary System

LAS	- Logical Compare Accumulator with Storage
	- Loop Actuating Signal
LASER	- Light Amplification by Stimulated Emission of Radiation
LASHUP	- Land-Air Synergic Homogeneous Ultra-Processor
LASL	- Los Alamos Scientific Laboratory
LASO	- Lower Altitude Search Option
LASS	- Library Automated Service Systems
	- Line Amplifier and Super Sync Mixer
LASSO	- Landing and Approach System, Spiral-Oriented
LASV	- Low Altitude Supersonic Vehicle (formerly SLAM)
LAT	- Latitude
	- Local Apparent Time
LAVD	- Laboratory of High Temperatures and Pressures (USSR)
LAW	- Light Anti-Tank Weapon
	- Local Air Warning
LAWSO	- Lockheed Anti-Submarine Warfare Systems Organization
LAYDET	- Layer Detection
LB	- Launch Bunker
	- Light Bombardment
	- Line Buffer
	- Local Battery
	- Pound
LB-FT	- Pound-Foot
LBH	- Lyman-Birge Hopfield System
LB-IN	- Pound-Inch
LBM	- Load Buffer Memory
	- Logic Buss Monitor
	- Lunar Breaking Module
LBNSY	- Long Beach Naval Shipyard
LBP	- Length Between Perpendiculars
LBS	- Launch Blast Simulator
LBT	- Low BIT Test
LC	- Inductance-Capacitance
	- Landing Craft
	- Launch Complex
	- Launch Countdown
	- Launching Control
	- Lead Covered
	- Letter Contract
	- Level Control
	- Liaison, Cargo
	- Library of Congress Classification
	- Light Case
	- Line Carrying
	- Line Connection
	- Line of Communications

LC

LC	- Line of Contact
	- Logistics Command
	- Lower Carbon
	- Lower Case
L/C	- Letter Contract
	- Loop Check
L-C	- Line of Communication
LCA	- Landing Craft Assault
	- Launch Control Area
	- Load Carrying Ability
	- Logistics Control Area
LCAL	- Lower Conformance Altitude
LCAO	- Linear Combination of Atomic Orbitals
LCAVAT	- Landing Craft and Amphibious Vehicle Assignment Table
LCB	- Launch Control Building
	- Longitudinal Position of Center of Buoyancy
LCC	- Landing Craft Control
	- Launch Control Center
	- Launch Control Console
	- Local Communications Complex
	- Local Communications Console
LCDTL	- Load-Compensated Diode Transistor Logic
LCE	- Launch Complex Engineer
	- Launch Complex Equipment
	- Load Circuit Efficiency
LCF	- Launch Control Facility
	- Longitudinal Position of Center of Flotation
LCFS	- Launch Control Facility Simulator
LCG	- Logistics Control Group
	- Longitudinal Position of Center of Gravity
LCH	- Load Channel
LCHR	- Launcher
LCI	- Landing Craft, Infantry
LCL	- Less than Carload
	- Lifting Condensation Level
	- Local
	- Localizer
	- Lower Control Limit
LCM	- Landing Craft Mechanized
	- Least Common Multiple
LCN	- Landing Craft, Navigation
LCNT	- Link Celestial Navigation Trainer
LCO	- Launch Control Officer
	- Launch Control Operation
LCP	- Loading Control Program
LCPL	- Landing Craft, Personnel, Large

LCR	- Inductance-Capacitance-Resistance
LCS	- Landing Control System
	- Landing Craft, Support
	- Lateral Control System
	- Launch Control System
	- Loudness Contour Selector
LCSB	- Launch Control Support Building
LCSR	- Landing Craft, Swimmer Reconnaissance
LCSS	- Launch Control System Simulator
LCT	- Landing Craft, Tank
	- Latest Closing Time
	- Local Civil Time
LCTB	- Launch Control Training Building
LCU	- Landing Craft, Utility
LCV	- Landing Craft, Vehicle
LCVP	- Landing Craft, Vehicle and Personnel
LD	- Launching Division
	- Lethal Dose
	- Level Discriminator
	- Lift-Drag Ratio
	- Line of Departure
	- List of Drawings
	- Logic Driver
	- Low Drag
	- Lunar Docking
L/D	- Length to Diameter
	- Lift over Drag
LDA	- Line Driving Amplifier
	- Locate Drum Address
LDB	- Lexington Development Branch
	- Light Distribution Box
	- Load B-Register
LDC	- Load Complement of Decrement in Index Register
	- Local 10-Address Counter
	- Local Defense Center
LDG	- Landing
	- Lexington Design Group
LDGE	- LEM Dummy Guidance Equipment
LDI	- Load Indicators
LDK	- Lower Deck
LDO	- Launch Division Officer
LDP	- Language Data Processing
	- Local Data Package
	- Low Density Polyethylene
LDQ	- Load the MQ
LDR	- Low Data Rate

■ LDRI

LDRI	– Low Data Rate Input
LDT	– Level Detector
	– Linear Differential Transformer
LE	– Laboratory Evaluation
	– Launch Emplacement
	– Launch Escape
	– Launching Equipment
	– Leading Edge
	– Lifting Eye
	– Light Equipment
	– Low Explosive
L/E	– Launch Escape (LE preferred)
LEAD	– Lens Electronic Automatic Design
LEAP	– Leading Edge Airborne PANAR
	– Lift-Off Elevation and Azimuth Programmer
LEAR	– Logistics Evaluation and Review
LEARSYN	– Lear Synchro
LEB	– Lower Equipment Bay
LEC	– Launch Escape Control
	– Lockheed Electronics Company
LECA	– Launch Escape Control Area
LECS	– Launching Equipment Checkout Set
LED	– Laboratory Evaluation Department
LEDC	– Low Energy Detonating Cord
LEET	– Limiting Equivalent Exposure Time
LEG	– Logistical Expediting Group
LEID	– Low Energy Ion Detector
LEJ	– Longitudinal Expansion Joint
LELU	– Launch Enable Logic Unit
LEM	– Launch Escape Motor
	– Lunar Excursion Module
LENGIRD	– Leningrad Group for the Study of Jet Propulsion (USSR)
LEO	– Librating Equidistant Observer
	– Low Earth Orbit
LEP	– Lowest Effective Power
LEPS	– Launch Escape Propulsion System
LEPT	– Long-Endurance Patrolling Torpedo
LES	– Launch Enable System
	– Launch Escape System
	– Loop Error Signal
LESA	– Lunar Exploration System for Apollo
LESC	– Launch Escape System Control
LESCS	– Launch Escape Stabilization and Control System (Obsolote)
LESS	– Launch Escape System Simulator
	– Least Cost Estimating and Scheduling System
LET	– Launch Escape Tower

LET	- Linear Energy Transfer
	- Live Environmental Testing
LETCS	- Launch Escape Tower Canard System
LEU	- Launch Enable Unit
LEV	- Lunar Escape Vehicle
LEX	- Land Exercise
LF	- Launch Facility
	- Line Feed
	- Load Factor
	- Low Frequency (band 5)
L/F	- Launch Facility
LFC	- Laminar Flow Control
	- Low Frequency Correction
LFE	- Laboratory For Electronics
LFLM	- Lowest Field Level of Maintenance
LFM	- Landing Force Manual
	- Low-Powered Fan Marker
LF/MF	- Low Frequency, Medium Frequency
LFO	- Low-Frequency Oscillator
LFP	- Late Flight Plan
LFPRL	- Lewis Flight Propulsion Research Laboratory
LFR	- Low Flux Reactor
LFRD	- Lot Fraction Reliability Deviation
LFS	- Launch Facility Simulator
	- Loop Feedback Signal
LFT	- Left Half Indicators, Off Test
LFTU	- Landing Force Training Unit
LG	- Landing Gear
	- Landing Ground
	- Light Gun
	- Loop Gain
LGA	- Light-Gun Amplifier
LGE	- LEM Guidance Equipment
LGG	- Light-Gun Pulse Generator
LGL	- Logical Left Shift
LGN	- Line Gate Number
LGR	- Logical Right Shift
LH	- Left Hand
	- Long Haul
L/H	- Long Haul
LH_2	- Liquid Hydrogen
LHA	- Local Hour Angle
LHDC	- Lateral Homing Depth Charge
LHE	- Liquid Helium
LHFEB	- Left Hand Forward Equipment Bay
LHSC	- Left Hand Side Console

■ LHW

LHW	- Left Hand Word
LI	- Letter of Intent
	- Level Indicator
	- Litton Industries
	- Location Identifier
	- Lot Item
LIAP	- Leningrad Institute of Aviation Instrument Building (USSR)
LICOF	- Land Line Communication Facility
LICOR	- Lightening Correlation
LIDAR	- Light Equivalent of Radar
LIFE	- Lear Integrated Flight Equipment
LIFMOP	- Linearly Frequency-Modulated Pulse
LIFO	- Last In, First Out
LII	- Flight-Testing Institute (USSR)
LIL	- Lunar International Laboratory
LILOC	- Light Line Optical Correlation
LIM	- Limiting
LIMDAT	- Limiting Date
LIMFAC	- Limiting Factors
LINAC	- Linear Accelerator
LINCOS	- Lingua Cosmica
LINFT	- Linear Foot
LINS	- Lightweight Inertial Navigation System
LINUS	- Lexington Integrated Utility System
LIP	- Launch In Process
	- Light-Gun Input Program
LIS	- Loop Input Signal
LISP	- List Processor
LITR	- Low-Intensity Test Reactor
LITVC	- Liquid Injection Thrust Vector Control
LIVE	- Lunar Impact Vehicle
LJ	- Little Joe
LL	- Landline
	- Left Label
	- Light Line
	- Lincoln Laboratory
	- Live Load
	- Loudness Level
	- Low Level
	- Lower Limit
L/L	- Landline
	- Latitude/Longitude
LLI	- Land Landing Impact
	- Long Lead Item
LLL	- Low-Level Logic
LLM	- Lunar Landing Mission

LO

LLM	- Lunar Landing Module (see LEM)
LLOS	- Landmark Line of Sight
LLP	- Live Load Punch
	- Lunar Landing Program
LLR	- Load Limiting Resistor
LLRR	- Lowest Level Remove-Replace
LLRV	- Lunar Landing Research Vehicle
LLS	- Long Left Shift
	- Lunar Logistics System
LLSV	- Lunar Logistics System Vehicle
LLTT	- Landline Teletype
LLV	- Lunar Landing Vehicle (see LEM)
	- Lunar Logistics Vehicle
LM	- Landmark
	- Land Mine
	- Lumen (see also L)
L/M	- List of Material
L&M	- Light and Medium Vehicles Office
LMAL	- Langley Memorial Aeronautical Laboratory
LME	- Launch Monitor Equipment
LMED	- Light Military Electronics Department
LMFR	- Liquid Metal Fuel Reactor
LMFRE	- Liquid Metal Fuel Reactor Equipment
LMG	- Light Machine Gun
	- Liquid Methane Gas
LMI	- Logistics Management Institute
LML	- Lookout Mountain Laboratories
LMM	- Locator Middle Marker
LMO	- Lens-Modulated Oscillator
LMR	- Lowest Maximum Range
LMSC	- Lockheed Aircraft Corporation Missile and Space Company
LMT	- Local Mean Time
	- Log Mean Temperature
LMTD	- Log Mean Temperature Difference
LMU	- Line Monitor Unit
LN	- Liquid Nitrogen
	- Logarithm, Natural
LN_2	- Liquid Nitrogen
LNE	- Land Navigation Equipment
LNT	- Launch Network Test
	- Left Half Indicators, On Test
LNVT	- Launch Network Verification Test
LO	- Launch Operations
	- Letter Order
	- Level Off
	- Level Originator

■ LO

LO	- Liaison Officer
	- Lift-Off
	- Local Oscillator
	- Lock-On
	- Lubricating Oil
	- Lubrication Order
	- Lunar Orbiter
L/O	- Lift-Off
LO_2	- Liquid Oxygen
LOA	- Leave of Absence
	- Length Overall
	- Light Observation Aircraft
LOAC	- Low Accuracy
LOAP	- List of Applicable Publications
LOAS	- Lift-Off Acquisition System
LOB	- Launch Operations Building
	- Line of Balance
LOBAR	- Long Base Line Radar
LOC	- Launch Operations Center (formerly LOD; now KSC)
	- Launch Operations Complex
	- Limited Operation Clearance
	- Line of Communication
	- Localizer
LOCC	- Launch Operations Control Center
LOCS	- Librascope Operations Control System
LOCTRACS	- Lockheed Tracking and Control System
LOD	- Launch Operations Directorate
	- Launch Operations Division (now Launch Operations Center)
LODESTAR	- Logically Organized Data Entry, Storage and Recording
LOF	- Local Oscillator Frequency
LOFAR	- Low Frequency Acquisition and Ranging
	- Low Frequency Analysis and Recording
LOFTI	- Low Frequency Trans-Ionospheric Satellite
LOG	- Logarithm
LOGAIRNET	- Logistics Air Network
LOGAN	- Logical Language
LOGBALNET	- Logistics Ballistics Network
LOGCOMD	- Logistical Command
LOGIPAC	- Logical Processor and Computer
LOGR	- Logistical Ratio
LOH	- Light Observation Helicopter
LOI	- Lunar Orbit Injection
LOLA	- Lunar Orbit and Landing Approach
LOM	- Locator Outer Marker
	- Lunar Orbital Mission
LOMA	- Life Office Management Association

LOMAR	- Logistics, Maintenance and Repair
LONG	- Longitude
LOP	- Launch Operations Panel
	- Line of Position (also PL)
	- Lunar Orbit Plane
LOPAR	- Low Power Acquisition Radar
LOPC	- Lunar Orbital Photo Craft
LOP-GAP	- Liquid Oxygen Petrol, Guided Aircraft Protectile (British)
LOPP	- Lunar Orbital Photographic Project
LO-QG	- Locked Oscillator-Quadrature Grid
LOR	- Lunar Orbit Rendezvous
LORAC	- Long Range Accuracy
LORAD	- Long Range Active Detection
LORADAC	- Long Range Active Detection and Communications System
LORAN	- Long Range Navigation
LORAPH	- Long Range Passive Homing
LORC	- Lockheed Radio Command
LOREC	- Long Range Earth Current Communications
LORL	- Large Orbital Research Laboratory
LORV	- Low Observable Re-Entry Vehicle
LOS	- Line of Sight
	- Loop Output Signal
	- Loss of Signal
	- Lunar Orbiter Spacecraft
LOTS	- Launch Optical Trajectory System
	- Logistics Over the Shore
LOV	- Limit of Visibility
LOX	- Liquid Oxygen
LOZ	- Liquid Ozone
LP	- Landplane
	- Launch Pad
	- Linear Programming
	- Liquified Petroleum
	- Liquid Propellant
	- Local Procurement
	- Long Persistance
	- Long-Playing
	- Lower Panel
	- Low-Pass
	- Low Point
	- Low Pressure
L&P	- Lunar and Planetary
LPA	- Log Periodic Antenna
LPARM	- Liquid-Propellant Applied Research Motor
LPC	- Linear Power Controller
	- Lockheed Propulsion Company

■ LPD

LPD	- Local Procurement District
	- Log-Periodic Dipole
	- Low Performance Drone
LPDA	- Log-Periodic Dipole Array
LPF	- Leucocytosis-Promoting Factor
	- Low Pass Filter
LPG	- Liquified Petroleum Gas
LPGE	- LEM Partial Guidance Equipment
LPH	- Landing Platform Helicopter
LPLM	- Lowest Planned Level of Maintenance
LPM	- Lines per Minute
	- Long Particular Meter
LPO	- Lincoln Project Office
	- Local Purchase Order
LPR	- Late Position Report
	- Liquid Propellant Rocket
	- Long Playing Rocket
LPRD	- Launch Program Requirement Document
LPSD	- Logically Passive Self-Dual
LPT	- Low Pressure Test
LPV	- Lightproof Vent
LPW	- Lumens per Watt
LR	- Laboratory Request
	- Line Relay
	- Liquid Rocket
	- Load Ratio
	- Long Range
	- Lunar Rendezvous
L/R	- Locus of Radius
LRBM	- Long Range Ballistic Missile
LRBR	- Long Range Ballistic Rocket
LRC	- Langley Research Center
	- Lewis Research Center
	- Load Ratio Control
LRD	- Long Range Data
LRDP	- Long Range Development Program
LRE	- Liquid Rocket Engine
LRI	- Left-Right Indicator
	- Long Range Interceptor
	- Long Range Radar Input
LRIM	- Long Range Input Monitor
LRIP	- Long Range Impact Point
LRIR	- Low Resolution Infrared Radiometer
LRL	- Lawrence Radiation Laboratory
LRLM	- Lower Reject Limit Median
LRM	- Lunar Reconnaissance Module

LRM	- Lunar Rendezvous Mission
LRP	- Long Range Plans
LRPGD	- Long Range Proving Ground Division
LRR	- Long Range Radar
	- Long Range Rocket
LRRP	- Lowest Required Radiated Power
LRS	- Long Range Schedule
	- Long Right Shift
LRT	- Last Resort Target
LRTM	- Low Range Training Mission
LRU	- Line Replaceable Unit
LS	- Late Scramble
	- Launch Simulator
	- Launch Site
	- Least Significant
	- Left Sign
	- Life Systems
	- Limited Standard
	- Limit Switch
	- Logistics Support
	- Long Shot
	- Loudspeaker
	- Low Speed
L&S	- Launch and Service
LSA	- Large Search Area
LSB	- Launcher Support Building
	- Launch Service Building
	- Least Significant BIT
	- Lower Sideband
LSC	- Linear Shaped Charge
LSCE	- Launch Sequence and Control Equipment
LSD	- Landing Ship, Dock
	- Launch Systems Data
	- Least Significant Digit
	- Life Systems Division (now Crew Systems Division)
	- Linkage System Diagnostics
	- Low Speed Data
	- Lysergic Acid Diethylamine
LSE	- Launch Sequencer Equipment
	- Life Science Experiment
	- Life Support Equipment
LSI	- Lear Siegler, Incorporated
LSL	- Logistics System Laboratory
LSM	- Launcher Status Multiplexer
	- Logistic Support Manager
LSN	- Linear Sequential Network

■ LSO

LSO	- Landing Signal Officer
	- Launch Safety Officer
	- Life Systems Officer
LSOC	- Logistical Support Operations Center
LSP	- Low Speed Pointer
LSR	- Load Shifting Register
	- Local Sunrise
LSS	- Launcher Status-Summarizer
	- Life Support System
	- Limited Storage Site
	- Logistics Support Squadron
LSSF	- Limited Service Storage Facility
LST	- Landing Ship, Tank
	- Left Store
	- Local Sidereal Time
	- Local Standard Time
LSU	- Line Selector Unit
LSV	- Landing Ship, Vehicle
LSX	- Large Scale Exercise
LT	- Laboratory Test
	- Landing Team
	- Language Translation
	- Launch Tube
	- Logic Theory
	- Lot Time
	- Low Tension
	- Low Torque
L/T	- Line Telecommunications
LTA	- LEM Test Article
	- Lighter Than Air
LTAL	- Lower Transition Altitude
LTBO	- Linear Time Base Oscillator
LTC	- Launch Vehicle Test Conductor
LTDS	- Launch Trajectory Data System
LTDT	- Langley Transonic Dynamics Tunnel
LTFRD	- Lot Tolerance Fraction Reliability Deviation
LTI	- Light Transmission Index
LTL	- Less Than Truckload
LTM	- Leave Trapping Mode
	- Line Type Modulator
LTO	- Lot Time Order
LTP	- Linear Time Plot
	- Low Trip Point
LTPS	- Lincoln Tube Process Specification
LTR	- Left Test Register
LTS	- Lateral Test Simulator

LTS	- Launch Telemetry Stations
	- Launch Tracking System
	- Liftoff Transmission Subsystem
	- Lunar Touchdown System
LTTM	- Launch Tube Test Missile
LTV	- Launch Test Vehicle
	- LEM Test Vehicle
	- Ling-Tempco-Vought, Incorporated
L-T-V	- Ling-Tempco-Vought, Incorporated
LUB	- Lubrication
LUCID	- Language for Utility Checkout and Instrumentation Development
	- Language Used to Communicate Information System Design
LUCOM	- Lunar Communication System
LUF	- Lowest Usable Frequency
LUHF	- Lowest Usable High Frequency
LUMAS	- Lunar Mapping System
LUPWT	- Langley Unitary Plan Wind Tunnel
LUS	- Load, Update, Subset
LUSI	- Lunar Surface Inspection
LUSTER	- Lunar Dust and Earth Return
LUT	- Launch Umbilical Tower
LV	- Launch Vehicle
	- Local Vertical
	- Low Voltage
L/V	- Launch Vehicle
LVCD	- Least Voltage Coincidence Detection
LVD	- Low Voltage Drop
LVDT	- Linear Variable Differential Transformer
LVH	- Landing Vehicle Hydrofoil
LVM	- Launch Vehicle Monitor
LVMI	- Leningrad Military-Mechanical Institute (USSR)
LVO	- Launch Vehicle Operations
LVOD	- Launch Vehicle Operations Division
LVP	- Low Voltage Protection
LV&P	- Launch Vehicles and Propulsion
LVR	- Low Voltage Release
LVS	- Launch Vehicle Simulator
	- Low Velocity Scanning
LVSE	- Launch Vehicle Systems Engineer
LVSG	- Launch Vehicle Study Group
LVSSTS	- Launch Vehicle Safety System Test Set
LVT	- Landing Vehicle Tracked
LVT(A)	- Landing Vehicle Tracked (Armored)
LVTC	- Launch Vehicle Test Conductor
LVTE	- Landing Vehicle, Tracked, Engineer

- **LVTH**

LVTH	- Landing Vehicle, Tracked, Howitzer
LVTR	- Low VHF Transmitter-Receiver
LW	- Late Warning
	- Launch Window
	- Limited Warfare
LWASV	- Lightweight Air-to-Surface Vessel
LWB	- Long Wheelbase
LWCS	- Limited War Capabilities Study
LWD	- Launch Window Display
LWL	- Limited War Laboratory
	- Load Water Line
	- Low Water Line
LWM	- Load Water Mark
	- Low Water Mark
LWO	- Limited Warning Operations
LWOP	- Leave Without Pay
LWP	- Limited War Plan
LXA	- Load Index from Address
LXD	- Load Index from Decrement
LZ	- Landing Zone
	- Load Zone
LZT	- Local Zone Time

M

M	- Mach
	- Magnaflux
	- Medium
	- Mega
	- Meter
	- Milli
	- Missile
	- Mode
	- Model
	- Moment
M-	- Mobilization
	- Mockup
MA	- Machine Accountant
	- Main Alarm
	- Maritime Administration
	- Marshalling Area
	- Mauchly Associates
	- Memory Address
	- Mercury-Atlas
	- Message Assembler
	- Milliampere
	- Missile Assembly
	- Mission Accomplished
	- Modify Address
M-A	- Mercury-Atlas
MAA	- Master Army Aviator
	- Mathematical Association of America
	- Missile Assembly Area
MAAG	- Military Assistance Advisory Group
MAAM	- Medium Antiaircraft Missile
MAAMA	- Middletown Air Materiel Area
MAB	- Missile Assembly Building
MABLE	- Miniature Autonetics Base Line Equipment

■ MABRON

MABRON	- Marine Air Base Squadron (MABS preferred)
MABS	- Marine Air Base Squadron
MAC	- Maintenance Advisory Committee
	- Maintenance Allocation Chart
	- Major Air Command
	- McDonnell Aircraft Corporation
	- Mean Aerodynamic Chord
	- Missile Advisory Committee
	- Monitor and Control
	- Motion Analysis Camera
	- Multi-Analyzer Configuration
MACCS	- Manufacturing Accountability Cost Control System
MACG	- Marine Air Control Group
MACL	- Minimum Acceptable Compliance Level
MACON	- Maintenance Analysis Control
MACS	- Marine Air Control Squadron
	- Medium Altitude Communications Satellite
MACSS	- Medium Altitude Communications Satellite System
MAD	- Magnetic Airborne Detector
	- Magnetic Anomaly Detection
	- Maintenance, Assembly, Disassembly
	- Marine Aviation Detachment
	- Michigan Algorithmic Decoder
	- Minimum Approach Distance
	- Multiple Access Device
	- Multiple-Aperature Device
	- Multiply and Add
MADAEC	- Military Application Division of the Atomic Energy Commission
MADAM	- Multi-Purpose Automatic Data Analysis Machine
MADCAR	- Management Data Charting and Review
MADDAM	- Macro-Module and Digital Differential Analyzer Machine
MADDIDA	- Magnetic Drum Digital Differential Analyzer
MADE	- Minimum Airborne Digital Equipment
	- Multi-Channel Analog-to-Digital Data Encoder
MADIS	- Millivolt Analog-Digital Instrumentation System
MADRAC	- Malfunction Detection and Recording System
MADRE	- Magnetic Drum Receiving Equipment
	- Martin Automatic Data Reduction Equipment
MADT	- Micro-Alloy Diffused Transistor
MADW	- Military Air Defense Warning Network
MAE	- Mean Absolute Error
	- Missile Airborne Equipment
	- Missile Assembly Equipment
MAECON	- Mid-America Electronics Conference
MAERU	- Mobile Ammunition Evaluation and Reconditioning Unit

MAET	- Missile Accident Emergency Team
MAF	- Manpower Authorization File
	- Mixed Amine Fuel
MAFB	- Malmstrom Air Force Base
MAFD	- Minimum Acquisition Flux Density
MAFR	- Merged Accountability and Fund Reporting
MAG	- Magneto
	- Magnetic
	- Magnetron
	- Marine Aircraft Group
	- Military Advisory Group
MAGAMP	- Magnetic Amplifier
MAGI	- Multi-Array Gamma Irradiator
MAGIC	- MIDAC Automatic General Integrated Computation
	- Modified Action Generated Input Control
MAGLATCH	- Magnetic Latch
MAGLOC	- Magnetic Logic Computer
MAGMOD	- Magnetic Modulator
MAGNETTOR	- Magnetic Modulator
MAGTRAC	- Magnetic Amplifier, Transistorized, Automatic Target Tracker
MAI	- Military Assistance Institute
	- Sergo Ordzhonikidze Moscow Aviation Institute (USSR)
MAID	- Maintenance Automatic Integration Director
	- Monrobot Automatic Internal Diagnosis
MAIDS	- Multipurpose Automatic Inspection and Diagnostic System
MAIN	- Medical Automation Intelligence
	- Military Authorization Identification Number
MAINT	- Maintenance
MAJAC	- Monitor, Anti-Jam, and Control
MAJCON	- Major Air Command Controlled Units
MAL	- Material Allowance List
MALF-OUT	- Malfunction or Failure of Component
MALLAR	- Manned Lunar Landing and Return
MAM	- Military Air Movement
MAMB	- Missile Assembly and Maintenance Building
MAMBO	- Minuteman Assembly-Maintenance Building, Ogden
MAMIE	- Minimum Automatic Machine for Interpolation and Extrapolation
MAMOS	- Marine Automatic Meteorological Observing Station
MAMRON	- Marine Aircraft Maintenance Squadron
MAMS	- Missile Assembly and Maintenance Shop
	- Missile Assistance Maintenance Structure
	- Modern Army Maintenance System
MAN	- Microwave Aerospace Navigation
MANIAC	- Mechanical and Numerical Integrator and Computer

■ MANIP

MANIP	- Manual Input
MANOP	- Manual of Operations
MANTRAC	- Manual Angle Tracking Capability
MAOT	- Military Assistance Observer Team
MAP	- Manifold Absolute Pressure
	- Message Acceptance Pulse
	- Military Assistance Program
	- Missile Assignment Program
	- MORT (Master Operational Recording Tape) Assembly Program
	- Mutual Assistance Program (now DIAC)
MAPAG	- Military Assistance Program Advisory Group
MAPCHE	- Mobile Automatic Programmed Checkout Equipment
MAPP	- Masking Parameter Printout
MAPROS	- Maintain Production Schedule
MAPS	- Master Activation Phasing Schedule
	- Multivariate Analysis and Prediction of Schedules
MAR	- Memory Address Register
	- Multifunction Array Radar
MARAD	- Maritime Administration
MARC	- Material Accountability/Recoverability Code
MARCOR	- Marine Corps
MARCORB	- Marine Corps Base
MARCORCAMPDET	- Marine Corps Camp Detachment
MARCORMAN	- Marine Corps Manual
MARI	- Motivator and Response Indicator
MARISP	- Maritime Strike Plan
MARITCOM	- Maritime Commission
MARLIS	- Multiple Aspect Relevance Linkage Information System
MARS	- Machine Retrieval System
	- Manned Astronautical Research Station
	- Marine Aircraft Repair Squadron
	- Master Attitude Reference System
	- Military Affiliate Radio System
	- Mobile Atlantic Range Stations
	- Multi-Aperature Reluctance Switch
MARSA	- Military Authority Assumes Responsibility for Separation of Aircraft
MART	- Maintenance Analysis Review Technique
	- Missile Automatic Radiation Tester
MARTAC	- Martin Automatic Rapid Test and Control
MARTC	- Marine Air Reserve Training Command
MARTCOM	- Marine Air Reserve Training Command
MARTD	- Marine Air Reserve Training Detachment
MARTI	- Maneuverable Re-Entry Technology Investigation
MARTINI	- Massive Analog Recording Technical Instrument for Nebulous Indications

MARV	- Maneuvering Anti-Radar Vehicle
MAS	- Military Agency for Standardization
MASA	- Military Automotive Supply Agency
MASCOT	- Meteorological Auxiliary Sea Current Observation Transmitter
	- Motorola Automatic Sequential Computer Operated Tester
MASER	- Microwave Amplification by Stimulated Emission of Radiation
MASP	- Material Analysis Specification Plan
MASRU	- Marine Air Support Radar Unit
MASS	- Marine Air Support Squadron
	- Modern Army Supply System
MASSDAR	- Modular Analysis Speedup, Sampling, and Data Reduction System
MASSPO	- Manned Spaceflight Support Project Office
MAST	- Magnetic Annular Shock Tube
	- Missile Automatic Supply Technique
MASTIF	- Multi-Axis Spin Test Inertia Facility
MAT	- Manned Anti-Tank Rocket
	- Micro-Alloy Transistor
	- MORT (Master Operational Recording Tape) Address Table
MATABE	- Multiple-Weapon Automatic Target and Battery Evaluator
MATC	- Military Air Transport Command
MATCON	- Microwave Aerospace Terminal Control
	- Military Air Traffic Control
MATCU	- Marine Air Traffic Control Unit
MATD	- Mine and Torpedo Detector
MATE	- Modular Automatic Test Equipment
MATRA	- Societe Generale de Mecanique, Aviation, Traction
MATRS	- Miniature Airborne Telemetry Receiving Station
MATS	- Major Assembly Test Set
	- Military Air Transport Service
MATTS	- Multiple Airborne Target Trajectory System
MAUDE	- Morse Automatic Decoder
MAULT	- Manual/Automatic Ultrasonic Laboratory Test
MAV	- Manpower Authorization Voucher
MAVAR	- Modulating Amplifier Using Variable Resistance
MAW	- Marine Aircraft Wing
	- Medium Anti-Tank Weapon
	- Medium Assault Weapon
MAWCS	- Mobile Air Weapons Control System
MAX	- Maximum
MAXSECOM	- Maximum Security Communications
MAYDAY	- m'aidez (help me) Radio-Telephone International Distress Call
MAZH	- Missile Azimuth Heading

■ MAZO

MAZO	- Missile Azimuth Orientation
MB	- Main Battery
	- Maneuver Box Tracking
	- Memory Buffer
	- Millibar
	- Missile Bomber
	- Modified By
	- Munitions Board
M-B	- Make-Break
MBB	- Make-Before-Break
MBC	- Manual Battery Control
	- Maximum Breath Capacity
	- Maximum Breathing Capability
MBE	- Missileborne Equipment
MBI	- May Be Issued
MBK	- Multiple Beam Klystron
MBL	- Model Breakdown List
MBOS	- Missile Base Operations Supervisor
MBR	- Intercontinental Ballistic Missile (USSR)
	- Marker Beacon Receiver
	- Mechanical Buffer Register
MBRUU	- May Be Retained Until Unserviceable
MBRV	- Maneuverable Ballistic Re-Entry Vehicle
MBS	- Magnetron Beam Switching
	- Main "Bang" Suppressor
	- Missile Bunker Storage
	- Multiple Business System
MBT	- Mobile Boarding Team
MBTS	- Missile Battery Test Set
MC	- Magnetic Core
	- Manual Control
	- Marginal Checking
	- Marine Corps
	- Master Change
	- Master Control
	- Material Code
	- Medical Corps (now AMEDS)
	- Megacycle
	- Message Composer
	- Midcourse Correction
	- Military Characteristics
	- Military Code
	- Military Computer
	- Millicurie
	- Missile Car
	- Missile Code

MC	- Missile Control
	- Momentary Contact
	- Multiple Contact
M&C	- Monitor and Control
MCA	- Main Console Assembly
	- Manufacturing Chemists' Association
	- Material Control Area
	- Military Construction, Army
MCAAF	- Marine Corps Auxiliary Air Facility
MCAAS	- Marine Corps Auxiliary Air Station
MCAB	- Marine Corps Air Base
MCAD	- Military Contracts Administration Department
MCAF	- Marine Corps Air Facility
	- Military Construction, Air Force
MCAS	- Marine Corps Air Station
	- Minuteman Configuration Accountability System
MCAT	- Missile Container for Air Transport
MCB	- Marine Corps Base
	- Mobile Construction Battalion
MCBETH	- Military Computer Basic Environment for Test Handling
MCBF	- Mean Cycles Between Failures
MCBR	- Multiple-Carriage Bomb Rack
MCC	- Maintenance Control Center
	- Maintenance of Close Contact
	- Manual Control Center
	- Master Control Console
	- Mercury Control Center
	- Military Coordinating Committee
	- Minuteman Change Committee
	- Missile Change Committed
	- Missile Checkout Console
	- Missile Command Coder
	- Missile Control Center
	- Mission Control Center
	- Modified Close Control
	- Movement Control Center
	- Multiple Computer Complex
MCCA	- Missile Change Committed (Action)
MCCB	- Minuteman Change Commitment Board
MCCG	- Mission Command and Control Group
MCD	- Manufacturing Construction Document
	- Marginal Checking and Distribution
	- Marine Corps District
MCE	- Manpower and Cost Estimate
	- Maximum Capability Envelope
	- Military Characteristics Equipment

■ MCEB

MCEB	- Marine Corps Equipment Board
	- Military Communications-Electronics Board
MCEC	- Marine Corps Education Center
MCF	- Missile Compatibility Firing
MCG	- Master Control Gauge
	- Microwave Command Guidance
	- Mobile Command Guidance
MCGS	- Microwave Command Guidance System
MCI	- Marine Corps Institute
MCIS	- Multiple Corridor Identification System
MCL	- Master Change Log
	- Mid-Canada Line
MCM	- Magnetic Core Memory
	- Master Change Memorandum
	- Missile Carrying Missile
MCN	- Master Change Notice
	- Mid-Course Navigation
MCO	- Marine Corps Order
MCON	- Military Construction Program
MCP	- Master Change Proposal
	- Military Construction Program
	- Miscellaneous Change Proposal
MCPM	- Marine Corps Personnel Manual
MCR	- Marine Corps Representative
	- Marine Corps Reserve
	- Master Change Record
	- Military Compact Reactor
	- Minuteman Change Request
MCRD	- Marine Corps Reserve District
MCRTC	- Marine Corps Reserve Training Center
MCS	- Maintenance Control Section
	- Management Control System
	- Marine Corps Schools
	- Mast Connection System
	- Master Control System
	- Megacycles per Second
	- Microwave Carrier Supply
	- Military Communication Stations
	- Mine Countermeasures Support
	- Motor Circuit Switch
MCSC	- Marine Corps Supply Center
MCSD	- Marine Corps Supply Depot
MCSO	- Marine Corps Special Orders
MCSRP	- Management Control Systems Research Project
MCSS	- Military Communications Satellite System
MCTB	- Marine Corps Technical Bulletin

MCTG	- Model Change Training Guide
MCTS	- Master Central Timing System
MCU	- Medium Close-Up
	- Miniature Command Unit
MCUG	- Military Computer Users Group
MCW	- Modulated Continuous Wave
MCXM	- Marine Corps Exchange Manual
MD	- Magnetic Drum
	- Main Drum
	- Manual Data
	- Marine Detachment
	- Mass Density
	- Master Dimension
	- Maximum Design Meter
	- Mean Deviation
	- Messages per Day
	- Military Decision
	- Mine Disposal
	- Missile Division
	- Movement Directive
M/D	- Messages per Day
	- Modulator-Demodulator
MDA	- Minimum Decision Altitude
MDAA	- Mutual Defense Assistance Act
MDAP	- Military Defense Assistance Program
	- Mutual Defense Assistance Program
MDB	- Microsecond Delay Blasting
MDC	- Main Display Console (formerly MDP)
	- Master Direction Center
	- Missile Development Center
	- Missile Direction Center
	- Movement Designator Code
MDCAC	- Manufacturing Department Change Analysis Commitment
MDCC	- Master Data Control Console
MDCS	- Master Digital Command System
MDDA	- Mechanisburg Defense Depot Activity
MDE	- Mission Display Equipment
	- Modular Design of Electronics
MDF	- Main Distributing Frame
	- Manual Direction Finder
	- Mild Detonating Fuze
MDI	- Magnetic Detection Indicator
	- Manual Data Input
	- Miss Distance Indicator
MDIF	- Manual Data Input Function
MDIRT	- Miss Distance Indicator Radioactive Test

■ MDIS

MDIS	- Manual Data Input System
MDIU	- Manual Data Insertion Unit
MDLC	- Materiel Development and Logistic Command
MDMS	- Miss Distance Measuring System
MDP	- Main Display Panel (now Main Display Console)
MDR	- Manual Data Room
	- Mission Data Reduction
	- Multi-Channel Data Recorder
MDS	- Malfunction Detection System
	- Manual Data Supervisor
	- Master Development Schedule
	- Meteoroid Detection Satellite
	- Minimum Discernible Signal
	- Multiple Deployment System
MDSA	- Manual Data Supervisor Assistant
MDSR	- Manufacturing Development Service Request
MDSS	- Mission Data Support System
MDT	- Manual Data Technician
	- Mean Down Time
	- Mountain Daylight Time
	- Munitions Disposal Technician
MDU	- Message Decoder Unit
	- Mine Disposal Unit
	- Missile Design Unit
MDV	- Map and Data Viewer
MDW	- Military District of Washington
ME	- Manufacturing Engineering
	- Mechanical Efficiency
	- Middle East
	- Multi-Engine
MEA	- Middle East Airlines
	- Minimum Enroute Altitude
MEAL	- Master Equipment Authorization List
MEAR	- Maintenance Engineering Analysis Record
MEATR	- Materials, Engineering, and Advanced Test Reactor
MEB	- Marine Expeditionary Brigade
MEC	- Manual Emergency Controls
	- Microwave Electronics Corporation
	- Milgo Electronics Corporation
MECO	- Main Engine Cutoff
MECR	- Maintenance Engineering Change Report
MED	- Manufacturing Engineering Document
MEDAL	- Micro-Mechanized Engineering Data for Automated Logistics
MEDDA	- Mechanized Defense Decision Anticipation
MEDIA	- Magnavox Electronic Data Image Apparatus

MEDIA	- Missile Era Data Integration Analysis
MEDIC	- Medical Electronic Data Interpretation and Correlation
MEDIUM	- Missile Era Data Integration—Ultimate Method
MEDLARS	- Medical Literature Analysis and Retrieval System
MEDT	- Military Equipment Delivery Team
MEE	- Minimum Essential Equipment
	- Mission Essential Equipment
MEETA	- Maximum Improvement in Electronics Effectiveness Through Advanced Techniques
MEF	- Marine Expeditionary Force
MEG	- Megohm
MEGO	- Megohm
MEGV	- Megavolt
MEGW	- Megawatt
MEGWH	- Megawatt-Hour
MEI	- Maintenance Engineering Inspection
MEIM	- Minuteman Engineering Instruction Manual
MEK	- International Electrochemical Commission (USSR)
MEL	- Master Equipment List
MELH	- Missile Elevation Heading
MELI	- Master Equipment List Identification
MELVA	- Military Electronic Light Valve
MEM	- Mars Excursion Module
MEMS	- Missile Equipment Maintenance Set
MEN	- Multiple Event Network
MEND	- Medical Education for National Defense
MEO	- Maintenance Engineering Order
	- Major Engine Overhaul
MEOSAB	- Missile Explosive Ordnance Safety Advisory Board
MEP	- Mean Effective Pressure
	- Moon-Earth-Plane
MERCAST	- Merchant Ship Broadcast
MERL	- MIT (Massachusetts Institute of Technology) Electronic Research Laboratories
MERT	- Maintenance Engineering Review Team
MERU	- Milli-Earth Rate Unit
MES	- Mission Event Sequence
M&ES	- Manufacturing and Engineering Support
MESROM	- Materials-Evaluation Subcaliber Rocket Motor
MEST	- Munitions Evaluation and Standardization Team
MET	- Meteorological
	- Missile Electrical Technician
	- Mission Environment Tape
METCO	- Meteorological Coordinating Committee
METEOR	- Manned Earth-Satellite Terminal Evolving from Earth-to-Orbit Ferry Rockets

■ METJET

METJET	- Meteorological Sounding Rocket, Ramjet-Powered
METO	- Maximum Engine Take-Off
	- Meteorological Officer
	- Middle East Treaty Organization
METROC	- Meteorological Rocket
METU	- Marine Electronic Technical Unit
MEU	- Marine Expeditionary Unit
MEV	- Million Electron Volts
MEW	- Microwave Early Warning
MEWS	- Microwave Early Warning System
MF	- Main Feed
	- Medium Frequency (band 6)
	- Microfarad
	- Millifarad
	- Motor Field
	- Munitions Facility
M&F	- Male and Female
MFB	- Munitions Facility Building
MFC	- Manual Frequency Control
MFCT	- Major Fraction Thereof
MFD	- Magnetic Frequency Detector
MFE	- Make Factory Equipment
MFED	- Maximum Flat Envelope Delay
MFEQ	- Mechanical Facilities and Equipment
MFG	- Major Functional Group
MFIC	- Military Flight Information Center
MFIT	- Modified Flight Intersection Tape
MFL	- Missile Firing Laboratory
MFP	- Minimal Flight Path
MFR	- Memorandum for Record
	- Military Field Representative
MFS	- Military Flight Service
	- Missile Firing Station
	- Multifunction Sensor
MFSC	- Missile Flight Safety Center
MFSFU	- Matt-Finish Structural Facing Units
MFSG	- Missile Firing Safety Group
MFSK	- Multiple Frequency Shift Keying
MFSO	- Missile Flight Safety Officer
MFSOA	- Missile Flight Safety Officer Assistant
MFSOC	- Missile Flight Safety Officer Console
MFSS	- Missile Flight Safety System
MFT	- Major Fraction Thereof
	- Mechanized Flame Thrower
MFTP	- Munitions Facility Test Program
MFV	- Main Fuel Valve

MG	- Machine Gun
	- Master Gage
	- Message Generator
	- Middle Gimbal
	- Military Government
	- Milligram
	- Motor-Generator
	- Multi-Gage
MGA	- Middle Gimbal Angle
	- Middle Gimbal Axis
MGC	- Manual Gain Control
	- Missile Guidance and Control
MGCC	- Missile Guidance and Control Computer
MGCIS	- Marine Ground Control Intercept Squadron
MGCR	- Maritime Gas Cooled Reactor
MGD	- Magnetogasdynamics
MGE	- Maintenance Ground Equipment
	- Missile Ground Equipment
MGF	- Marine Garrison Force
MGG	- International Physical Year (USSR)
	- Memory Gate Generator
MGMC	- Multiple Gun Motor Carriage
MGP	- Maintenance Ground Point
MGRS	- Military Grid Reference System
MGS	- Message Generation Study
	- Missile Guidance Set
	- Missile Guidance System
MGSE	- Maintenance Ground Support Equipment
	- Mobile Ground Support Equipment
MGSE-ECM	- Maintenance Ground Support Equipment—Environmental Controls and Mechanisms
MGST	- Military Geography Specialist Team
MH	- Magnetic Head
	- Magnetic Heading
	- Millihenry
M-H	- Minneapolis-Honeywell
M&H	- Mechanical and Hydraulic
MHA	- Modified Handling Authorized
MHCP	- Mean Horizontal Candlepower
MHD	- Magnetohydrodynamics
MHE	- Materials Handling Equipment
MHF	- Medium High Frequency
MHS	- Magnetic Heading System
MHW	- Mean High Water
MI	- Manned Interceptor
	- Manual Input

MI

MI	- Mile
	- Military Intelligence
	- Miller Integrator
	- Minimum Impulser
	- Miscellaneous Input
	- Missed Interceptor
	- Moment of Inertia
	- Movements Identification
	- Mutual Inductance
MIA	- Metal Interface Amplifier
	- Minimum Intercept Altitude
	- Mouse in Able
MIAC	- Material Identification and Accounting Code
MIADS	- Minot Air Defense Sector
MIAN	- V. A. Steklov Institute of Mathematics (USSR)
MIAPD	- Mid-Central Air Procurement Division
MIAR	- MIDAS (Missile Defense Alarm System) Intercept Assembly Register
MIB	- Manual Input Buffer
MIC	- Monitoring, Identification, and Correlation
MICA	- Macro Instruction Compiler Assembler
MICON	- Military Construction Program
MICOS	- Master Information Control Station
MICR	- Magnetic Ink Character Recognition
MICRAM	- Micro-Miniature Individual Component Reliably Assembled Modules
MICROACE	- Micro-Miniature Automatic Checkout Equipment
MICRO-MIN	- Micro-Miniaturize
MICU	- Message Injection Control Unit
MIDAC	- Michigan Digital Automatic Computer
MIDAR	- Microwave Detection and Ranging
MIDARM	- Micro-Dynamic Angle and Rate Monitoring System
MIDAS	- Missile Defense Alarm System (system of satellites)
	- Missile Intercept Data Acquisition System (equipment manufactured by Cubit Corporation)
MIDEASTFOR	- Middle East Force
MIDIZ	- Mid-Canada Identification Zone
MIDOP	- Missile Doppler
MIDOT	- Multiple Interferometer Determination of Trajectory
MIE	- Mobile Inspection Equipment
MIEETAT	- Maximum Improvement in Electronic Effectiveness Through Advanced Techniques
MIFE	- Minimum Independent Failure Element
MIFI	- Moscow Engineering-Physics Institute (USSR)
MIFSA	- Missile In-Flight Safety Approval
MIG	- Metal Inert Gas

MIG	- Mikoyan and Gurevich (Russian aircraft designers)
	- Miniature Integrating Gyroscope
MII	- Manned Interceptor Integration
	- Military Intelligence Interpreter
MIIT	- Manned Interceptor Integration Team
MIKE	- Microphone
MIL	- Military
	- Missile Industry Liaison
MILA	- Merritt Island Launch Area
MILCEST	- Military Communications Electronic System Technology
MILDD	- Military-Industry Logistics Data Development
MILECON	- Military Electronics Convention
MIL HDBK	- Military Handbook
MILPAS	- Miscellaneous Information Listing Program Apollo Spacecraft
MILS	- Missed Intercept—Late Scramble
	- Missile Impact Location System
MILS/PAC	- Missile Impact Location System, Pacific
MIL SPEC	- Military Specification
MILSTAMP	- Military Standard Transportation and Movement Procedures
MIL STD	- Military Standard
MILSTRIP	- Military Standard Requisitioning and Issue Procedure
MILTAG	- Military Technical Assistance Group
MIM	- Maintenance Instructions Manual
MIMO	- Man In, Machine Out
MIN	- Minimum
	- Minute
MINAC	- Minuteman Action Committee
MIND	- Magnetic Integrator Neuron Duplicator
MINEAC	- Miniature Electronic Autocollimator
MINIAPS	- Miniature Accessory Power Supply
MINICOM	- Minimum Communications
MININAN	- Minimal Doppler Navigator
MINITAS	- Miniature True Air Speed Computer
MINITRACK	- Minimum-Weight Tracking System (now STADAN)
MINLANT	- Mine Warfare Force, Atlantic
MINPAC	- Mine Warfare Force, Pacific
MINS	- Management Information Network System
	- Miniature Inertial Navigation System
MINSAT	- Minimum Safe Air Travel
MINSY	- Mare Island Naval Shipyard
MINT	- Material Identification and New Item Control Technique
MINWR	- Minimum Weapon Radius
MIO	- Military Identification Officer

■ MIP

MIP	- Manual Input Processing
	- Manual Input Program
	- Material Improvement Project
	- Methods Improvement Program
	- Minimum Impulse Pulse
	- Missile Impact Prediction
	- Model Implementation Plan
MIPE	- Magnetic Induction Plasma Engine
	- Modular Information Processing Equipment
MIPIR	- Missile Precision Instrumentation Radar
MIPR	- Military Interdepartmental Procurement Request (Army)
	- Military Interdepartmental Purchase Request (Air Force)
MIPS	- Missile Impact Prediction System
	- Missile Information Processing System
MIR	- Manual Input Room
	- Memory Input Register
M&IR	- Manufacturing and Inspection Records
MIRAK	- Minimum Rocket (German)
MIRAN	- Missile Ranging
MIRF	- Multiple Instantaneous Response File
MIRR	- Material Inspection Receiving Report
MIRROS	- Modulation Inducing Reactive Retrodirective Optical System
MIRTRAK	- Martin Infrared Tracker
MIRV	- Multiple Independent Re-Entry Vehicle
MIS	- Man in Space
MISC	- Miscellaneous
MISER	- Manned Interceptor SAGE Evaluation Routine
MISHAP	- Missile High-Speed Assembly Program
MISMA	- Major Item Supply Management Agency
MISP	- Man-in-Space Program
	- Manned Interceptor Simulation Program
MISR	- Major Items Status Report
MISRAN	- Missile Range
MISS	- Man in Space Soonest
	- Miniature SOFAR System
	- Minuteman Integrated Schedules and Status
	- Missile Intercept Simulation System
MISSD	- Minuteman Integrated Schedules and Status Data System
MISSOPH	- Man in Space Sophisticated
MIST	- Multi-Input Standard Tape
MIST-FOAL	- Multi-Stage Force Allocation
MISTRAM	- Missile Trajectory Measurement System
MIT	- Manual Inputs-Tracks
	- Massachusetts Institute of Technology
	- Master Instruction Tape

MIT	- Minimum Individual Training
	- Miscellaneous Tool
	- Movements Identification Technican
MITE	- Master Instrumentation Timing Equipment
	- Missile Integration Terminal Equipment
	- Multiple Input Terminal Equipment
MITIL	- Massachusetts Institute of Technology Instrumentation Laboratory
MITMS	- Military/Industry Technical Manual Specifications
MITOC	- Missile Intercommunications Technical Operational Communications
MITRE	- Massachusetts Institute of Technology Research and Engineering Corporation
MIU	- Malfunction Insertion Unit
	- Message Injection Unit
	- Mobile Inspection Unit
MIU/FCO	- Mobile Inspection Unit/Functional Checkout
MIX	- Magnavue Information Extractor
MK	- Mark
MKS	- Meter-Kilogram-Second
MKSA	- Meter-Kilogram-Second, Absolute
ML	- Machine Language
	- Mainland
	- Maintenance Laboratory
	- Manual Lean
	- Material List
	- Military Liaison
	- Millilambert
	- Milliliter
	- Missile Launcher
	- Mission Life
	- Modify and Load
	- Mold Line
M/L	- Manual Local
M&L	- Modify and Load
MLC	- Mesh Level Control
	- Military Liaison Committee
MLCAEC	- Military Liaison Committee to the Atomic Energy Comission
MLCB	- Missile Launch Control Blockhouse
MLCC	- Mobile Launch Control Center
MLD	- Median Lethal Dose
	- Minimum Line of Detection
MLE	- Mobile Launcher Equipment
MLF	- Multi-Lateral Force
MLFC	- Moses Lake Flight Center

■ MLG

MLG	- Main Landing Gear
MLI	- Minimum Line of Interception
MLL	- Manned Lunar Landing
	- Manned Lunar Launching
MLLP	- Manned Lunar Landing Program
MLLR	- Manned Lunar Landing and Return (MALLAR preferred)
MLLW	- Mean Lower Low Water
MLO	- Metal Layout
MLP	- Machine Language Program
MLR	- Main Line of Resistance
MLRV	- Missile Less Re-Entry Vehicle
MLS	- Medium Long Shot
	- Model Lunar Service
MLT	- Metal Layout Template
	- Mission Life Test
MLW	- Mean Low Water
MM	- Middle Marker
	- Millimeter
	- Minuteman
	- Missile Master
M/M	- Master Model
	- Maximum and Minimum
MMC	- Maximum Material Condition
	- Midcourse Measurement Correction
	- Mortar Motor Carriage
MMD	- Master Makeup and Display
MMEF	- Mobile Missile Emplacement Facility
MMF	- Magnetomotive Force
	- Micromicrofarads
MMG	- Medium Machine Gun
MMH	- Monomethylhydrazine
MMH/FH	- Maintenance Manhours per Flight Hour
MMIT	- Man-Machine Interrogation Technique
MMM	- Mars Mission Module
	- Missile Maintenance Mechanic
	- Multi-Mission Module
MMO	- Minuteman Ordnance
MMP	- Multiplexed Message Processor
MMPC	- Mobilization Material Procurement Capability
MMR	- Mobilization Materiel Requirement
MMRBM	- Mobile Mid-Range Ballistic Missile
MMRS	- Manned Military Recovery System
MMS	- Missile Maintenance Squadron
MMSA	- Mining and Metallurgical Society of America
MMSS	- Manned Maneuverable Space System
MMSTP	- Master Missile System Training Program

MMT	- Missile Maintenance Technician
MMTLN	- Map Margin Top Line
MMTT	- Mobile Minuteman Train Test
MMU	- Midcourse Measurement Unit
	- Millimass Unit
MN	- Magnetic North
MNBA	- Minimum Normal Burst Altitude
MNEE	- Mission Non-Essential Equipment
MNOR	- Missile Not Operationally Ready
MNORM	- Missile Not Operationally Ready for lack of Maintenance (see MOCM)
MNORP	- Missile Not Operationally Ready for lack of Parts (see MOCP)
MNTV	- Mercury Network Test Vehicle
MO	- Manual Output
	- Manufacturing Order
	- Master Oscillator
	- Medical Officer
	- Meteorology Officer
	- Miscellaneous Operation
	- Mobile Station
	- Molecular Orbit
	- Movement Orders
M/O	- Manned and Operational
M&O	- Maintenance and Operation
MOA	- Made on Assembly
	- Medium Observation Aircraft
MOADS	- Montgomery Air Defense Sector
MOAMA	- Mobile Air Materiel Area
MOARS	- Mobilization Assignment Reserve Section
MOB	- Missile Order of Battle
	- Mobile Operations Building
MOBIDIC	- Mobile Digital Computer
MOBL	- Macro Oriented Business Language
MOBOT	- Mobile Remote-Controlled Robot
MOBRASOP	- Mobilization Requirements in Support of the Army Strategic Objectives Plan
MOBU	- Mobilization Base Units
MOC	- Magnetic Optic Converter
	- Master Operations Console
	- Master Operational Controller
	- Master Operations Control
	- Missile Operation Center
	- Mission Operation Computer
	- Mission Operational Controller
MOCC	- Master Operations Control Center
MOCM	- Missile Out of Commission for Maintenance (see MNORM)

■ MOCOM

MOCOM	- Mobile Command
MOCP	- Missile Out of Commission for Parts (see MNORP)
MOCR	- Mission Operations Control Room
MOD	- Medical Officer of the Day
	- Ministry of Defence (British)
	- Minuteman Operating Directive
	- Miscellaneous Obligation Document
	- Model
	- Modification
	- Modulator
MODAC	- Mountain System Digital Automatic Computer
MODEM	- Modulate-Demodulate
	- Modulator-Demodulator
MODERN	- Modulator-Demodulator (MODEM preferred)
MODICON	- Modular Dispersed Control
MODS	- Manned Orbital Development Station (NASA)
	- Military Orbital Development Station (Air Force)
MOE	- Measure of Effectiveness
	- Model Operational Environment
	- Mythical Operational Environment
MOGAS	- Motor Gasoline
MOHO	- Mohorovicic Discontinuity
MOI	- Make on Installation
MOK	- International Commission on Illumination
MOL	- Machine Oriented Language
	- Manned Orbiting Laboratory (Air Force)
MOLAB	- Mobile Laboratory
MOLC	- Multiple Operational Launch Complex
MOLECOM	- Molecularized Digital Computer
MOM	- Mark XII Output and Monitoring System
	- Missile Operations Manager
MOMP	- Mid-Ocean Meeting Place
MON	- Monitoring
MOO	- Missile Operations Officer
MOOSE	- Man Out of Space Easiest
MOP	- Model Operational Plan
	- Modular Operating Procedure
	- MORT (Master Operational Recording Tape) Plotter
MOPA	- Master Oscillator Power Amplifier
MOPAR	- Master Oscillator Power Amplifier Radar
MOPR	- Manner of Performance Rating
MOPS	- Maintenance and Operation Paging System
	- Military Operation Phone System
	- Missile Operating Phone Set
	- Missile Operations
MOPTAR	- Multi-Object Phase Tracking and Ranging System

MOPTS	- Mobile Photographic Tracking System
MOR	- Maintenance Operations Report
	- Mars Orbital Rendezvous
	- Memory Output Register
	- Missile Operationally Ready
	- MORT (Master Operational Recording Tape) Recording
MOREST	- Mobile Arrested Landing Equipment
MORL	- Manned Orbital Research Laboratory (NASA)
MORS	- Military Operations Research Symposium
MORT	- Master Operational Recording Tape
	- Missile Operational Readiness Test
MOS	- Management Operating System
	- Metal Oxide Semiconductor
	- Military Occupational Speciality
	- Missile on Site
MOSAIC	- Mobile System for Accurate ICBM Control
MOSL	- Manned Orbital Space Laboratory
MOSS	- Manned Orbital Space Station
MOST	- Manufacture of 709 System Tapes
MOTS	- Mobile Test Set
MOTU	- Mobile Optical Tracking Unit
MOUSE	- Minimum Orbital Unmanned Satellite, Earth
MOV	- Main Oxidizer Valve
	- Manned Orbiting Vehicle
MOW	- Mission Operations Wing
MOWS	- Manned Orbital Weapon System
MP	- Magnetic Bearing (USSR)
	- Manifold Pressure
	- Manual Proportional
	- Manufacturing Plan
	- Medium Pressure
	- Melting Point
	- Missile Possessed
	- Multiplier Phototube
	- Multipointing
	- Multipole
MPA	- Miniature Pendulum Accelerometer
MP&A	- Machine Process and Assembly
MPACS	- Management Planning and Control System
	- Manual Proportional Attitude Control System
MPB	- Maintenance Parts Breakdown
	- Material Price Book
MPC	- Maximum Permissible Concentration
	- Missile Production Center
	- MORT (Master Operational Recording Tape) Processor Control

■ MPC

MPC	– Multiple-Purpose Communications
MPCM	– Manual Proportional Control Mode
MPD	– Make per Drawing
	– Maximum Permissible Dosage
	– Missile Products Division
	– Missile Profile Drawing
	– Missile Pulse Detector
MPDM	– Maintenance Planning Data Manual
MPE	– Maximum Permissible Exposure
	– Mechanized Production of Electronics
MPG	– Methods and Procedures Group
	– Microwave Pulse Generator
	– Miles per Gallon
MPH	– Miles per Hour
MPHPS	– Miles per Hour per Second
MPI	– Mean Point of Impact
MPIC	– Message Processing Interrupt Count
MPL	– Maintenance Parts List
MPLE	– Multi-Purpose Long Endurance
MPM	– Miles per Minute
MPO	– Maximum Power Output
	– Mercury Project Office
	– Military Planning Officer
MPP	– Maintainability Program Plan
	– Materials Preparation Program
	– Most Probable Position
MPPM	– Mission Prediction and Performance Module
MPPPG	– Manual Problem Production Planning Group
MPPS	– Master Program Preparation Section
MPR	– Malfunction Problem Report
	– Maximum Potential Representation
	– Multiply and Round
MPRI	– Manual Probe Request Indicator
MPS	– Materiel Planning Study
	– Miles per Second
MPSC	– Military Personnel Security Committee
MPSP	– Military Personnel Security Program
MPTE	– Multi-Purpose Test Equipment
MPTO	– Methods and Procedures Technical Order
MPTR	– Mobile Position Tracking Radar
MPTS	– Multi-Purpose Tool Set
MPX	– Multiplex
MQ	– Multiplier Quotient
MQR	– Multiplier Quotient Register
MR	– Machine Records
	– Machine Rifle

MRL ■

MR	- Manufacturing Record
	- Master Reset
	- Master Routing
	- Material Review
	- Medium Range
	- Memorandum Receipt
	- Mercury-Redstone
	- Message Repeat
	- Micro-Miniature Relay
	- Military Requirement
	- Milliroentgen
	- Missed Recognition
	- Monitor Recorder
	- Morning Report
M/R	- Map Reading
	- Maximum Range
M&R	- Maintenance and Repair
MRA	- Machine Records Activity
	- Material Review Area
	- Material Review Authorization
MRAL	- Material Readiness Authorization List
MRB	- Material Review Board
	- Missile Review Board
	- Modification Requirements Board
MRBM	- Medium Range Ballistic Missile
MRC	- Master Routing Control
	- Military Representative Committee
MRCF	- Missile Recycle Facility
MRCN	- Minuteman Requirement Control Number
MRCP	- Mobile Radar Control Post
MRCR	- Measurement Requirement Change Request
MRD	- Manufacturing Research Directive
	- Military Reference Data
MRDF	- Maritime Radio Direction Finding
MRE	- Mid Range Estimate
	- Missile Recycle Equipment
MRF	- Missile Recycle Facility
MRG	- Model Reprogramming Group
MRH	- Mechanical Recording Head
MR/HR	- Milliroentgen per Hour
MRI	- Medium Range Interceptor
	- Midwest Research Institute
	- Miscellaneous Radar Input
	- Monopulse Resolution Improvement
MRICC	- Missile and Rocket Inventory Control Center
MRL	- Master Repair List

■ MRL

MRL	- Multiple Rocket Launcher
MRML	- Medium Range Missile Launcher
MRMR	- Mobilization Reserve Material Requirement
MRN	- Meteorological Rocket Network
MRNC	- Meteorological Rocket Network Committee
MRO	- Maintenance, Repair, and Operation
	- Maintenance Report Order
	- Material Release Order
	- Movement Report Office
MRPM	- Military Revolutions per Minute
MRR	- Maintenance, Repair, and Retrofix
	- Material Rejection Report
MRRC	- Material Requirements Review Committee
MRS	- Manned Reconnaissance Satellite
	- Material Returned to Stores
	- Memo Routing Slip
	- Mobilization Requirement Study
MRSR	- Manufacturing Research Service Request
MRS V	- Maneuverable Recoverable Space Vehicle
MRT	- Military Rated Thrust
	- Mobile Radar Target
	- Mobility R&D Testing
MRU	- Machine Records Unit
	- Message Retransmission Unit
	- Microwave Relay Unit
	- Mobile Radio Unit
MRV	- Maneuvering Re-Entry Vehicle
MS	- Machine Screw
	- Machine Steel
	- Magnetic Storage
	- Main Stage
	- Mapping Supervisor
	- Margin of Safety
	- Mark Sense
	- Master Switch
	- Medium Shot
	- Meterological Systems
	- Millisecond
	- Military Specification
	- Military Standard
	- Missile Station
	- Mission Sequencer
	- Most Significant
M/S	- Magnetostruction
	- Main Stage
	- Master Schedule

M&S	- Maintenance and Supply
MSA	- Master Schedule Authorization
	- Mineralogical Association of America
	- Missile System Analyst
	- Mutual Security Agency
MSAT	- Missile System Analyst Technician
MSATF	- Missile Site Activation Task Force (see SATAF)
MSB	- Missile Storage Bunker
	- Missile Support Base
	- Most Significant BIT
MSBS	- Mer-Sol Bolistiques Strategiques
MSC	- Management Systems Corporation
	- Manned Spacecraft Center
	- Medical Service Corps
	- Meteorological Satellite Center
	- Minesweeper, Coastal
MSCE	- Mobile Systems Checkout Equipment
MSCF	- Millions of Standard Cubic Feet
MSCOP	- Missile System Checkout Program
MSCP	- Mean Spherical Candlepower
	- Missile System Checkout Programmer
MSD	- Marine Signal Detachment
	- Mean Solar Day
	- Minimum Safe Distance
	- Missile System Development
	- Mission Systems Data
	- Most Significant Digit
MSE	- International Telecommunications Union (USSR)
	- Minus Sense
MSEC	- Millisecond
MSEF	- Missile System Evaluation Flight
MSEM	- Mission Status and Evaluation Module
MSEO	- Military System Exercising Organization
MSF	- Manned Space Flight
	- Mobile Striking Force
MSFC	- Marshall Space Flight Center
MSFL	- Manned Space Flight Laboratory
MSFN	- Manned Space Flight Network
MSFP	- Manned Space Flight Program
MSFS	- Manned Space Flight System
MSG	- Mapper Sweep Generator
	- Mapping Supervisor Gap Filler
	- Message
	- Miscellaneous Simulation Generator
MSGR	- Mobile Support Group
MSHB	- Minimum Safe Height of Burst

- **MSI**

MSI	- Manned Satellite Inspector
	- Master Specification Index
	- Mine Sweeper, In Shore
	- Missile Subsystem Integration
MSIS	- Manned Satellite Inspection System
MSL	- Mean Sea Level
MSLCOM	- Missile Command
MSLS	- Maneuverable Satellite Landing System
MSM	- Manufacturing Shop Manual
MSMS	- Mutual Security Military Sales
MSMV	- Mono-Stable Multi-Vibrator
MSN	- Master Serial Number
MSO	- Mine Sweeper, Ocean
MSP	- Mutual Security Program
MSR	- Main Supply Route
MSS	- Make Suitable Subsitution
	- Manufacturers Standardization Society
	- Mean Solar Second
	- Military Supply Standard
	- Mission Status Summary
	- Model Shape Survey
MSSA	- Military Subsistence Supply Agency (obsolete)
MSSCC	- Military Space Surveillance Control Center
MSSMS	- Munitions Section, Strategic Missile Squadron
MST	- Mean Solar Time
	- Military Science Training
	- Mountain Standard Time
MSTO	- Military System Training Organization
MSTP	- Manual System Training Program
MSTS	- Military Sea Transportation Service
	- Missile Static Test Site
MSV	- Meteor Simulation Vehicle
MSVD	- Missile and Space Vehicle Department
MSVO	- Missile and Space Vehicle Office
MT	- Machine Translation
	- Magnetic Tape
	- Magnetic Tube
	- Maintenance Technician
	- Maximum Torque
	- Mechanical Time
	- Mechanical Translation
	- Mechanical Transport
	- Megaton
	- Megatron
	- Metric Ton
	- Microsyn Torquer

MT	- Military Transport
	- Missed Target
	- Missile Test
	- Mode Transducer
	- Motor Transport
	- Moving Target
	- Multiple Transfer
MTA	- Missile Transfer Area
MTAC	- Mathematical Tables and Other Aids to Computation
MTB	- Maintenance of True Bearing
	- Motor Torpedo Boat
MTBF	- Mean Time Between Failures
MTC	- Master Tape Control
	- Memory Test Computer
	- Missile Test Center
	- Mission and Traffic Control
MTCC	- MATS Traffic Control Center
MTCF	- Mean Time to Catastrophic Failure
MTD	- Manufacturing Technical Directive
	- Mean Temperature Difference
	- Mobile Training Detachment
	- Multiple Target Deception
MTDR	- Maintainability Test and Demonstration Requirement
	- Malfunction Tear Down Report
MTDS	- Marine Tactical Data System
	- Missile Trajectory Data System
MTE	- Maximum Tracking Error
	- Missile Targeting Equipment
	- Missile Test Engineer
	- Multi-System Test Equipment
MTECP	- Maintenance Test Equipment Certification Procedure
MTECR	- Maintenance Test Equipment Certification Requirement
MTF	- Mechanical Time Fuze
	- Mississippi Test Facility
MTG	- Microsyn Torque Generator
	- Multiple-Trigger Generator
MTH	- Magnetic Tape Handler
MTI	- Materials Technology Institute
	- Metal Treating Institute
	- Moving Target Indicator
MTL	- Master Tape Loading
MTLP	- Master Tape Loading Program
MTM	- Material Transfer Memorandum
	- Mechanical Test Model
	- Mission Test Module
MTMA	- Military Traffic Management Agency (obsolete)

■ MTO

MTO	- Manufacturing Technical Order
	- Missile Test Operator
	- Mission, Task, Objective
	- Mississippi Test Operations
MTON	- Measurement Ton
MTP	- Mobilization Training Program
MTR	- Magnetic Tape Recorder
	- Master Tool Record
	- Materials Testing Reactor
	- Mean Time to Restore
	- Missile Tracking Radar
	- Motor
	- Moving Target Reactor
	- Multiple Tracking Range
MT&RC	- Marine Training and Replacement Command
MTRE	- Magnetic Tape Recorder End
	- Missile Test and Readiness Evaluation
MTRI	- Missile Test Range Instrumentation
MTRS	- Magnetic Tape Record Start
MTS	- Magnetic Tape System
	- Master Timing System
	- Mechanical Test Set
	- Message Traffic Study
	- Missile Test Station
	- Motor-Operated Transfer Switch
MTSE	- Manufacturing Test Support Equipment
MTSQ	- Mechanical Time, Super Quick
MTSS	- Military Test Space Station
MTT	- Mobile Training Team
MTTF	- Mean Time to Failure
MTTFF	- Mean Time to First Failure
MTTR	- Mean Time to Repair
MTU	- Magnetic Tape Unit
	- Mechanical Test Unit
	- Missile Training Unit
	- Mobile Training Unit
MTUOP	- Mobile Training Unit Out for Parts
MTVAL	- Master Tape Validation
MTW	- Mobile Training Wing
MU	- Millimicron
	- Mobile Unit
	- Mockup
	- Multiple Unit
M/U	- Mockup
	- Multiple Use
MUC	- Muchea, Australia (remote site)

MUF	- Maximum Usable Frequency
MUGSE	- Mobile Unit—Ground Support Equipment
MULS	- Mobile Unit Launch Site
MURA	- Midwestern Universities Research Association
MURG	- Machine Utilization Report Generator
MUSA	- Multiple-Unit Steerable Antenna
MUSB	- Mobile Unit Support Base
MUSCM	- Missile Unit Simulated Combat Mission
MUSF	- Mobile Unit Support Facility
MUSS	- Missile Unit Support System
	- Mobile Unit Support System
MUV	- Mobile Underwater Vehicle
MUX	- Multiplex
MV	- Mean Value
	- Millivolt
	- Muzzle Velocity
MVB	- Multivibrator
MVC	- Manual Volume Control
MVD	- Map and Visual Display
MVS	- Magnetic Voltage Stabilizer
MW	- Medium Wave
	- Microwave
	- Milliwatt
	- Molecular Weight
M/W	- Microwave
MWDDEA	- Mutual Weapons Development Data Exchange Agreement
MWDP	- Mutual Weapons Development Program
MW(E)	- Megawatt, Electrical
MWG	- Manpower Working Group
	- Meteorological Working Group
	- Music Wire Gage
MWH	- Megawatt-Hour
MWL	- Milliwatt Logic
MWO	- Modification Work Order
MWP	- Maximum Working Pressure
	- Membrane Waterproofing
MWR	- Mean Width Ratio
MWS	- Microwave Station
MWSG	- Marine Wing Service Group
MW(T)	- Megawatt, Thermal
MWV	- Maximum Working Voltage
MX	- Matrix
	- Missile Experimental
	- Multiplex
MXD	- Multiple Transmitter Duplicator
MYMSG	- Reference My Message

N

N	- National Form
	- Navigation
	- Navy
	- Nitrogen
	- North
	- Novelty
	- Nylon
NA	- National Acme
	- Naval Aircraft
	- Naval Aviator
	- Normal Alarm
	- Not Applicable
	- Not Appropriate
	- Not Appropriated
	- Not Available
N/A	- Next Assembly
	- Not Applicable
	- Not Available
NAA	- National Aeronautic Association
	- North American Aviation, Incorporated
NAACD	- North American Aviation Columbus Division
NAADC	- North American Air Defense Command
NAAL	- North American Aerodynamics Laboratory
NAALA	- North American Aviation Los Angeles
NAAPS	- Nozzle Actuator Auxiliary Power Supply
NAARD	- North American Aviation Rocketdyne Division
NAAS	- Naval Auxiliary Air Station
NAASC	- North American Aviation Science Center
NAASD	- North American Aviation Space Division
NAAT	- Naval Air Advanced Training
NAB	- National Acoustics Board
	- National Association of Broadcasters
	- Naval Air Base

NAB	- Naval Amphibious Base
	- Navigational Aid to Bombing
NABAC	- National Association for Bank Audit, Control, and Operation
NABD	- Naval Advance Base Depot
NABTC	- Naval Air Basic Training Command
NABUG	- National Association of Broadcast Unions and Guilds
NAC	- National Agency Check
	- North Atlantic Council
	- Northeast Air Command
NACA	- National Advisory Committee for Aeronautics (now NASA)
NACCAM	- National Coordinating Committee for Aviation Meteorology
NACE	- National Advisory Committee for Electronics
	- National Association of Corrosion Engineers
NAD	- Naval Air Depot
	- Naval Ammunition Depot
NADAC	- National Damage Assessment Center
NADAR	- North American Data Airborne Recorder
NADC	- Naval Air Development Center
	- Naval Ammunition Depot, Crane
NADEFCOL	- NATO Defense College
NADEVCEN	- Naval Air Development Center
NADGE	- NATO Air Defense Ground Environment
NADMC	- Naval Air Development and Material Center
NADOP	- North American Defense Operational Plan
NADU	- Naval Air Development Unit
NAE	- National Academy of Engineering (proposed)
NAEC	- National Aviation Education Council
NAECON	- National Aerospace Electronics Conference
NAEF	- Naval Aeronautical Engineering Facility
NAES	- Naval Air Experiment Station
NAESU	- Naval Aviation Engineering Service Unit
	- Navy Airborne Electronics Service Unit
NAF	- Naval Aircraft Factory
	- Naval Air Facility
	- Naval Avionics Facility
NAFEC	- National Aviation Facilities Experimental Center
NAFI	- Naval Avionics Facility, Indianapolis
NAFLI	- Natural Flight Instrument System
NAG	- Naval Advisory Group
	- Naval Astronautics Group
NAGARD	- NATO Advisory Group for Aeronautical Research and Development
NAI	- Northrop Aircraft, Incorporated
NAICP	- Nuclear Accident and Incident Control Plan
NAIF	- No A-1 Equipped Fighters

■ NAIO

NAIO	- Naval Air Intelligence Office
NAIOP	- Navigational Aid Inoperative for Parts
NAITF	- Naval Air Intercept Training Facility
NAL	- National Astronomical League
NAM	- National Association of Manufacturers
	- Naval Air Museum
NAMAP	- Northern Air Material Area, Pacific
NAMATCEN	- Naval Air Material Center
NAMC	- Naval Air Material Center
	- North Atlantic Military Committee
NAMDDU	- Naval Air Mine Defense Development Unit
NAMFI	- NATO Missile Firing Installation
NAMISCEN	- Naval Missile Center
NAMISTESTCEN	- Naval Air Missile Test Center
NAMPPF	- Nautical Air Miles per Pound of Fuel
NAMRU	- Naval Medical Research Unit
NAMS	- Naval Advent Management Staff
NAMT	- Naval Aircraft Maintenance Trainer
NAMTC	- Naval Air Missile Test Center
NAMTRAGRUP	- Naval Air Maintenance Training Group
NAND	- Naval Ammunition and Net Depot
NANEP	- Naval Air Navigation Electronics Project
NANWEP	- Navy Numerical Weather Prediction
NAO	- Naval Aviation Observer
NAOA	- Naval Aviation Observer Aerology
NAOB	- Naval Aviation Observer Bombardier
NAOC	- Naval Aviation Observer Controller
NAOI	- Naval Aviation Observer Intercept
NAON	- Naval Aviation Observer Navigation
NAOR	- Naval Aviation Observer Radar
NAOSP	- North Atlantic Ocean Stations Program
NAOT	- Naval Aviation Observer Tactical
NAOTS	- Naval Aviation Ordnance Test Station
NAP	- Naval Air Priorities
	- Naval Aviation Pilot
	- Nuclear Auxiliary Power
NAPA	- National Association of Purchasing Agents
NAPE	- National Association of Power Engineers
NAPOG	- Naval Airborne PRESS Operations Group
NAPU	- Nuclear Auxiliary Power Unit
NAPUS	- Nuclear Auxiliary Power Unit System
NAQF	- North Atlantic Quality Figures
NAR	- National Association of Rocketry
	- No Action Required
	- Non-Advanced Readiness
	- Numerical Analysis Research

NARAD	- Naval Air Research and Development
NARASPO	- Navy Regional Airspace Officer
NARATE	- Navy Automatic Radar Test Equipment
	- Northrop Automatic Radar Test System
NARCOM	- North Atlantic Relay Communication System
NARDA	- Naval Air Research and Development Activities
NAREC	- Naval Research Laboratory Electronic Computer
NARF	- Naval Air Reserve Facility
	- Nuclear Aircraft Research Facility
NARM	- National Association of Relay Manufacturers
NARS	- Naval Air Rescue Service
NARTB	- National Association of Radio and Television Broadcasters
NARTC	- Naval Air Rocket Test Center
NARTS	- Naval Air Rocket Test Station
NARTU	- Naval Air Reserve Training Unit
NARUC	- National Association of Railroad and Utilities Commissions
NAS	- National Academy of Sciences
	- National Aircraft Standard (prepared by National Aerospace Standards Committee)
	- Naval Air Station
	- Naval Amphibious School
NASA	- National Aeronautics and Space Administration (formerly NACA)
	- Naval Aircraft Safety Activity
NASA-AEC	- National Aeronautics and Space Administration and Atomic Energy Commission
NASAO	- National Association of State Aviation Officials
NASARR	- North American Search and Ranging Radar
NASC	- National Aeronautics and Space Council
	- National Aerospace Standards Committee
	- NATO Supply Center
	- Navy Aviation Safety Center
NASCO	- National Academy of Sciences Committee on Oceanography
NASCOM	- NASA World-Wide Communications Network
NASD	- Naval Air Supply Depot
NASIR	- Nuclear Amplification by Stimulated Isomer Radiation
NASMO	- NATO Starfighter Manufacturing Organization
NASRR	- North American Search and Ranging Radar
NASS	- National Aids Support System
	- Navy Advent Ship Station
NAST	- Navy Advent Ship Terminal
NASW	- National Association of Science Writers
NASWF	- Naval Air Special Weapons Facility
NAT	- National Air Transport
NATA	- National Aviation Trades Association
NATB	- Naval Air Training Base

■ NATC

NATC	- National Air Taxi Conference
	- Naval Air Test Center
	- Naval Air Training Center
	- Naval Air Training Command
NATCO	- National Air Transport Coordinating Committee
NATCOM	- National Communications Symposium
NATEC	- Naval Airship Training and Experimental Command
NATESTCEN	- Naval Air Test Center
NATF	- Naval Air Test Facility
NATIV	- North American Test Instrument Vehicle
NATMC	- National Advanced Technology Management Conference
NATO	- North Atlantic Treaty Organization
NATOPS	- Naval Air Training and Operating Procedures Standardization Program
NATS	- Naval Air Transport Service
NATSF	- Naval Air Technical Services Facility
NATT	- Naval Air Technical Training
NATTC	- Naval Air Technical Training Center
NATTS	- Naval Air Turbine Test Station
NATTU	- Naval Air Technical Training Unit
NATU	- Naval Aircraft Torpedo Unit
NAUS	- National Airspace Utilization System
NAV	- Naval
	- Navigation
	- Navy
NAVA	- National Audio-Visual Association
NAVACT	- Naval Activity
NAVAID	- Navigation Aid
NAVAR	- Navigation Radar
NAVARHO	- Navigation and Radio Homing
NAVBMC	- Naval Ballistic Missile Committee
NAVCAT	- Naval Career Appraisal Team
NAVCM	- Navigation Countermeasures and Deception
NAVCOMMSTA	- Naval Communications Station
NAVCOMMSYS	- Naval Communications System
NAVCOMMU	- Naval Communication Unit
NAVDAC	- Navigational Data Assimilation Center
NAVEXOS	- Publication Issued by Executive Office of the Secretary of the Navy
NAVFAC	- Naval Facility
NAVFORNORAD	- Naval Forces, North American Air Defense Command
NAVGMU	- Naval Guided Missile Unit
NAVIC	- Naval Information Center
NAVID	- Navigational Aid
NAVORD	- Naval Ordnance
NAVPERS	- Publication Issued by Bureau of Naval Personnel

NAVSAT	- Navigational Satellite
NAVSHIPS	- Publication Issued by Bureau of Ships
NAVSHIPY	- Naval Shipyard
NAVSPASUR	- Naval Space Surveillance System
NAVSTA	- Naval Station
NAVWAG	- Naval Warfare Analysis Group
NAVWEPS	- Publication Issued by Bureau of Naval Weapons
NAWAF	- Navy with Air Force
NAWAR	- Navy with Army
NAWAS	- National Attack Warning System
NAWC	- Naval War College
NAWL	- North American Iterative Weighted Least Squares
NAWM	- Naval Air Weapons Meeting
NAWS	- National Aviation Weather System
NAXSTA	- Naval Air Experimental Station
NB	- Narrow Band
	- Naval Base
	- No Bias
N/B	- Narrow Band
NBA	- Naval Radio Station, Canal Zone
NBAA	- National Business Aircraft Association
NBC	- Nuclear-Biological-Chemical
NBDL	- Narrow Band Data Line
NBER	- National Bureau of Economic Research
NBFM	- Narrow Band Frequency Modulation
NBL	- Naval Biological Laboratory
NBMG	- Navigation Bombing and Missile Guidance System
NBO	- Naval Bureau of Ordnance
NBS	- National Bureau of Standards
	- New British Standard
NBS-A	- National Bureau of Standards—Atomic
NBSR	- National Bureau of Standards Reactor
NBTS	- New Boston Tracking Station
NC	- National Course
	- Network Controller
	- Nitrocellulose
	- No Change
	- No Connection
	- Non-Conformance
	- Normally Closed
	- Northrop Corporation
	- Nose Cone
	- Nuclear Capability
N/C	- No Change
	- Normally Closed
	- Nose Cone

■ NCA

NCA	- Northwest Computing Association
NCAE	- National Conference on Airborne Electronics
NCAN	- Nationwide Combat Alert Network
NCAR	- National Center for Atmospheric Research
	- Non-Conformance Analysis Report
NCB	- National Conservation Bureau
	- Naval Communications Board
NCC	- NORAD Control Center
NCCS	- National Command and Control System
NCEL	- Naval Civil Engineering Laboratory
NCF	- Naval Communications Facility
NCM	- Noncorrosive Metal
	- Northern Cruise Master
NCO	- Noncommissioned Officer
NCOIC	- Noncommissioned Officer in Charge
NCR	- National Cash Register Company
	- Non-Combat Ready
	- Non-Compliance Report
	- Non-Conformance Report
NCRP	- National Committee on Radiation Protection
NCS	- Naval Communications Station
	- Naval Communications System
	- Network Control Station
NCSHA	- Naval Communications System Headquarters Activity
NCTC	- Naval Communications Training Center
NCU	- Naval Communications Unit
	- Nozzle Control Unit
N/C/W	- Not Compliable With
ND	- Naval District
	- Navy Department
	- No Drawing
	- Nuclear Dam
NDA	- Nuclear Development of America
NDAC	- National Damage Assessment Center
	- National Defense Advisory Commission
NDAP	- National Damage Assessment Program
NDASG	- National Damage Assessment Steering Group
NDB	- Non-Directional Beacon
NDCCC	- National Defense Communications Control Center
NDEA	- National Defense Education Act
NDEI	- National Defense Education Institute
NDFS	- Non-Dwelling Floor Space
NDR	- Network Data Reduction
NDRC	- National Defense Research Committee
NDRO	- Non-Destructive Readout
NDS	- Nuclear Detection Satellite

NDT	- Non-Destructive Testing
NDTA	- National Defense Transportation Association
NE	- Northeast
NEAC	- Northeast Air Command
NEAR	- National Emergency Alarm Repeater
NEAT	- National Cash Register Electronic Autocoding Technique
NEC	- National Electrical Code
	- National Electronics Conference
NECPA	- National Emergency Command Post Afloat
NECS	- National Electrical Code Standards
NED	- Navigation Error Data
NEDA	- National Electronics Distributors Association
NEEP	- Nuclear Electronics Effects Program
NEES	- Naval Engineering Experiment Station
NEF	- National Extra Fine
NEFA	- Northeast Frontier Agency
NEFD	- Noise-Equivalent Flux Density
NEG	- Negative
NEGDF	- Naval Emergency Ground Defense Force
NEL	- Naval Electronics Laboratory
	- Naval Explosives Laboratory
NELIAC	- Naval Electronics Laboratory International Algebraic Compiler
NELMCOM	- Navy Eastern Atlantic and Mediterranean Command
NEMA	- National Electrical Manufacturers Association
NEN	- New England Nuclear Corporation
NEO	- Northeastern Office
NEP	- Noise-Equivalent Power
NEPA	- Nuclear Energy for Propulsion of Aircraft
NEPD	- Noise Equivalent Power Density
NEREM	- Northeast Electronics Research and Engineering Meeting
NERV	- Nuclear Emulsion Recovery Vehicle
NERVA	- Nuclear Engine for Rocket Vehicle Applications
NESC	- National Electric Safety Code
	- Nuclear Engineering and Science Congress
NESTS	- Non-Electric Stimulus Transfer System
NET	- Net Equivalent Temperature
	- Noise Equivalent Temperature
	- Not Earlier than
NETF	- Nuclear Engineering Test Facility
NETR	- Nuclear Engineering Test Reactor
NETS	- Network Techniques
NEW	- Navy Early Warning
NEWS	- Naval Electronic Warfare Simulator
NF	- Nanofarad
	- National Fine

NF

NF	- Near Face
	- Neighborhood Final Fade
	- Noise Factor
	- Noise Figure
	- Nose Fuze
NFC	- Not Favorably Considered
NFL	- No Fire Line
NFR	- No Further Requirement
NFS	- Not on Flying Status
NFSL	- No Fighter Suitably Located
NFT	- Non-Functional Test
NG	- Narrow Gage
	- National Guard
	- Naval Gunfire
	- Nitroglycerin
N&G	- Navigation and Guidance
NGA	- National Glider Association
NGAUS	- National Guard of the United States (NGUS preferred)
NGB	- National Guard Bureau
NGF	- Naval Gun Factory
	- Naval Gun Fire
NGL	- No-Gimbal-Lock
NGLO	- Naval Gun Fire Liaison Officer
NGO	- National Gas Outlet
NGR	- National Guard Regulation
NGUS	- National Guard of United States
NH	- Nonhygroscopic
NHA	- Next Higher Assembly
NHH	- Neither Help nor Hinder
NHO	- Navy Hydrographic Office
NHPO	- NATO HAWK Production Organization
NHTS	- New Hampshire Tracking Station
NI	- Noise Index
	- Non-Interceptor
	- Numerical Index
NIAT	- Non-Indexable Address Tag
	- Scientific Research Institute of Aviation Technology (USSR)
NIB	- Non-Interference Basis
NIC	- National Inventors Council
	- Negative Impedance Converter
	- Not in Contract
NICAD	- Nickel Cadmium
NICB	- National Industrial Conference Board
NICP	- National Inventory Control Point
	- Nuclear Incident Control Plan
NIER	- National Industrial Equipment Reserve

NME ■

NIFA	- Not-in-File Area
NIFS	- Nuclear Influence Fuzing System
NIFTE	- Neon Indicator Flashing Test Equipment
NIG	- Naval Inspector General
NIGS	- Non-Inertial Guidance Set
NIH	- National Institutes of Health
	- Not Invented Here
NIN	- National Information Network
NIP	- Naval Intercept Project
	- Nominal Impact Point
NIPR	- National Industrial Plant Reserve
NIR	- Nose Fuze Impact Rocket
NIRAP	- Naval Industrial Reserve Aircraft Plant
NI&RT	- Numerical Index and Requirements Table
NIRTS	- New Integrated Range Timing System
NIS	- Not In Stock
NJCC	- National Joint Computer Committee (now AFIPS)
NL	- Night Letter
	- No Limit
N/L	- No Ledger
	- No Limit
NLF	- Nearest Landing Field
NLFT	- No Load Frame Time
NLG	- Nose Landing Gear
NLM	- National Library of Medicine
	- Noise Level Monitor
NLO	- Naval Liaison Officer
NLOGM	- Naval Liaison Officer for Guided Missiles
NLR	- Noise Load Ratio
NLRB	- National Labor Relations Board
NLS	- Non-Linear Systems
NLT	- Not Later Than
NLUS	- Navy League of United States
NM	- Nautical Mile
	- Naval Magazine
	- No Message
	- Nonmetallic
NMAA	- National Machine Accountants Association
NMB	- National Mediation Board
NMC	- National Meteorological Center
	- Naval Missile Center
NMCCS	- National Military Command and Control System
NMCS	- National Military Command System
	- Navy Mine Countermeasures Station
NMDL	- Navy Mine Defense Laboratory
NME	- National Military Establishment

NME

NME	- Naval Military Establishment
NMF	- Naval Missile Facility
NMFHAWAREA	- Naval Missile Facility, Hawaiian Area
NMFPA	- Naval Missile Facility, Point Arguello
NMFPM	- Naval Missile Facility, Point Mugu
NMI	- Nautical Mile
NMIC	- National Missile Industry Conference
NMILA	- NASA Merritt Island Launch Area
NMO	- Navy Management Office
	- Normal Manual Operation
NMPS	- Nautical Miles per Second
NMR	- National Military Representative
	- Naval Missile Range
	- Nuclear Magnetic Resonance
NMRDC	- National Nuclear Rocket Development Center
NMRI	- Naval Medical Research Institute
NMRL	- Naval Medical Research Laboratory
NMRM	- Nuclear Magnetic Resonance Measurement
NMSSA	- NATO Maintenance Supply Service Agency
NMSSS	- NATO Maintenance Supply Service System
NMTBA	- National Machine Tool Builders Association
NMTC	- Naval Missile Test Center
NNR	- Northern NORAD Region
NO	- Naval Observatory
	- Normally Open
	- Number
N/O	- Normally Open
NOA	- New Obligation Authority
	- Not Otherwise Authorized
	- Scientific Testing Airfield (USSR)
NOACT	- Naval Overseas Air Cargo Terminal
NOB	- Naval Operating Base
	- Number of Bursts
NOC	- Notice of Change
NOCM	- Nuclear Ordnance Commodity Manager
NOCO	- Noise Correlation
NOD	- Notice of Discrepancy
NODAC	- Naval Ordnance Data Automation Center
NODC	- National Oceanographic Data Center
NODEL	- Not to Delay
NODI	- Notice of Delayed Items
NOE	- Notice of Exception
NOEV	- NORAD Operational Evaluation
NOF	- Naval Operating Facility
	- Network Operations and Facilities
	- Nitrosyl Fluoride

NOFORN	- Not Releasable to Foreign Nationals – Special Handling Required
NOIBN	- Not Otherwise Indexed By Name
NOL	- Naval Ordnance Laboratory
NOLC	- Naval Ordnance Laboratory, Corona
NOLO	- No Live Operator
NOLTESTFAC	- Naval Ordnance Laboratory Test Facility
NOM	- Nominal
NOMA	- National Office Management Association
NOMAD	- Navy Oceanographic and Meteorological Automatic Device
NOMHL	- Naval Ordnance Materials Handling Laboratory
NOMSS	- National Operational Meteorological Satellite System
NOMTF	- Naval Ordnance Missile Test Facility
NOO	- Naval Oceanographic Office
NOOOA	- NORAD Office of Operational Analysis
NOP	- Naval Ordnance Plant
	- Near Object Probe
	- No Operation
N-OP	- Non-Operational
NOR	- No Order Required
NORAD	- North American Air Defense Command
NORAD COC	- North American Air Defense Command Combat Operations Center
NORAD CPX	- North American Air Defense Command Command Post Exercise
NORC	- National Opinion Research Center
	- Naval Ordnance Research Computer
NORD	- Naval Ordnance
NORIS	- North Island
NORLANTAACS	- North Atlantic Airways and Air Communications Service
NORM	- Not Operationally Ready—Maintenance
NORPAC	- Naval Overhaul and Repair, Pacific
NORS	- Not Operationally Ready—Supply
NORTAM	- Northrop Terminal Attrition Model
NORTHAG	- North European Army Group
NORVIPS	- Northrop Voice Interruption Priority System
NOSO	- Naval Ordnance Supply Office
NOSS	- National Orbiting Space Station
	- Nimbus Operational Satellite System
NOTAL	- Not To, Nor Needed By All
NOTAM	- Notice to Airmen
NOTIP	- Northern Tier Integration Project
NOTS	- Naval Ordnance Test Station
NOU	- Naval Ordnance Unit
NOUS	- Naval Order of the United States

■ NP

NP	- Name Plate
	- National Pipe
	- Nonpropelled
NPA	- National Petroleum Association
NPC	- NASA Procurement Circular
NPD	- Navy Procurement Directive
	- Nuclear Power Division
NPE	- Naval Pilot Evaluation
NPF	- Naval Powder Factory
NPFO	- Nuclear Power Field Office
NPG	- Naval Proving Ground
NPIC	- Naval Photographic Interpretation Center
NPL	- National Physics Laboratory
NPN	- Negative-Positive-Negative
N-P-N	- Negative-Positive-Negative
NPO	- Naval Purchasing Office
	- Navy Post Office
NPOLA	- Naval Purchasing Office, Los Angeles
NPP	- Naval Propellant Plant
NPPS	- Naval Publications and Printing Service
NPPSBO	- Naval Publications and Printing Service Branch Office
NPPSO	- Naval Publications and Printing Service Office
NPR	- Noise Power Ratio
	- Nuclear Power Reactor
NPSH	- Net Positive Suction Head
NPT	- National Taper Pipe
NR	- Navigational Radar
	- Noise Ratio
	- Nonreactive
	- Nonrecoverable
	- NORAD Region
	- No Requirements
	- Not Required
N/R	- Not Responsible
NRA	- Naval Reserve Association
	- Network Resolution Area
	- Non-Recurrence Action
	- No Repair Action
NRAO	- National Radio Astronomy Observatory
	- Naval Regional Accounts Office
NRB	- Naval Reactor Branch
NRC	- National Referral Center
	- National Research Corporation
	- National Research Council
	- Naval Retraining Command
NRCC	- National Research Council of Canada

254

NSB

NRD	- National Range Division (AFNRD preferred)
	- National Range Document
	- Negative Resistance Diode
NRDC	- National Defense Research Committee
NRDL	- Naval Radiological Defense Laboratory
NRDS	- Nuclear Rocket Development Station
NRDSCG	- Naval Research and Development Satellite Communications Group
NREC	- National Resources Evaluation Center
NRFI	- Not Ready for Issue
NRI	- National Radio Institute
	- Not Requiring Identification
	- Number of Records Ignored
NRIP	- Number of Rejected Initial Pickups
NRL	- Naval Research Laboratory
NROTC	- Naval Reserve Officers Training Corps
NRP	- Non-Registered Publication
NRS	- National Range Support
NRT	- Non-Radiating Target
	- Normal Rated Thrust
NRTS	- National Reactor Test Station
	- Not Repairable This Station
NRX	- NERVA Reactor, Experimental
NRZ	- Non-Return-to-Zero
NRZC	- Non-Return-to-Zero Change
NRZL	- Non-Return-to-Zero Level
NRZM	- Non-Return-to-Zero Mark
NS	- National Special
	- National Standard
	- Naval Shipyard
	- Naval Station
	- Near Side
	- Non Standard
	- No Scramble
	- No Stock
	- Nuclear Ship
	- Nuclear Systems
NSA	- National Security Act
	- National Security Agency
	- National Shipping Authority
	- National Standards Association
	- Navy Stock Account
	- Nuclear Science Abstracts
NSALO	- National Security Agency Liaison Office
NSAS	- Non-Scheduled Air Services
NSB	- Naval Submarine Base

■ NSC

NSC	- National Safety Council
	- National Security Council
	- National Simulation Council
	- Navigational Star Catalogue
	- Navy Supply Center
	- Numerical Sequence Code
NSD	- Naval Supply Depot
	- Network Status Display
	- Nonself-Destroying
NSDC	- Naval Special Devices Center
NSE	- North Steaming Error
	- Nuclear Support Equipment
NS&E	- Nuclear Science and Engineering
NSF	- National Science Foundation
NSFD	- Notice of Structural or Functional Deficiency
NSG	- Naval Security Group
NSGN	- Noise Generator
NSI	- Non-Satellite Identification
	- Nonstandard Item
NSIA	- National Security Industrial Association
NSIC	- Nuclear Strike Information Center
NSIF	- Near Space Instrumentation Facility (obsolete)
NSL	- Naval Supersonic Laboratory
	- Northrop Space Laboratory
	- Nuclear Safety Line
NSM	- Network Space Monitor
NSMSES	- Naval Ship Missile Systems Engineering Station (also USNSMSES)
NSN	- No Stock Number
NSP	- Nonstandard Part
	- Nose Shipping Plug
NSPE	- National Society of Professional Engineers
NSPO	- Nuclear Systems Project Office
NSRB	- National Security Resources Board
NSRDS	- National Standard Reference Data System
NSRP	- National Search and Rescue Plan
NSS	- National Stockpile Site
	- Navy Shore Station
NSSCC	- National Space Surveillance Control Center
NSSS	- Naval Space Surveillance System
NST	- Network Support Team
	- Nonslip Thread
	- Not Sooner Than
NSTA	- National Science Teachers Association
NT	- Nontight
	- Not Tested

NT	- No Tool
NTA	- Nuclear Test Aircraft
NTC	- Naval Training Center
	- Negative Temperature Coefficient
NTCAVAL	- Notice of Availability
NTDC	- Naval Training Devices Center
NTDS	- Naval Tactical Data System
NTE	- Navy Technical Evaluation
	- Not to Exceed
NTI	- No Travel Involved
NTIOC	- No Travel Involved for Officer Concerned
NTL	- No Time Lost
NTO	- Nitrogen Tetroxide
NTOL	- Normal Take-Off and Landing
NTP	- Normal Temperature and Pressure
	- Number of Theoretical Plates
NTPC	- National Technical Processing Center
NTS	- Naval Torpedo Station
	- Naval Transportation Service
	- Nevada Test Site
	- Not To Scale
NTSA	- National Technical Service Association
NTSC	- National Television System Committee
NTU	- Number of Transfer Units
NTWS	- Non-Track While Scan
NTX	- National Teletypewriter Exchange
	- Naval Teletype Exchange
	- Navy Target Exchange System
NUAU	- Nuclear Authorization
NUDET	- Nuclear Detection
NUDETS	- Nuclear Detection System
	- Nuclear Detonation Detection and Reporting System
NUFP	- Number of Uncorrected Flight Plans
NULACE	- Nuclear Liquid Air Cycle Engine
NUOS	- Naval Underwater Ordnance Station
NUSL	- Naval Underwater Sound Laboratory
NUSUM	- Numerical Summary Message
NV	- Not Valid
N-V	- Northrop-Ventura
NVAL	- Not Available
NVB	- Navigational Base
NVE	- Non-Visual Eyepiece
NVPO	- Nuclear Vehicle Projects Office
NVR	- No Voltage Release
NW	- Northwest
NWC	- National War College

■ NWC

NWC	- Naval War College (NAWC preferred)
	- Not Worth Considering
NWDS	- Number of Words
NWDSEN	- Number of Words per Entry
NWFAL	- Nation-Wide Fallout
NWL	- Naval Weapons Laboratory
NWO	- Nuclear Weapons Officer
NWP	- Nuclear Weapons Plant
NWPOG	- Numerical Weather Prediction Operational Grid
NWS	- Naval Weapons Station
NWSSG	- Nuclear Weapons System Safety Group
NWSY	- Naval Weapons Station, Yorktown
NWT	- Nonwatertight
NYADS	- New York Air Defense Sector
NYAP	- New York Assembly Program
NYAS	- New York Academy of Science
NYD	- Navy Yard
NYOD	- New York Ordnance District
NYPFO	- New York Procurement Field Office
NYR	- Not Yet Required
	- Not Yet Returned
NZ	- Nike-Zeus
NZTJWG	- Nike-Zeus Target Joint Working Group
NZTV	- Nike-Zeus Target Vehicle

O	- Observation
	- Orange
	- Output
	- Oxygen
OA	- Office of Applications
	- Omniantenna
	- On or About
	- Operations Analyst
	- Originating Agency
	- Output Axis
	- Overall
O/A	- Omnirange Antenna
	- On or About
	- Output Axis
OAB	- Ordnance Assembly Building
OAC	- Oceanic Area Control
	- Ordnance Ammunition Command
OAFB	- Offutt Air Force Base
OAFIE	- Office of Armed Forces Information and Education
OAG	- Official Airline Guide
OAI	- Office of Aeronautical Intelligence
	- OR Accumulators to Indicators
OAIDE	- Operational Assistance and Instructive Data Equipment
OAL	- Operational Applications Laboratory
	- Ordnance Aerophysics Laboratory
OAM	- Office of Aerospace Medicine
OAMA	- Ogden Air Materiel Area (OOAMA preferred)
OAME	- Orbital Attitude and Maneuvering Electronics
OAMS	- Orbital Attitude and Maneuvering System
OAO	- Orbiting Astronomical Observatory
OAR	- Office of Aerospace Research
	- Office of Analysis and Review
	- Operations Analysis Report

■ OARC

OARC	- Office of Air Research Automatic Computer
OART	- Oakland Army Terminal
	- Office of Advanced Research and Technology
OAS	- Operational Announcing System
	- Organization of American States
OASD	- Office of the Assistant Secretary of Defense
OASIS	- Operational Automatic Scheduling Information System
OASM	- Office of Aerospace Medicine
OASMS	- Ordnance Ammunition Surveillance and Maintenance School
OASN	- Office of the Assistant Secretary of the Navy
OASNM	- Office of the Assistant Secretary of the Navy for Material
OASNR&D	- Office of the Assistant Secretary of the Navy for Research and Development
OAT	- Outside Air Temperature
	- Overall Test
OATC	- Oceanic Air Traffic Center
	- Overseas Air Traffic Control
OAV	- Operational Aerospace Vehicle
OB	- Obsolete
	- Operational Base
	- Order of Battle
	- Output Buffer
OBD	- Omnibearing Distance
OBHR	- Office of Biotechnology and Human Research
OBI	- Omnibearing Indicator
OBP	- Office of Bioscience Programs
OBS	- Omnibearing Selector
OBW	- Observation Window
OC	- Object Class
	- On Call
	- On Center
	- On Course
	- Operating Characteristic
	- Operations Conductor
	- Ordnance Corps
	- Overcurrent
O/C	- Open Circuit
O&C	- Operation and Checkout
	- Operations and Control Building
OCA	- Oceanic Control Area
	- Operational Control Authority
OCADS	- Oklahoma Air Defense Sector
OCAF	- Office, Chief of Aerospace
OCAFF	- Office, Chief of Army Field Forces
OCAMA	- Oklahoma City Air Materiel Area

OD ■

OCB	- Oil Circuit Breaker
	- Operations Coordinating Board
	- Override Control BITs
OCC	- Operational Control Center
	- Ordnance Command Converter
OCCM	- Office of Commercial Communications Management
OC/CP	- Operations Conductor, Command Post
OCD	- Office of Civil Defense
	- Ordnance Classification of Defects
OCDM	- Office of Civil Defense Mobilization (obsolete)
OCDU	- Optics Coupling Display Unit
OCE	- Office, Chief of Engineers
	- Officer Conducting the Exercise
OCEANS	- Omnibus Conference on Experimental Aspects of Naval Missile Range Spectroscopy
OCF	- Operational Control Facility
	- Owens-Corning Fiberglas Corporation
OCh	- Octane Number (USSR)
OCINFO	- Office of the Chief of Information
OCL	- Operational Check List
	- Operational Control Level
OCLI	- Optical Coating Laboratory, Incorporated
OCLUS	- Outside Continental Limits of United States
OCN	- Office, Chief of Naval Operations
OCNS	- Oklahoma City NORAD Sector
OCO	- Office, Chief of Ordnance
	- Open-Close-Open
O&CO	- Operation and Checkout (O&C preferred)
O&C/O	- Operation and Checkout (O&C preferred)
OCONUS	- Outside Continental United States
OCP	- Orbital Control Program
OCR	- Office of Coordinating Responsibility
	- Optical Character Recognition
	- Order Control Record
	- Overhead Component Requirement
OCRD	- Office, Chief of Research and Development
OCS	- Office of Cataloguing and Standardization
	- Office of Communications Systems
	- Officer Candidate School
	- Optical Character Scanner
OCSA	- Office, Chief of Staff, Army
OCSIGO	- Office, Chief Signal Officer
OD	- Officer of the Day
	- On Dock
	- One Day
	- Operations Directive

■ OD

OD	- Operations Director
	- Ordnance Department
	- Original Drawing
	- Outside Diameter
O&D	- Ordering and Distribution
ODA	- Operational Data Analysis
	- Operational Design and Analysis
ODAP	- Operation Data Analysis Program
ODB	- Operational Design Branch
	- Output Display Branch
ODCM	- Office of Defense and Civilian Mobilization
ODD	- Operations Development Department
ODDA	- Office of Deputy Director for Administration
ODDP	- Office of Deputy Director for Programs
ODDRD	- Office of Deputy Director for Research and Development
ODDRE	- Office of Director of Defense Research and Engineering
ODDS	- Office of Deputy Director for Systems
ODG	- Operational Design Group
ODH	- Operations Directive Handbook
ODI	- Official Documents Index
	- Operational Development Inspection
ODM	- Office of Defense Mobilization
	- One-Day Mission
	- Orbital Determination Module
ODN	- Own Doppler Nullifier
ODP	- Operational Design Proposal
	- Operational Development Phase
	- Operational Development Program
	- Organic Development Problem
	- Output-to-Display Parity Error
ODR	- Operational Design Resolution
ODS	- Operations Directorate Station
ODT	- Office of Defense Transportation
	- Operational Development Team
ODTF	- Operational Development Test Facility
ODU	- Output Display Unit
ODVAR	- Orbit Determination and Vehicle Attitude Reference
OEC	- Office of Electronics and Control
OEEC	- Organization for European Economic Cooperation
OEG	- Operations Evaluation Group
OEI	- Overall Efficiency Index
OEL	- Organizational Equipment List
OEM	- Original Equipment Manufacturer
OEP	- Office of Emergency Planning
OER	- Officer Efficiency Report
	- Operational Equipment Requirement

OERC	- Optimum Earth Re-Entry Corridor
OES	- Operations and Equipment Section
OETP	- Operations Experimental Test Plan
OF	- Optical Form
	- Outside Face
O/F	- Orbital Flight
	- Oxidizer to Fuel
OFACS	- Overseas Foreign Aeronautical Communications Station
OFB	- Operational Facilities Branch
OFC	- Operational Flight Control
OFG	- Optical Frequency Generator
OFHC	- Oxygen-Free High Conductivity
OFIR	- Oceanic Flight Information Region
OFMC	- Operational Fixed Microwave Council
OFO	- Office of Flight Operations
OFP	- Operating Force Plan
	- Oscilloscope Face Plane
OFR	- Operational Failure Report
	- Ordering Function Register
OFSM	- Operational Flight Safety Monitor
OFT	- Operational Flight Trainer
OFTMS	- Output Format Table Modification Submodule
OG	- OR Gate
	- Outer Gimbal
OGA	- Outer Gimbal Axis
OGAMA	- Ogden Air Materiel Area (OOAMA preferred)
OGE	- Operating Ground Equipment
OGG	- Organic Geochemistry Group of the Geochemical Society
OGM	- Outside Gage Marks
OGO	- Orbiting Geophysical Observatory
OGR	- Outgoing Repeater
OGRC	- Office of Grants and Research Contracts
OGSE	- Operational Ground Support Equipment
OGSESS	- Operational Ground Support Equipment Systems Specification
OGT	- Outgoing Trunk
OH	- On Hand
	- Operational Hardware
OHDETS	- Over-Horizon Detection System
OI	- Oil Insulated
	- Operating Instructions
OIA	- Optics Inertial Analyzer
OIB	- Official Information Base
	- Operations Integration Branch
OIC	- Officer in Charge
OICC	- Officer in Charge of Construction
OICO	- Office of Integration and Checkout

■ OID

OID	- Order Initiated Distribution
OIDPS	- Overseas Intelligence Data Processing System
OIFC	- Oil Insulated Fan-Cooled
OIG	- Office of the Inspector General
OIR	- Office of Industrial Relations
O&IR	- Operation and Inspection Record
OIS	- Office of Information Services
	- Operational Instrumentation System
OISC	- Oil Insulated Self-Cooled
OIST	- Operator Integration Shakedown Test
OIYaI	- Joint Institute for Nuclear Research (USSR)
OJ	- Open-Joisted
OJT	- On-the-Job Training
OK	- All Right
OL	- Open Loop
	- Operating Location
	- Overlap
O/L	- On-Line
	- Operations and Logistics
O&L	- Operations and Logistics
OLA	- Office of Legislative Affairs
OLF	- Orbital Launch Facility
	- Outlying Field
OLM	- On-Line Monitor
OLO	- Orbital Launch Operation
OLS	- Optical Landing System
OLSS	- Overseas Limited Storage Site
OLU	- Outgoing Line Unit
OLV	- Orbital Launch Vehicle
OLVP	- Office of Launch Vehicles and Propulsion
OM	- Officer Messenger
	- Orbit Modification
	- Ordnance Mission
	- Outer Marker
O&M	- Operation and Maintenance
OMA	- Office of Military Assistance
	- Operation and Maintenance, Army
OMB	- Outer Marking Beacon
OMD	- Ocean Movement Designator
OMEGA	- Operations Model Evaluation Group, Air Force
OMF	- Operation and Maintenance of Facilities
O&MF	- Operation and Maintenance Facilities
OMGUS	- Office of Military Government, United States
OMI	- Optical Measurement Instrument
OMIS	- Operational Management Information System
OML	- Outside Mold Line

OMLIT	- One-Man Live Interception Test
OMNITENNA	- Omnirange Antenna
OMP	- Output Makeup
OMR	- Operational Modification Report
O&MR	- Operations and Management Research
OMRD	- Operations Management Research Department
OMS	- Office of Meteorological Systems
	- Operational and Mathematical Specification
	- Organizational Maintenance Shop
	- Output Multiplex Synchronizer
OMSA	- Ordnance Missile Support Agency
OMSF	- Office of Manned Space Flight
OMTS	- Organizational Maintenance Test Station
OMU	- Operating Mockup
	- Optical Measuring Unit
ONC	- Operational Navigation Charts
ONERA	- Office National d'Etudes et de Recherches Aeronautiques
ONI	- Office of Naval Intelligence
ONM	- Office of Naval Material
ONMINST	- Office of Naval Material Instruction
ONOF	- Office of Network Operations and Facilities
ONOZ	- Oil Nozzle
ONR	- Office of Naval Research
ONRS	- Office of National Range Support
ONS	- Office of Nuclear Systems
O/O	- Off Ocean
OOA	- Open Ocean Area
OOAMA	- Ogden Air Materiel Area
OOD	- Officer of the Day (OD preferred)
	- Officer of the Deck
OOL	- Operator-Oriented Language
OON	- United Nations (USSR)
OOPS	- Off-Line Operating Simulator
OOR	- Office of Ordnance Research
OOSC	- Out-of-Sight Controller
OOSCC	- Out-of-Sight Control Center
OOSS	- Overseas Operational Storage Site
OP	- Observation Post
	- Operating Procedure
	- Operations Plan
	- Ordnance Pamphlet
	- Ordnance Personnel
	- Output
O-P	- Oppenheimer-Phillips Process
O&P	- Operations and Procedures
OPADEC	- Optical Partial Decoy

■ OPAL

OPAL	- Optical Platform Alignment Linkage
OPC	- Office of Program Coordination
OPCOM	- Operations-Communications
OPCON	- Operation and Control System
	- Optimizing Control
OPCONCEN	- Operations Control Center
OPCONCENTER	- Operations Control Center
OPD	- Office of Programs Director
	- Operational Plans Division
OPDAR	- Optical Direction and Ranging
OPDEVFOR	- Operational Development Forces
OPERA	- Ordnance Pulses Experimental Research Assembly
OPFAD	- Outer-Perimeter Fleet Air Defense
OPI	- Office of Primary Interest
	- Oil Pressure Indicator
OPL	- Outpost Line
OPLAN	- Operation Plan
OPLR	- Outpost Line of Resistance
OPM	- Office of Procurement and Materiel
	- Office of Production Management
	- Operations per Minute
	- Ordnance Proof Manual
OPMS	- Outplant Procurement Manufacturing Specification
OPNAV	- Publication Issued by Office of Chief of Naval Operations
OPO	- Outside Production Order
	- Outside Purchase Order
OPOC	- Onboard Pilot-Observer Camera
OPORD	- Operation Order
OPP	- Octal Print Punch
	- Operator Preparation Program
	- Outside Production Part
OP&PB	- Oceanographic Plans and Policy Board
OPPE	- Office of Plans and Program Evaluation
OPPG	- Office of Propulsion and Power Generation
OPR	- Office of Primary Responsibility
	- Operational Program Requirement
	- Operations Procedure
	- Optional Parts Request
OPRD	- Office of Production Research and Development
OPRM	- Office of Program Review and Resources Management
OPRR	- Outside Production Requirement Record
OPS	- Operational Station
	- Operations
	- Outside Production Service
OPSAS	- Office of Program Support and Advanced Systems

OPT	- Operations and Telling
	- Optical
	- Optics
OPTAG	- Optical Pickoff Two-Axis Gyroscope
OPTAR	- Optical Automatic Ranging
OPTEVFOR	- Operational Test and Evaluation Force
OQC	- Outside Quality Control
OR	- Operationally Ready
	- Operations Requirement
	- Operations Research
	- Ordering Register
	- Out of Range
	- Outside Request
O/R	- Operationally Ready
O&R	- Operations and Regulation
	- Overhaul and Repair
ORA	- Office of Research Analysis
	- OR to Accumulator
	- Output Register Address
ORACLE	- Oak Ridge Automatic Computer and Logical Engine
ORATE	- Ordered Random Access Talking Equipment
ORBIT	- ORACLE Binary Internal Translator
	- Orbit Ballistic Impact and Trajectory
ORCON	- Organic Control
ORD	- Operational Ready Date
	- Operations Requirement Document
	- Optical Rotary Dispersion
	- Ordnance
ORDCIT	- Ordnance, California Institute of Technology
ORDIR	- Omnirange Digital Radar
ORDRAT	- Ordnance Dial Reader and Translator
ORDVAC	- Ordnance Variable Automatic Computer
ORE	- Operational Readiness Evaluation
ORH	- Operations Requirement Handbook
ORI	- Operational Readiness Inspection
	- Operations Research, Incorporated
ORIE	- Operational Radiation Instrumentation Equipment
ORINS	- Oak Ridge Institute of Nuclear Studies
ORIT	- Operational Readiness Inspection Test
ORL	- Orbital Research Laboratory
ORNL	- Oak Ridge National Laboratory
ORO	- Operations Research Office (now RAC)
ORPICS	- Orbital Rendezvous Positioning, Indexing, and Coupling System
ORR	- Omnidirectional Radar Range
	- Omnidirectional Radio Range

■ ORRT

ORRT	– Operational Readiness and Reliability Test
ORS	– Operations Research Society
	– OR to Storage
ORSA	– Operations Research Society of America
ORT	– Operational Readiness Test
	– Operational Readiness Training
ORT/CTL	– Operational Readiness Training—Combat Training Launch
ORTE	– Operational Readiness Training Equipment
ORTUAG	– Organized Reserve Training Unit, Vessel Augmentation
ORTUEL	– Organized Reserve Training Unit, Electronics
ORTUPS	– Organized Reserve Training Unit, Port Security
ORTUR	– Organized Reserve Training Unit, Rescue Coordination Center
ORV	– Ocean Range Vessel
	– Orbital Re-Entry Vehicle
OS	– Oil Switch
	– On Station
	– Operational Spare
	– Operation Snapper
	– Operations Section
	– Ordnance Specification
	– Oversize
OSA	– Office of the Secretary of the Army
	– Optical Society of America
OSAF	– Office of the Secretary of the Air Force
OSB	– Orbital Solar Observation
	– Ordnance Supply Bulletin (obsolete)
OSC	– On-Site Commander
	– Oscillator
OSCAR	– Optimum Survival Containment and Recovery
	– Orbiting Satellite Carrying Amateur Radio
OSD	– Office of the Secretary of Defense
	– Office Services Division
	– Operating System Description
	– Ordnance Supply Depot (obsolete)
	– Over Short and Damage Report
OSE	– Office of Systems Engineering
	– Operational Support Equipment
	– Orbital Sequence of Events
OSEDS	– Operational Support Equipment Design Specification
OSFM	– Office of Spacecraft and Flight Missions
OSI	– Office of Special Investigations
	– Office of Strategic Information
	– Operational Status Indicator
	– OR Storage to Indicators

OSL	- Orbiting Space Laboratory
	- Outstanding Leg
OSN	- Office of the Secretary of the Navy
OSO	- Office of Space Operations
	- Operations Support Office
	- Orbiting Solar Observatory
OSP	- ODAP Starter Program
	- Office of Scientific Personnel
	- Offshore Procurement
	- On-Station Position
	- Operating System Plan
	- Orbital Support Plan
OSPE	- Organizational Spare Parts and Equipment
OSR	- Office of Scientific Research (AFOSR preferred)
	- Office of Security Review
	- Operational Scanning Recognition
	- Operational Support Requirement
	- Output Shift Register
OSRD	- Office of Scientific Research and Development
OSS	- Ocean Surveillance System
	- Office of Space Sciences
	- Office of Space Systems
	- Office of Strategic Services
	- Operational Storage Site
	- Orbital Space Station
OSSA	- Office of Space Science and Applications
OSSS	- Orbital Space Station Study
OST	- All-Union Standard (USSR)
	- On Site Test
	- Operational Suitability Test
	- Operational System Test
OSTD	- Ordnance Standard
OSTF	- Operational Silo Test Facility
OSTI	- Office of Scientific and Technical Information
OSTP	- Operational System Test Program
OSURF	- Ohio State University Research Facility
OSV	- Ocean Station Vessel
	- Office of Space Vehicles
OSWAC	- Ordnance Special Weapons-Ammunition Command
OSWD	- Office of Special Weapons Development
OSWG	- Optical Systems Working Group
OS&Y	- Outside Screw and Yoke
OT	- Observer-Target
	- Oil Tight
	- One Time
	- Operating Time

■ OT

OT	– Organization Table
	– Overlap Technician
	– Overlap Telling
	– Overtime
O/T	– Organization Table
O-T	– Observer-Target
OTAC	– Ordnance Tank and Automotive Command
OTAD	– Oversea Terminal Arrival Date
OTAN	– Organisation du Traite de l'Atlantique du Nord (NATO)
OTB	– Operational Training Base
OTC	– Objective Time and Cost Control
	– Office of Transport and Communications
	– Officer in Tactical Command
	– Operational Test Center
	– Ordnance Technical Committee
OTD	– Orbital Test Direction
	– Orbital Test Directive
OTDA	– Office of Tracking and Data Acquisition
OTE	– Operational Test Equipment
	– Operator Training Exercise
OTEP	– Operational Test and Evaluation Plan
OTF	– Optimum Traffic Frequency
OTI	– Office of Technical Information
	– Optics Technology, Incorporated
OTIA	– Ordnance Technical Intelligence Agency
OTJ	– On the Job
OTM	– Operational Technical Manual
OTN	– Operational Teletype Network
OTO	– Optical Tracker Operator
O to O	– Out to Out
OTP	– Operational Test Procedure
OTRAC	– Oscillogram Trace Reader
OTS	– Office of Technical Services
	– Officer Training School
	– Overlap Technician Supervisor
	– Overlap Telling and Surveillance
OTT	– Office of Transportation and Traffic
	– One-Time Tape
OTU	– Ogden Test Unit
	– Operational Test Unit
	– Operational Training Unit
OTV	– Operational Test Vehicle
OTWR	– Oblique Tape-Wound Refrasil
OUD	– Operational Use Date
OUTRAN	– Output Translator
OVE	– On Vehicle Equipment

OVERS	- Orbital Vehicle Re-Entry Simulator
OVM	- On Vehicle Material
	- Orbiting Velocity Meter
OVSP	- Overspeed
OW	- Order Wire
	- Outer Wing
OWC	- Ordnance Weapons Command (obsolete)
OWF	- Optimum Working Frequency
	- Orbital and Weightless Flight
OWG	- Oil, Water, Gas
OX	- Oxide
	- Oxidizer
OXY	- Oxygen
OYCV	- Optimum Yaw Control Vertical
OZ	- Ounce
OZARC	- Ozone-ARCAS
OZ-IN	- Ounce-Inch

P

P	- Pad
	- Page
	- Paper
	- Parity
	- Phosphorus
	- Pitch
	- Plate
	- Pole
	- Port
	- Pressure
	- Probe
	- Proton
	- Prototype
PA	- Pad Abort
	- Pending Availability
	- Performance Analysis
	- Picatinny Arsenal
	- Power Amplifier
	- Precision Angle
	- Pressure Angle
	- Programmed Arithmetic
	- Project Authorization
	- Public Address
P/A	- Pilotless Aircraft
	- Planetary Atmosphere
	- Polar to Analog
	- Pressure Actuated
P&A	- Priorities and Allocations
	- Procurement and Assignment
PAAC	- Program Analysis Adaptable Control
PAAFB	- Patrick Auxiliary Air Force Base
PAALR	- PICE Assembly and Load Routine
PABX	- Private Automatic Branch Exchange (same as PBX only PBX is manual)

PAC	- Pacific Air Command
	- Pacific Ocean
	- Place Complement of Address in Index Register
	- Planning Approval Committee
	- Program Advisory Committee
	- Program Allocation Checker
	- Pursuant to Authority Contained in
PACAACS	- Pacific Area Airways and Air Communications Service
PACAF	- Pacific Air Forces
PACC	- PERT Associated Cost Control
	- Product Administration Contract Control
PACCS	- Post Attack Command and Control System
PACE	- Packaged CRAM Executive
	- Performance and Cost Evaluation
	- Precision Analog Computing Equipment
	- Pre-Flight Acceptance Checkout Equipment
	- Prelaunch Automatic Checkout Equipment
	- Program Analysis Control and Evaluation
PACE/LV	- Pre-Flight Acceptance Checkout Equipment—Launch Vehicle
PACER	- Program of Active Cooling Effects and Requirements
PACE/SC	- Pre-Flight Acceptance Checkout Equipment—Spacecraft
PACFLT	- Pacific Fleet
PACIR	- Practical Approach to Chemical Information Retrieval System
PACK	- Packing and Allocation for a COMPOOL Kaleidoscope
PACM	- Pulse Amplitude Code Modulation
PACMISRAN	- Pacific Missile Range (now AFWTR)
PACOM	- Pacific Command
PACOR	- Passive Correlation and Ranging
PACT	- Pay Actual Computer Time
	- Print Active Computer Tables
	- Production Action Control Technique
	- Programmed Automatic Circuit Tester
	- Project for the Advancement of Coding Techniques
PACTA	- Packed Tape Assembly
PAD	- Pilotless Aircraft Division
	- Port of Aerial Debarkation
	- Power Amplifier Driver
	- Propellant Actuated Device
PADAR	- Passive Airborne Detection and Ranging
	- Passive Detection and Ranging
PADL	- Pilotless Aircraft Development Laboratory
PADLOC	- Passive Detection and Location of Countermeasures
PADRE	- Portable Automatic Data Recording Equipment
PADS	- Passive-Active Data Simulation
	- Precision Aerial Display System

■ **PAE**

PAE	- Port of Aerial Embarkation
PAF	- Pacific Air Forces (PACAF preferred)
PAFB	- Patrick Air Force Base
PAFS	- Primary Air Force Speciality
PAFSC	- Primary Air Force Speciality Code
PAG	- Program Advisory Group
PAI	- Place Accumulator in Indicators
	- Production Acceptance Inspection
PAINCO	- Paraglider Inflation Complete
PAIRC	- Pacific Air Command
PAL	- Pacific Aerospace Library
	- Programmed Application Library
PALC	- Point Arguello Launch Complex
PALS	- Permissive Action Link Systems
PAM	- Pad Abort Mission
	- Property Accountability Manual
	- Pulse Amplitude Modulation
PAMAC	- Parts and Material Accountability Control
PAM-FM	- Pulse Amplitude Modulation—Frequency Modulation
PAMI	- Personnel Accounting Machine Installation
PAMO	- Port Air Material Office
PAMUSA	- Post-Attack Mobilization of the United States Army
PANAR	- Panoramic Radar
PAO	- Parts Assembly Order
	- Procurement and Accountability Order
	- Public Affairs Office
PAP	- Pacific Automation Products
PAR	- Parameter
	- Planning Accountability Records
	- Precision Approach Radar
	- Preventive Aircraft Repair
	- Program Appraisal and Review
	- Progressive Aircraft Rework
	- Publication Analysis Report
	- Pulse Acquisition Radar
PARADE	- Passive-Active Range Determination
PARAMI	- Parsons Active Ring-Around Miss Distance Indicator
PARAMIS	- Parsons Passive Miss Distance Indicating System
PARAMP	- Parametric Amplifier
PARC	- Progressive Aircraft Repair Cycle
PARD	- Pilotless Aircraft Research Division
PARDOP	- Passive Ranging Doppler
PARIS	- Pulse Analysis-Recording Information System
PARSECS	- Program for Astronomical Research and Scientific Experiments Concerning Space
PARSET	- Precision Askania Range System of Electronic Timing

PARSIP - Point Arguello Range Safety Impact Predictor
PARSYN - Parametric Synthesis
PART   - Parametric Target System
PARTAC - Precision Askania Range Target Acquisition and Control
PARTEI - Purchasing Agents of the Radio, Television, and Electronics Industries
PARTNER - Proof of Analog Results Through a Numerical Equivalent Routine
PAS   - Passed to the Adjacent Sector
        - Phase Address System
        - Primary Alerting System
PASEP   - Passed Separately
PAT   - Parametric Artificial Talker
        - Patent
        - Pattern Analysis Test
        - Performance Acceptance Test
        - Platoon Anti-Tank
        - Plenum Air Tread
        - Polaris Acceleration Test
        - Programmer Aptitude Tester
PAT(A) - Plenum Air Tread, Amphibious
PATH   - Performance Analysis and Test Histories
PATRIC - Pattern Recognition Interpretation and Correlation
PAV   - Phase Angle Voltmeter
        - Position and Velocity
PAVE   - Position and Velocity Extraction
PAVT   - Position and Velocity Tracking
PAW   - Powered-All-the-Way
PAX   - Place Address in Index Register
        - Private Automatic Exchange (Inter-Office)
PAZ   - Antiatomic Defense (USSR)
PB   - Packard Bell
        - Painted Base
        - Pilotless Bomb
        - Plotting Board
        - Plug Board
        - Process Bulletin
        - Publications Bulletin
        - Pull Box
        - Push Button
P/B   - Piggy Back
PBAA   - Polybutydiene Acrylic Acid
PBAC   - Program Budget Advisory Committee
PBC   - Pitch Bank Compensation
PBCC   - Packard Bell Computer Corporation
PBF   - Power Burst Facility

■ PBG

PBG	– Program and Budget Guidance
PBHP	– Pounds per Brake Horsepower
PBI	– Particle Background Investigation
	– Production Backup Item
PBLD	– Progressive Base Line Dimensioning
PBO	– Plotting Board Operator
PBP	– Plotting Board Plot
	– Push Button Panel
PBR	– Precision Bombing Range
PBRE	– Pebble Bed Reactor Experiment
PBS	– Physiological Biomedical System
PBT	– P-BIT Test
PBX	– Plastic-Bonded Explosives
	– Private Branch Exchange (Same as PABX only PABX is Automatic)
PC	– Parameter Checkout
	– Part Card
	– Personnel Carrier
	– Phase Coherent
	– Photo-Cell
	– Pitch Circle
	– Plant Clearance
	– Plug Cock
	– Point of Curve
	– Printed Circuit
	– Program Coordination
	– Program Counter
	– Pulsating Current
	– Pulse Compression
	– Purchasing and Contracting
P/C	– Pitch Control
	– Planning Change
	– Polar to Cartesian
	– Power Conversion
P-C	– Polar to Cartesian
	– Power Conversion
P&C	– Procurement and Contracts
	– Purchasing and Contracting
PCA	– Parts Control Area
	– Polar-Cap Absorption
	– Production Control Area
PCAM	– Punch Card Accounting Machine
PCAT	– Procedures for the Control of Air Traffic
PCB	– Pre-Change Board
PCC	– Planning Coordination Committee
	– Point of Compound Curve

PCC	- Processor Control Cards
PCCC	- Participating College Correspondence Course
PCCDS	- Patrol Craft Combat Direction System
PCCN	- Part Card Change Notice
PCCP	- Preliminary Contact Change Proposal
PCCS	- Positive Control Communication System
	- Program Control Correspondence Station
PCD	- Power Control and Distribution
	- Procurement Control Document
PCE	- Parameter Checkout Engineer
	- Peripheral Control Element
PC&E	- Parts Control and Expediting
PCEA	- Pacific Coast Electrical Association
PCF	- Pounds per Cubic Foot
	- Power Cathode Follower
	- Pulse-to-Cycle Fraction
PCH	- Patrol Craft, Hydrofoil
PCIN	- Production Change Incorporation Notice
PCL	- Positive Control Line
PCLA	- Power Control Linkage Assembly
PCM	- Passive Countermeasure
	- Photo Contact Master
	- Pitch Control Motor
	- Pulse Code Modulation
	- Pulse Code Modulator
	- Punched Card Machine
PCMD	- Pulse Code Modulation Digital
PCME	- Pulse Code Modulation Event
PCM/FSK/AM	- Pulse Code Modulation/Frequency Shift Keying/Amplitude Modulation
PCMS	- Punched Card Machine System
PCN	- Pacific Communications Network
	- Prelaunch Channel Number
	- Procedure Change Notice
	- Production Change Number
PCO	- Prime Contracting Officer
	- Procuring Contracting Officer
	- Production Control Office
PCP	- Processor Control Program
	- Program Change Proposal
	- Pulse Comparator
PCPS	- Program Change Packages
PCPCN	- Part Card Procurement Change Notice
PCPL	- Proposed Change Point Line
PCR	- Production Control Record
	- Programmer in Charge of Records

PCR

PCR	– Project Cost Report
PCS	– Pacific Command Ship
	– Permanent Change of Station
	– Permanent Change of Status
	– Physical-Chemical System
	– Position, Course, Speed
	– Probability of Crew Safety
PCSP	– Program Communications Support Program
PCT	– Pitch Centering Torquer
PCTF	– Power Conversion Test Facility
PCTR	– Physical Constants Testing Reactor
PCU	– Pound Centigrade Unit
	– Power Control Unit
	– Pressurization Control Unit
	– Print Control Unit
P/C/W	– Previously Complied With
PD	– Passive Detection
	– Per Diem
	– Pitch Diameter
	– Point Detonating
	– Position Description
	– Potential Difference
	– Power Distribution
	– Preliminary Design
	– Prime Depot
	– Priority Directive
	– Probability of Detection
	– Process Drawing
	– Production Design
	– Production Division
	– Programs Director
	– Project Directive
	– Pulse Duration
PDA	– Patient Data Automation
	– Precision Drive Axis
	– Preliminary Drawing Approval
	– Probability Distribution Analyzer
	– Proposed Development Approach
	– Pump Drive Assembly
PDB	– Program Design Branch
PDC	– Percent Defective Chart
	– Probability of Detection and Conversion
PDD	– Premodulation Processor—Deep Space Data
PDF	– Point Detonating Fuze
PDG	– Passive Defense Group
	– Production Development Group

PDG	- Programs Development Group
PDGW	- Principle Directorate of Guided Weapons (British)
PDI	- Pilot Direction Indicator
PDL	- Passed Down the Line
PDM	- Pulse Duration Modulation
PDMAMS	- Product Design Minuteman Airborne Mechanical System
PDM-FM	- Pulse Duration Modulation—Frequency Modulation
PDMMS	- Product Design Minuteman Mechanical System
PDO	- Publications Distribution Office
PDP	- Program Development Plan
	- Programmed Data Processor
	- Project Definition Phase
PDPC	- Position Display Parallax Corrected
PDR	- Precision Depth Recorder
	- Preliminary Design Review
	- Priority Data Reduction
	- Processed Data Recorder
	- Program Discrepancy Report
	- Program Document Requirement
	- Program Drum Recording
	- Power Directional Relay
PDRF	- Passive Defense Recovery Force
PDS	- Paradynamic Sketch
	- Parameter Drift Screening
	- Personnel Data System
	- Power Distribution System
	- Production Data Sheet
	- Pyrotechnic Devices Simulator
PDSD	- Point Detonating Self-Destroying
PDSQ	- Point Detonating Super Quick
PDT	- Pacific Daylight Time
PDU	- Pressure Distribution Unit
PDV	- Premodulation Processor—Deep Space Voice
PDX	- Place Decrement in Index Register
PE	- Performance Evaluation
	- Permanent Echo
	- Personnel Error
	- Photo-Electric
	- Planning Estimate
	- Plastic Explosive
	- Positive Expulsion
	- Potential Energy
	- Probable Error
	- Program Element
	- Project Engineer
	- Purchased Equipment

■ P&E

P&E	– Planning and Estimating
	– Procurement and Expedition
	– Propellant and Explosive
	– Pyrotechnic and Explosive
P-E	– Perkin-Elmer Company
PEAC	– Photoelectric Autocollimator
PEB	– Program Element Breakdown
PEC	– Production Equipment Code
	– Program Environmental Control
PECAN	– Pulse Envelope Correlation Air Navigation
PECE	– Proposed Engineering Change Estimate
PECM	– Preliminary Engineering Change Memorandum
PECOS	– Program Environmental Checkout System
PECR	– Process Equipment Certification Requirement
PED	– Personnel Equipment Data
	– Production Engineering Document
PEDRO	– Pneumatic Energy Detector with Remote Optics
PEGE	– Program for Evaluation of Ground Environment
PEH	– Planning Estimate Handbook
PEI	– Preliminary Engineering Inspection
	– Prince Edward Island
PEJ	– Premolded Expansion Joint
PEM	– Production Engineering Measures
	– Program Element Monitor
PEMA	– Procurement of Equipment and Missiles, Army
PEN	– Program Error Note
PENNSTAC	– Pennsylvania State Automatic Computer
PEP	– Peak Envelope Power
	– Physiological Evaluation of Primates
	– Plant Engineering Procedure
	– Political and Economic Planning
	– Princeton Experimental Package
	– Program Evaluation Procedure
	– Program Evaluation Program
PEPP	– Professional Engineers in Private Practice
PEPSS	– Programmable Equipment for Personnel Subsystem Simulation Facility
PER	– Phase Engineering Report
PERG	– Production Equipment Redistribution Group
PERI	– Production Equipment Redistribution Inventory
PERMINVAR	– Permeability Invariable
PERP	– Perpendicular
PERS	– Preliminary Engineering Reports
PERT	– Preliminary Flight Readiness Test
	– Program Evaluation and Review Technique
PERT-NAP	– Program Evaluation and Review Technique—Network Automatic Plotting

PERU	– Production Equipment Records Unit
PET	– Performance Evaluation Test
	– Peripheral Equipment Test
	– Polyethylene Terephthalate
	– Position-Event-Time
	– Production Environmental Test
	– Production Experimental Test
	– Propulsion Experimental Test
PETN	– Pentaerythritoltetranitrate
PETP	– Polyethylene Terephthalate Capacitors
PETS	– Pacific Electronic Show
	– Position Equipment Task Summary
PEWG	– Program Evaluation Work Group
PF	– Pattern Flight
	– Performance Factor
	– Pneumatic Float
	– Point of Frog
	– Port Facilities
	– Power Factor
	– Preflight
	– Probability of Failure
	– Proximity Fuze
	– Pulse Frequency
P/F	– Pattern Flight
PFB	– Pre-Formed Beam
PFL	– Propulsion Field Laboratory
P/FLT	– Pattern Flight
PFM	– Power Factor Meter
	– Pulse Frequency Modulation
PFN	– Pulse Forming Network
PFP	– Probability of Failure, Performance
PFR	– Part Failure Rate
PFRT	– Preliminary Flight Rating Test
PFRTE	– Preliminary Flight Rating Test Engine
PFS	– Probability of Failure, Stress
PFT	– Portable Flame Thrower
	– Positive Flight Termination
PFV	– Probability of Failure, Vehicle
PG	– Pregnant Guppy (reference to Boeing 377)
	– Program Guidance
	– Proving Ground
	– Pulse Generator
	– Pyrolytic Graphic
P/G	– Programmer Group
PGA	– Pendulous Gyro Accelerometer
	– Pressure Garment Assembly

■ **PGANE**

PGANE	– Professional Group on Aeronautical and Navigational Electronics
PGAP	– Professional Group on Antennas and Propagation
PGB	– Borane Pyrographalloy
PGBTR	– Professional Group on Broadcast and Television Receivers
PGC	– Proving Ground Command
PGD	– Program for Geographical Display
PGEC	– Professional Group on Electronic Computers
PGEM	– Professional Group of Engineering Management
PGEWS	– Professional Group on Engineering Writing and Speech
PGHFE	– Professional Group on Human Factors in Electronics
PGME	– Professional Group on Medical Electronics
PGMIL	– Professional Group on Military Electronics
PGMSJ	– Professional Group of Mathematical Symbol Jugglers
PGMTT	– Professional Group on Microwave Theory and Techniques
PGPE	– Pre-Flight Ground Pressurization Equipment
PGPEP	– Professional Group on Product Engineering and Production
PGRF	– Pulse Group Repetition Frequency
PGRFI	– Professional Group of Radio Frequency Interference
PGS	– Power Generation System
PH	– Power House
PHA	– Prelaunch Hazard Area
PHD	– Phase Shift Driver
PHI	– Position Homing Indicator
PHIBLANT	– Amphibious Force, Atlantic
PHIBPAC	– Amphibious Force, Pacific
PHIBRON	– Amphibious Squadron
PHLAG	– Phillips Petroleum Load and Go System
PHLODOT	– Phase Lock Doppler Tracking System
PHM	– Phase Meter
PHO	– Philco Houston Operations
PHP	– Planetary Horizon Platform
	– Pounds per Horsepower
PHR	– Pounds per Hour
PI	– Personal Identification
	– Photo Interpreter
	– Planetary Interior
	– Point Initiating
	– Point Insulating
	– Point of Intersection
	– Position Indicator
	– Precision Instrument
	– Priviledged Information
	– Problem Inputs
	– Program Interrupter
	– Programmed Instruction

PI	- Program of Instrumentation
	- Proprietary Information
P-I	- Photogrammetric Instrumentation
P&I	- Production and Installation
PIA	- Place Indicators in Accumulators
	- Pre-Inspection Acceptance
	- Pre-Installation Acceptance
PIAPACS	- Psychophysical Information Acquisition, Processing, and Control System
PIAT	- Projector Infantry, Anti-Tank (British)
PIB	- Polar Ionosphere Beacon
	- Propellant Inspection Building
PIBAL	- Pilot Balloon
PIBD	- Point Initiating Base Detonating
PIBOL	- Pilot in Booster Loop
PIC	- Periodic Inspection Control
	- Photo Interpretation Console
	- Pilot-Integrated Cockpit
	- Procurement Information for Contracts
	- Program Initiations and Commitments
	- Pursuant to Instructions Contained in
PICE	- Programmable Integrated Control Equipment
PICOE	- Programmed Initiations, Commitments, Obligations, and Expenditures
PID	- Public Information Division
PIDEP	- Pre-Interservice Data Exchange Program
PIDR	- Parts Inventory and Disposition Request
PIE	- Pulse Interference Elimination
PIF	- Package Information Form
	- Pilot Information File
PIGA	- Pendulous Integrating Gyroscopic Accelerometer
PILAC	- Pulsed Ion Linear Accelerator
PILO	- Public Information Liaison Officer
PIM	- Precision Instrument Mount
	- Pulse Interval Modulation
PIMP	- Permissible Individual Maximum Pressure
PIN	- Position Indicator
	- Program Identification Number
PINT	- Interpretive Program
PIO	- Public Information Office
	- Public Information Officer
PIOUS	- Peripheral Integrated Off-Line Utility System
PIP	- Problem Input Preparation
	- Production Implementation Program
	- Program in Progress
	- Program Intergration Plan

■ PIP

PIP	– Pulsed Integrating Pendulum
PIPA	– Pulsed Integrating Pendulum Accelerometer
PIPER	– Pulsed Intense Plasma for Exploratory Research
PIPS	– Pulsed Integrating Pendulums
PIQSY	– Probes for the International Quiet Solar Year
PIR	– Personnel Information Report
	– Precision Inspection Request
	– Pressure Impact Arming Rocket
	– Publications Information Register
PIRA	– Personnel Industrial Relations Association
PIRD	– Program Instrumentation Requirements Document
PIREP	– Pilot Weather Report
PIRT	– Precision Infrared Tracking
	– Preliminary Infrared Triangulation System
PISAB	– Pulse Interference Separation and Blanking
PIT	– Peripheral Input Tape
	– Print Illegal and Trace
PIU	– Plug-In Unit
PIUMP	– Plug-In Unit Mounting Panel
PIV	– Peak Inverse Voltage
	– Post Indicator Valve
PJ	– Plasma Jet
	– Pulse Jet
PJR	– Power Jets Report
PKG	– Phonocardiogram
PL	– Parting Line
	– Personnel Laboratory
	– Phase Line
	– Pipe Line
	– Position Line (also LOP)
	– Post Landing
	– Private Line
	– Proportional Limit
	– Public Law
P/L	– Parts List
PLA	– Post Landing Attitude
PLACE	– Programming Language for Automatic Checkout Equipment
PLANNET	– Planning Network
PLAP	– Prelaunch, Launch, and Ascent Procedures
PLAT	– Pilot/LSO Landing Aid, Television
PLATO	– Programmed Logic for Automatic Teaching Operations
PLB	– Pull Button
PLC	– Power-Line Carrier
	– Prime Level Code
	– Propellant Loading Console
PLCU	– Propellant Loading Control Unit

PLD	- Payload
PLDC	- Preliminary List of Design Changes
PLI	- Preload Indicator
PLIM	- Post Launch and Instrumentation Message
PLM	- Prelaunch Monitor
	- Production Line Maintenance
PLO	- Pacific Launch Operation
	- Point of Local Operation
	- Project Line Organization
PLOO	- Pacific Launch Operations Office
PLOP	- Pressure Line of Position
PLPS	- Propellant Loading and Pressurization System
PLS	- Private Line Service
	- Propellant Loading System
PLSS	- Portable Life Support System
	- Prelaunch Status Simulator
PLT	- Procurement Lead Time
PLUS	- PERT—Life Cycle Unified System
PLUTO	- Pipe Line Under the Ocean (British)
	- Programmed Logic for Automatic Teaching
PM	- Pattern Maker
	- Permanent Magnet
	- Phase Modulation
	- Point Mugu
	- Post Meridian
	- Pounds per Minute
	- Preventive Maintenance
	- Procedures Manual
	- Production Memorandum
	- Production Monitoring
	- Pulse Modulation
	- Purchase Memorandum
P/M	- Panel Maintenance
PMA	- Primary Mental Ability
	- Program Modification and Adaptation Group
PMC	- Program Marginal Checking
PMCS	- Pulse-Modulated Communication System
PMD	- Program Monitory and Diagnosis
PME	- Photomagnetoelectric
	- Precision Measuring Equipment
PMEL	- Precision Measurement Equipment Laboratory
PMG	- Prediction Marker Generator
PMIS	- Program Measurement Information System
PMK	- Pitch Mark
PMM	- Pulse Mode Multiplex
PMMC	- Permanent Magnet Movable Coil

■ **PMMP**

PMMP	– Permissible Mean Maximum Pressure
PMO	– Program Management Office
	– Project Manufacturing Order
PMOS	– Physical Movement of Spacecraft
	– Primary Military Occupational Speciality
PMP	– Premodulation Processor
	– Program Management Plan
	– Protective Mobilization Plan
PMR	– Pacific Missile Range (now AFWTR)
	– Propellant Mass Ratio
PMRF	– Pacific Missile Range Facility
PMRFAC	– Pacific Missile Range Facility
PMRM	– Periodic Maintenance Requirements Manual
PMRR	– Pacific Missile Range Representative
PMRTF	– Pacific Missile Range Tracking Facility
PMS	– Physiological Monitoring System
	– Probability of Mission Success
	– Program Management Support
PMSS	– Precision Measuring Subsystem
PMT	– Photo-Multiplier Tube
	– Production Monitoring Test
	– Program Master Tape
PMTS	– Press-Mold-to-Shape
PMU	– Portable Memory Unit
PN	– Part Number
	– Pneumatic
	– Project Notification
P-N	– Positive-Negative
P/N	– Part Number
PND	– Premodulation Processor—Near Earth Data
PNMO	– Provided No Military Objection
PNP	– Positive-Negative-Positive
P-N-P	– Positive-Negative-Positive
PNPN	– Positive-Negative-Positive-Negative
PO	– Parking Orbit
	– Patent Office
	– Planetary Orbit
	– Polarity
	– Power Oscillator
	– Pressure Oscillation
	– Production Order
	– Project Office
	– Project Officer
	– Purchase Order
P&O	– Paints and Oil
POA	– Pacific Ocean Area

POA	- Probability of Acceptance
	- Purchased on Assembly
	- Purchase Order Authorization
POADS	- Portland Air Defense Sector
POAE	- Port of Aerial Embarkation
POC	- Planning Objective Coordinators
	- Privately Owned Conveyance
	- Production Operational Capability
	- Purchase Order Change
POCN	- Purchase Order Change Notice
POD	- Philadelphia Ordnance District
	- Port of Debarkation
	- Pre-Flight Operations Division
POE	- Port of Embarkation
POET	- Program Operation and Environment Transfer
POGO	- Polar Orbiting Geophysical Observatory
	- Program Optimizer for Bendix G-15 Operations
POI	- Parking Orbit Injection
	- Program of Instruction
POINTER	- Particle Orientation Interferometer
POL	- Petroleum, Oil, and Lubricant
	- Polarize
	- Problem Oriented Language
	- Procedure Oriented Language
POM	- Preparation for Overseas Movement
POMAR	- Position Operational Meteorological Aircraft Report
POMS	- Panel on Operational Meteorological Satellites
POP	- Pump Optimizing Program
	- Purchase Outside Production
POPI	- Post-Office Position Indicator (British)
POR	- Production Order Records
	- Production Order Request
PORCN	- Production Order Revision Change Notice
PORT	- Photo-Optical Recorder Tracker
POS	- Pacific Ocean Ship
	- Probability of Survival
POSI	- Personnel On-Site Integration
POSITRON	- Positive Electron
POSP	- Pacific Ocean Stations Program
POSS	- Passive Optical Satellite Surveillance
	- Photo-Optical Surveillance Subsystem
	- Prototype Optical Surveillance System
POSTER	- Post Strike Emergency Reporting
POSV	- Pilot Operated Solenoid Valve
POT	- Potentiometer
POTS	- Photo-Optical Terrain Simulator

POUCHE

POUCHE	- Program Distribution Service of the AICHE
POV	- Peak Operating Voltage
PP	- Pages
	- Panel Point
	- Partial Pressure
	- Peak to Peak
	- Power Package
	- Power Plant
	- Preprocessor
	- Pressure Proof
	- Print-Punch
	- Push Pull
P/P	- Point to Point
P&P	- Policies and Procedures
	- Procurement and Production
PPA	- Program Problem Area
PPB	- Provisioning Parts Breakdown
PPBG	- Preliminary Program and Budget Guidance
PPCO	- Phillips Petroleum Company
PPCR	- Production Planning Change Request
PPD	- Plot Plan Drawing
	- Program Planning Directive
PPDD	- Plan Position Data Display
PPE	- Premodulation Processing Equipment
	- Print-Punch Editor
PPEN	- Purchased Parts Equipment Notice
PPG	- Policies and Procedures Guide
	- Program Production Group
	- Propulsion and Power Generation
PPH	- Pounds per Hour
	- Pulses per Hour
PPH/LB	- Pounds per Hour per Pound
PPI	- Pictorial Position Indicator
	- Plan Position Indicator
PPIID	- Provisional Participation In Industrial Duties
PPIL	- Production Pre-Flight Inspection Letter
PPK	- Anti-G Suit (USSR)
PPL	- Power Plant Laboratory
	- Preferred Parts List
PPM	- Parts per Million
	- Periodic Permanent Magnet
	- Post Past Message
	- Pounds per Minute
	- Pulse Position Modulation
	- Pulses per Minute
PPMPC	- Pilot Parachute Mortar Pyrotechnic Cartridge

PPN	– Procurement Program Number
PPNR	– Production Part Number Record
PPP	– Peak Pulse Power
P&PP	– Pull and Push Plate
PPPP	– Proposed Partial Package Program
PPR	– Program Profile
PPRN	– Purchased Parts Requirement Notice
PPRR	– Purchased Parts Requirement Request
PPS	– Pounds per Second
	– Preliminary Propulsion System
	– Pulses per Second
PPSB	– Program Plan Study Branch
PPT	– Preprototype
	– Punched Paper Tape
PPTS	– Pre-Problem Training Situation
PPU	– Anti-G Equipment (USSR)
PQ	– Premium Quality
PQC	– Procurement Quality Control
PR	– Parachute Rigger
	– Pen Record
	– Periodic Report
	– Pipe Rail
	– Position Report
	– Prandtl Number
	– Priority Regulation
	– Procurement Regulation
	– Production Repair
	– Program Register
	– Program Requirement
	– Project Rover
	– Puerto Rico
	– Pulse Rate
	– Purchase Request
P&R	– Policies and Regulations
PRA	– Precession Axis
	– Production Repair Area
	– Program Requirements Authorization
	– Psychological Research Associates
PRAC	– Public Relations Advisory Committee
PRADOR	– PRF (Pulse Repetition Frequency) Ranging Doppler Radar
PRC	– Parts Release Card
	– Point of Reverse Curve
	– Price Redetermination Contract
PRD	– Program Requirements Data
	– Program Requirements Document
PRE	– Personnel Restraint Equipment

■ PRE

PRE	– Problem Reproducer Equipment
PREAMP	– Preamplifier
PRELORT	– Precision Long Range Tracking Radar
PREP	– Programmed Educational Package
PRESS	– Pacific Range Electromagnetic Signature Study
PRESSAR	– Presentation Equipment for Slow Scan Radar
PRESTO	– Program Reporting and Evaluation System for Total Operation
PRF	– Pulse Rate Frequency
	– Pulse Recurrence Frequency (Repetition preferred)
	– Pulse Repetition Frequency
PRG	– Program Requirements Group
PRH	– Program Requirements Handbook
PRI	– Photo Radar Intelligence
	– Pulse Rate Indicator
PRIDE	– Programmed Reliability in Design Engineering
PRIM	– Plume Radiation Intensity Measurement
PRIME	– Preparedness of Resources in Mission Evaluation
PRINT	– Pre-Edited Interpretive System
PRIOR	– Program for In-Orbit Rendezvous
PRIP	– Parts Reliability Improvement Program
PRIS	– Pacific Range Instrumentation Satellite
PRISE	– Program for Integrated Shipboard Electronics
PRISM	– Programmed Integrated System Maintenance
	– Program Reliability Information System for Management
PRL	– Personnel Research Laboratory
	– Predictions Program List
PRM	– Posigrade Rocket Motor
	– Pulse Rate Modulation
PRN	– Parts Requirement Notice
	– Pseudo-Random Noise
	– Pulse Ranging Navigation
PRNC	– Potomac River Naval Command
PRO	– Pen Recorder Output
	– Propagation Prediction Report
	– Public Relations Officer
PROBCOST	– Probabilistic Budgeting and Forward Costing
PROC	– Programming Computer
PROCAER	– Progetti Costruzioni Aeronautiche
PROFAC	– Propulsive Fluid Accumulator
PROMPT	– Project Management and Production Team
PROP	– Pilot Repair Overhaul Provisioning
	– Planetary Rocket Ocean Platform
PROSE	– Program System Example
PRP	– Position Report Printout
	– Pulse Recurrence Period (Repetition preferred)

PRP	– Pulse Repetition Period
PRR	– Parts Replacement Request
	– Production Revision Record
	– Production Revision Request
	– Pulse Recurrence Rate (Repetition preferred)
	– Pulse Repetition Rate
PR&RM	– Program Review and Resources Management
PRS	– Program Requirements Summary
	– Program Requirements Survey
PR/S	– Prestrike
PRSA	– Public Relations Society of America
PRT	– Pattern Recognition Technique
	– Personnel Research Test
PRU	– Photo Reconnaissance Unit
	– Physical Research Unit
PRV	– Peak Reverse Voltage
	– Pressure Reducing Valve
PS	– Packing Sheet
	– Parts Store
	– Passing Scuttle
	– Phase Shift
	– Pilot Simulator
	– Planetary Surface
	– Point of Switch
	– Potentiometer Synchro
	– Power Supply
	– Probability of Success
	– Process Specification
	– Production Store
	– Proof Shot
	– Pull Switch
	– Pulse Stretcher
P/S	– Parallel to Serial
	– Post Strike
	– Power Supply
P-S	– Parallel to Serial
P&S	– Port and Starboard
PSA	– Post-Separation Arming
	– Power and Servo Assembly
	– Power Servo Amplifier
	– Production Stores Area
	– Program Study Authorization
PSAC	– President's Science Advisory Committee
PSACN	– Process Specification Advance Change Notice
PSAS	– Program Support and Advanced Systems
PSBLS	– Permanent Space Based Logistics System

■ PSCEC

PSCEC	- Planning Status of Committed Engineering Changes
PSCRT	- Passive Satellite Communications Research Terminal
PSCS	- Pacific Scatter Communication System
PSD	- Phase Sensitive Demodulator
	- Power Spectral Density
	- Process Specification Departure
	- Program Support Directive
	- Program Support Document
	- Program System Description
	- Pulse Shape Descriminator
PSDF	- Propulsion System Development Facility
PSE	- Pulse Sense
PSF	- Pounds per Square Foot
PSG	- Personnel Subsystem Group
PSGE	- Photosynthetic Gas Exchanger
PSH	- Program Support Handbook
PSI	- Pacific Semiconductor, Incorporated
	- Plan Speed Indicator
	- Pounds per Square Inch
	- Proto-Synthex Index
PSIA	- Pounds per Square Inch Absolute
PSID	- Pounds per Square Inch Differential
	- Preparatory Systems Integration Demonstration
PSIG	- Pounds per Square Inch Gage
	- Program System Integration Group
PSK	- Phase Shift Keying
PSL	- Physical Science Laboratory
PSO	- Pad Safety Officer (now Pad Safety Supervisor)
	- Product Support Organization
PSP	- Parts Screening Program
	- Predictable System Performance
	- Program Support Plan
PSPP	- Proposed System Package Plan
PSPS	- Program Support Plans Summary
PSQ	- Personnel Security Questionnaire
PSR	- Periodic Summary Report
	- Pilot Supply Regulator
	- Plow Steel Rope
	- Problem Status Report
	- Progress Summary Report
PSRA	- Problem Status Report Analysis
PSRI	- Personnel Specialities and Record Inventory
PSRU	- Production Support Repair Unit
PSS	- Pad Safety Supervisor (formerly PSO)
	- Personnel Subsystem
	- Premature Separation Switch

PSS	– Programming Support System
PSSA	– Pilot Signal Selector Adaptor
PSSC	– Parachute Subsystem Sequence Controller
PST	– Pacific Standard Time
	– Point of Spiral Tangent
	– Production Sampling Test
PST&E	– Personnel Subsystem Test and Evaluation
PSTL	– Pressure Model Static and Transient Launch Configuration
PSV	– Photographic–Spatial Volume
	– Probability of Success of Vehicle
PSYOP	– Psychological Operations
PSYWAR	– Psychological Warfare
PT	– Paper Tape
	– Part
	– Partially Tested
	– Parts Tag
	– Parts Transfer
	– Part Time
	– Performance Test
	– Photo Template
	– Pint
	– Plotting Technician
	– Pneumatic Tube
	– Point
	– Point of Tangency
	– Potential Transformer
	– Pressure Test
	– Pressure Transducer
	– Program Time
	– Programmed Temperature Gas Chromatography
	– Propellant Transfer
	– Prototype
	– Pulse Timer
	– Punched Tape
P/T	– Prototype
P&T	– Personnel and Training
PTA	– Power Transfer Assembly
	– Primary Target Area
	– Program Time Analyzer
	– Property Transfer Authorization
	– Proposed Technical Approach
PTB	– Production and Test Branch
	– Program Test Branch
PTC	– Pitch Trim Compensator
	– Portable Temperature Controller
	– Positive Temperature Coefficient

■ PTC

PTC	- Pulse Time Code
PTCP	- Parameter Test Control Program
PTCR	- Pad Terminal Connection Room
	- Patent, Trademark, and Copyright Research Institute
P&TD	- Parts and Tool Disposition
PTDB	- Point Target Data Base
PTF	- Parts Transfer Form
	- Patrol Torpedo Boat Fleet
PTFE	- Polytetrafluorethylene
PTFMO	- Peacetime Force Materiel Objective
PTFMPO	- Peacetime Force Materiel Procurement Objective
PTFMR	- Peacetime Force Materiel Regulations
PTFT	- Production Temporary Facility Tool
PTG	- Program Test Group
PTH	- Phenylthiohydant
PTI	- Product Techniques, Incorporated
PTIR	- Precision Tool Inspection Record
PTL	- Primary Target Line
PTM	- Pancake Torquer Motor
	- Proof Test Model
	- Pulse Time Modulation
	- Pulse Time Multiplex
PTO	- Power Take-Off
PTOMAIN	- Project to Optimize Many Individual Numbers
PTOS	- Peacetime Operating Stock
PTP	- Parameter Test Program
PTPS	- Propellant Transfer Pressurization System
PT/PT	- Point to Point
PTR	- Parts Transfer Record
	- Polar to Rectangular
	- Position Track Radar
	- Processor Tape Read
PTS	- Parameter Test Setup
	- Propellant Transfer System
PTT	- Push to Talk
PTU	- PICE Terminal Unit
	- PICE Transfer Unit
PTV	- Parachute Test Vehicle
	- Propulsion Test Vehicle
PU	- Performance Unit
	- Pickup
	- Pluggable Unit
	- Power Unit
	- Propellant Utilization
	- Propulsion Unit
PUC	- Provided You Concur

PUCS	- Propellant Utilization Control System
PUFS	- Proposed Underwater Fire Control Feasibility Study
PUNC	- Program Unit Counter
PUP	- Program Unit Punch
PUTWS	- Put Word in String
PV	- Payload Vehicle
	- Physical Vulnerable
	- Plan View
P/V	- Peak to Valley
P&VE	- Propulsion and Vehicle Engineering
P&VE-ADM	- Propulsion and Vehicle Engineering—Administrative
P&VE-DIR	- Propulsion and Vehicle Engineering—Director
P&VE-E	- Propulsion and Vehicle Engineering—Vehicle Engineering
P&VE-F	- Propulsion and Vehicle Engineering—Advanced Flight Systems
P&VE-M	- Propulsion and Vehicle Engineering—Engineering Materials
P&VE-N	- Propulsion and Vehicle Engineering—Nuclear Vehicle Projects
P&VE-O	- Propulsion and Vehicle Engineering—Engine Management
P&VE-P	- Propulsion and Vehicle Engineering—Propulsion and Mechanics
P&VE-PC	- Propulsion and Vehicle Engineering—Program Coordination
P&VE-REL	- Propulsion and Vehicle Engineering—Reliability
P&VE-S	- Propulsion and Vehicle Engineering—Structures
P&VE-TS	- Propulsion and Vehicle Engineering—Technical and Scientific Staff
P&VE-V	- Propulsion and Vehicle Engineering—Vehicle Systems Integration
PVF	- Polyvinyl Formal
PVOR	- Precision VHF Omnidirectional Range
PVR	- Precision Voltage Reference
PVSE	- Primary Vehicle System Engineer
PVT	- Pressure-Volume-Temperature
PW	- Plain Washer
	- Program Word
	- Public Works
	- Pulse Width
P&W	- Pratt & Whitney
P&WA	- Pratt & Whitney Aircraft
PWC	- Pulse Width Coded
PWD	- Pulse Width Discriminator
PWE	- Prisoner of War Enclosure
	- Pulse Width Encoder
PWI	- Proximity Warning Indicator
PWM	- Pulse Width Modulation
PWO	- Public Works Office

■ PWP

PWP	- Plasticized White Phosphorus
	- Prelaunch Wind Profile
PWR	- Power
	- Pressurized Power Reactor
PWRS	- Prepositioned War Reserve Stock
PWS	- Private Wire Service (same as PLS)
PWT	- Propulsion Wind Tunnel
PXA	- Place Index in Address
PXD	- Place Index in Decrement
PYRO	- Pyrotechnic

Q-0	- Q-Zero Compiler
QA	- Quality Assurance
	- Quick Acting
QAD	- Quality Assurance Division
QADS	- Quality Assurance Data System
QAFM	- Quality Assurance Forms Guide Manual
QAM	- Quality Assurance Manual
QAOP	- Quality Assurance Operating Procedure
QAST	- Quality Assurance Service Test
QATP	- Quality Assurance Test Procedure
QAVC	- Quiet Automatic Volume Control
QC	- Quality Control
QCCR	- Quality Control Change Request
QCD	- Quality Control Directive
QCDSP	- Quality Control Directive Special
QCDSU	- Quality Control Directive Supplement
QCE	- Quality Control Engineering
QCL	- Quality Control Level
QCM	- Quality Control Manual
QCO	- Quality Control Officer
QCOP	- Quality Control Operating Procedure
QCPC	- Quality Control Property Clearance
QCPM	- Quality Control Procedures Manual
QCPP	- Quality Control Planning Procedures
QCPT	- Quality Control Pressure Test
QCR	- Quality Control/Reliability
QCRI	- Quality Control Reliability Investigator
QCS	- Quality Control Survey
QCSR	- Quality Control Service Request
QC&T	- Quality Control and Techniques
QCTE	- Quality Control Test Engineering
QCVTI	- Quality Control Verification Test Inspection
QD	- Quick Disconnect

■ **QDRI**

QDRI	- Qualitative Development Requirements Information
QE	- Quadrant Elevation
	- Quoted Exhibit
QEC	- Quick Engine Change
QF	- Quality Factor
	- Quick Firing
QFI	- Qualified Flight Instructor
QGV	- Quantized Gate Video
QI	- Quality Indices
QIP	- Query Interpretation Program
QIT	- Quality Information and Test System
Q/L	- Quick Look
QLIT	- Quick Look Intermediate Tape
QM	- Quadrature Modulation
	- Quartermaster
QMC	- Quartermaster Corps
QMCR	- Quartermaster Corps Regulations
QMDO	- Qualitative Material Development Objective
QMG	- Quartermaster General
QMI	- Qualification Maintainability Inspection
QMQB	- Quick-Make, Quick-Break
QMR	- Qualitative Material Requirement
QOD	- Quick-Opening Device
QOR	- Qualitative Operational Requirement
	- Quality Operational Requirement
QPC	- Quality Performance Chart
QPL	- Qualified Products List
QPP	- Quantized Pulse Position
QPR	- Qualitative Personnel Requirements
	- Quarterly Progress Report
QPRI	- Qualitative Personnel Requirements Information
QPS	- Qualified Processing Source
QQPRI	- Qualitative and Quantitative Personnel Requirements Information
QR	- Quick Reaction
	- Quick Response
	- Quotation Request
QRA	- Quality Reliability Assurance
QRC	- Quick Reaction Capability
QRCR	- Qualitative Reliability Consumption Report
QRDN	- Quality Requirement Discrepancy Notice
QREC	- Quartermaster Research and Engineering Center
QRS	- Qualification Review Sheet
QSSP	- Quasi-Solid-State Panel
QT	- Qualification Test
	- Quart

QTP	- Qualification Test Program
QTPR	- Quarterly Technical Progress Report
QTPT	- Qualification Test and Proof Test
QTY	- Quantity
QUAD	- Quadrant
QUAL	- Qualify
	- Quality
QUAP	- Questionnaire Analysis Program
QUASER	- Quantum Amplification by Stimulated Emission of Radiation
QVT	- Quality Verification Test

R

R	- Radium
	- Radius
	- Range
	- Rankine
	- Read
	- Reaumur
	- Reconnaissance
	- Red
	- Resistance
	- Resistor
	- Reverse
	- Reynolds Number
	- Right
	- Riser
	- Roentgen
	- Rubber
	- Rydberg Constant
RA	- Radar Altimeter
	- Read Amplifier
	- Reduction of Area
	- Register Allotter
	- Regular Army
	- Requesting Agency
R/A	- Radar Altimeter
	- Radius of Action
R&A	- Research and Analysis
RAAF	- Royal Australian Air Force
RABAL	- Radiosonde Balloon
RABAR	- Raytheon Advanced Battery Acquisition Radar
RABH	- Reported Altitude Block Height
RAC	- Research Analysis Corporation (formerly ORO)
	- Rework After Completion

RACE	- Rapid Automatic Checkout Equipment
	- Random Access Computer Equipment
	- Restoration of Aircraft to Combat Efficiency
RACEP	- Random Access and Correlation for Extended Performance
RACI	- Reported Altitude Change Indicator
RACON	- Radar Beacon
RAD	- Radiation Absorbed Dose
	- Research and Advanced Development
	- Research and Development
RADA	- Radioactive
	- Random Access Discrete Address
RADAC	- Rapid Digital Automatic Computing System
RADAN	- Radar Doppler Automatic Navigator
RADAR	- Radio Detection and Ranging
RADARC	- Radially Distributed Annular Rocket Combustion Chamber
RADAS	- Random Access Discrete Address System
RADAT	- Radar Data Transmission
RADATA	- Radar Automatic Data Transmission Assembly
RADATAC	- Radiation Data Acquisition Chart
RADC	- Radar Countermeasures and Deception
	- Rome Air Development Center
RADCM	- Radar Countermeasures and Deception
RADCOT	- Radial Optical Tracking Theodolite
RADDEF	- Radiological Defense
RADEM	- Random Access Delta Modulation
RADFO	- Radiological Fallout
RAD-HAZ	- Radiation Hazard
RADI	- Radiographic Inspection
RADIAC	- Radioactivity Detection, Identification, and Computation
RADIST	- Radar Distance Indicator
RADL	- Radiological
RADLAB	- Radiation Laboratory
RADLCEN	- Radiological Center
RADLDEF	- Radiological Defense
RADLDEFLAB	- Radiological Defense Laboratory
RADLMON	- Radiological Monitor
RADLO	- Radiological Defense Officer
RADLSAFE	- Radiological Safety
RADLSO	- Radiological Safety Officer
RADLSV	- Radiological Survey
RADLWAR	- Radiological Warfare
RADOME	- Radar Dome
RADOP	- Radar Operator
	- Radar/Optical
RADOT	- Real-Time Automatic Digital Optical Tracker
RADREL	- Radio Relay

■ RADRON

RADRON	- Radar Squadron
RADVS	- Radar Altimeter and Doppler Velocity Sensor
RADWAR	- Radiological Warfare
RAE	- Range, Azimuth, and Elevation
	- Right Arithmetic Element
	- Royal Aircraft Establishment
RAF	- Royal Air Force
RAFAR	- Radio Automated Facsimile and Reproduction
RAFD	- Rome Air Force Depot
RAFT	- Recomp Algebraic Formula Translator
	- Re-Entry Advanced Fuzing Test
RAG	- Replacement Air Group
	- Requirements Advisory Group
RAI	- Random Access and Inquiry
RAIC	- Radiological Accident and Incident Control
RAILS	- Remote Area Instrument Landing System
RAINBO	- Research and Instrumentation for National Bio-Science Operations
RALT	- Reported Altitude
RAM	- Radar Absorbing Material
	- Radiation Attenuation Measurement
	- Random Access Memory
	- Recovery Aids Material
	- Research Aviation Medicine
RAMA	- Rome Air Materiel Area (ROAMA preferred)
RAMAC	- Random Access Method of Accounting and Control
RAMARK	- Radar Marker
RAMD	- Random Access Memory Device
	- Receiving Agency Materiel Division
RAMIS	- Repair, Assemble, Maintain, Issue, and Supply
RAMP	- Radar Masking Parameters
	- Radiation Airborne Measurement Program
	- Raytheon Airborne Microwave Platform
RAMPART	- Radar Advanced Measurements Program for Analysis of Re-Entry Techniques
RAMPS	- Resource Allocation and Multi-Project Scheduling
RAN	- Read Around Number
RANCID	- Real and Not Corrected Input Data
RANCOM	- Random Communication Satellite
RAND	- Research and Development
RANDAM	- Random Access Indestructive Advanced Memory
RANDID	- Rapid Alpha-Numeric Digital Indicating Device
RAO	- Radio Astronomical Observatory
RAOB	- Radiosonde Observation
RAP	- Random Access Projector
	- Reliability Assurance Program

RAPCOE	- Random Access Programming and Checkout Equipment
RAPCON	- Radar Approach Control
RAPEC	- Rocket Assisted Personnel Ejection Catapult
RAPID	- Rocketdyne Automatic Processing of Integrated Data
	- Ryan Automatic Plot Indicator Device
RAPIT	- Record and Process Input Tables
RAPO	- Resident Apollo Project Office
RAPPI	- Random Access Plan Position Indicator
RAPR	- Radar Processor
RAR	- Radio Acoustic Ranging
	- Record and Report
RARE	- Ram-Air-Rocket Engine
RAREP	- Radar Weather Report
RAS	- Radar Advisory Service
RASC	- Rear Area Security Controller
RASCC	- Rear Area Security Control Center
RASER	- Radio Frequency Amplification by Stimulated Emission of Radiation
RASP	- Range Airframe Separation Program
	- Resource Allocation and Schedule Projection
RASPO	- Resident Apollo Spacecraft Project Office
RASTAC	- Random Access Storage and Control
RASTAD	- Random Access Storage and Display
RAT	- Rocket-Assisted Torpedo
RATAC	- Radar Analog Target Acquisition Computer
	- Remote Airborne Television Display of Ground Radar Coverage via TACAN
RATAN	- Radar and Television Aid to Navigation
RATC	- Rate-Aided Tracking Computer
RATCC	- Radar Air Traffic Control Center
RATE	- Remote Automatic Telemetry Equipment
RATEL	- Radio Telephone
RATG	- Radiotelegraph
RATO	- Rocket-Assisted Take-Off
RATS	- Rate and Track Subsystem
RATT	- Radio Teletype
RAVE	- Radar Acquisition Visual-Tracking Equipment
RAVEN	- Ranging and Velocity Navigation
RAWIN	- Radar Wind
RAWIND	- Radar Wind
RAWINDS	- Radar Wind Sounding
RAWINSONDE	- Radar Wind Sounding and Radiosonde
RAWOOP-SNAP	- Ramo-Wooldridge One-Pass Assembly Program
RAY-COM	- Raytheon Communications Equipment
RAYDAC	- Raytheon Digital Automatic Computer
RAYDIST	- Ray-Path Distance

■ RAYSISTOR

RAYSISTOR - Raytheon Resistor
RAYSPAN - Raytheon Spectrum Analyzer
RAY-TEL - Raytheon Telephone
RAZEL   - Range, Azimuth, and Elevation
RAZON   - Range and Azimuth Only
RB   - Radar Beacon
         - Reconnaissance Bomber
         - Relative Bearing
         - Renegotiation Board
         - Return to Bias
         - Road Buffer
         - Roller Bearing
R/B   - Radar Beacon
RBA   - Radar Beacon Antenna
RBDE   - Radar Bright Display Equipment
RBE   - Relative Biological Effectiveness
RBF   - Retract-Before-Fire Connector
RBFPP   - Rocket Booster Fuel Pod Pickup
RBI   - Rocketborne Instrumentation
RBOD   - Required Beneficial Occupancy Data
RBS   - Radar Bomb Scoring
         - Random Barrage System
RBSS   - Recoverable Booster Support System
RC   - Range Correction
         - Rate of Change
         - Register Containing
         - Reinforced Concrete
         - Reinstate Card
         - Remote Control
         - Resistance-Capacitance
         - Resolver Control
         - Reverse Current
         - Ribbon-Frame Camera
R/C   - Radio Command
         - Radio Control
         - Range Clearance
RCA   - Radio Corporation of America
RCAF   - Royal Canadian Air Force
RCAT   - Radio Controlled Aerial Target
RCC   - Radio Common Channels
         - Range Control Center
         - Recovery Control Center
         - Remote Combat Center
         - Remote Communications Complex
         - Remote Communications Console
         - Remote Control Complex
         - Rescue Coordination Center

RCC	- Rocket Combustion Chamber
	- Rough Combustion Cutoff
RCCPLD	- Resistance Capacitance Coupled
RCD	- Range Control Division
RCDC	- Radar Course-Directing Center
	- Radar Course-Directing Control
RCDMB	- Regional Civil and Defense Mobilization Boards
RCG	- Recovery Control Group
	- Reverberation Control of Gain
RCI	- Request for Change Information
RC/I	- Request for Change and/or Information
RCM	- Radar Countermeasure
	- Radio-Controlled Mine
RCN	- Royal Canadian Navy
RCO	- Radar Control Officer
	- Remote Control Office
	- Remote Control Oscillator
	- Rendezvous-Compatibility Orbit
RCP	- Recovery Command Post
	- Relative Corrector Program
	- Roll Centering Pickoff
RCS	- Radar Cross Section
	- Radio Command System
	- Reaction Control System
	- Recurrent Change of Station
	- Recurrent Change of Status
	- Re-Entry Control System
	- Remote Control System
	- Reports Control Symbol
	- Request for Change Studies
RCSS	- Random Communication Satellite System
RCT	- Regional Combat Team
	- Resolver Control Transformer
RCTL	- Resistor-Capacitor Coupled Transistor Logic
RCTSR	- Radio Code Test, Speed of Response
RCU	- Rocket Countermeasure Unit
RCV	- Relative Conductor Volume
RCVR	- Receiver
RCW	- Register Containing Word
RCZ	- Radiation Control Zone
RD	- Radar Data
	- Radar Display
	- Radiation Detection
	- Random Drift
	- Range Development
	- Reaction Division

■ RD

RD	- Readiness Date
	- Register Driver
	- Reply Delay
	- Required Date
	- Research and Development
	- Restricted Data
	- Revision Directive
	- Rocket Engine (USSR)
	- Root Diameter
R/D	- Revision Directive
R&D	- Research and Development
RDA	- Rapid Damage Assessment
	- Research and Development, Army
RDB	- Reference Data and Bias
	- Requirements and Design Branch
	- Research and Development Board
RDC	- Reply Delay Compensation
	- Request for Design Change
	- Request for Drawing Change
	- Reliability Data Central
	- Reliability Data Control
RDCRIT	- Read Criteria
RDECP	- Research and Development Engineering Change Proposal
RDF	- Radar Direction Finder
	- Radar Direction Finding
	- Radio Direction Finder
	- Radio Direction Finding
RDG	- Resolver Differential Generator
RDI	- Radio, Doppler Inertial
RDK	- Research and Development Kit
RDL	- Radiological Defense Laboratory
RDM	- Recording Demand Meter
RDMU	- Range-Drift Measuring Unit
RDN	- Rejection Disposition Notice
RDO	- Range Development Officer
	- Research and Development Objective
RDPC	- Radar Data Processing Center
RDR	- Read Drum
	- Rejection Disposition Report
	- Research and Development Report
RDS	- Read Select
RDTE	- Research, Development, Test, and Evaluation
RDT&E	- Research, Development, Test, and Evaluation
R/E	- Re-Entry
R&E	- Research and Engineering
REA	- Re-Entry Angle

REPT ■

REAC	- Reeves Electronic Analog Computer
READ	- Radar Echo Augmentation Device
READEF	- Reason for Deficiency
READI	- Rocket Engine Analyzer and Decision Instrument
READS	- Reno Air Defense Sector
REAP	- Rocket Engine Advancement Program
REB	- Radar Evaluation Branch
REBE	- Recovery Beacon Evaluation
REBECCA	- Radar Responder Beacon
REC	- Request for Engineering Change
RECAU	- Receipt Acknowledged and Understood
RECG	- Radioelectrocardiograph
RECIPE	- Recomp Computer Interspective Program Expeditor
RECOMP	- Recommended Completion
RECON	- Reconnaissance
	- Reliability and Configuration Accountability Control System
RECP	- Receptacle
RECSTA	- Receiving Station
RED	- Radio Equipment Department
REDOX	- Reduction and Oxidation
REDS	- Redundant Drogue System
REFCM	- Revised Engineering Firm Change Memorandum
REG	- Regulator
	- Rheoencephalograph
REGAL	- Range and Elevation Guidance for Approach and Landing
REGENT	- Reduce Geography in No Time
REIC	- Radiation Effects Information Center
REINS	- Radio-Equipped Inertial Navigation System
REIQ	- Refrigeration Installation Equipment
REL	- Radio Engineering Laboratories, Incorporated
REM	- Range Evaluation Missile
	- Roentgen Equivalent Man
REMAD	- Remote Magnetic Anomaly Detection
REMC	- Resin Encapsulated Mica Capacitor
REMG	- Radioelectromyograph
REMS	- Registered Equipment Management System
REO	- Regenerated Electrical Output
REON	- Rocket Engine Operation—Nuclear
REP	- Range Error Probable
	- Recovery and Evacuation Program
	- Rendezvous Exercise Pod
	- Roentgen Equivalent Physical
REPCAT	- Report Corrective Action Taken
REPP	- Request for Process Planning
REPPAC	- Repetitively Pulsed Plasma Accelerator
REPT	- Reference Engineering Photographic Template

- REQANS

REQANS	- Request Answer by
REQIBO	- Request Item be Placed on Back Order
RER	- Radiation Effects Reactor
RERL	- Residual Equivalent Return Loss
RES	- Restraint System
RESA	- Research Society of America
RESCAN	- Reflecting Satellite Communication Antenna
RESCAP	- Rescue Combat Air Patrol
RESCU	- Rocket-Ejection Seat Catapult Upward
RESDAT	- Restricted Data—Atomic Energy Act of 1954
RESOJET	- Resonant Pulse Jet
RESP	- Regulated Electrical Supply Package
RESREP	- Resident Representative
RESS	- Radar Echo Simulation System
	- Radio Echo Simulation Study
REST	- Re-Entry Environment and Systems Technology
RETEST	- Reinforcement Testing for System Training
RETL	- Rocket Engine Test Laboratory
RETMA	- Radio, Electronics, and Television Manufacturers Association (now EIA)
RETORC	- Research Torpedo Configuration
RETROF	- Retro Fire
REV	- Re-Entry Vehicle
REVEL	- Reverberation Elimination
REV/MIN	- Revolutions per Minute
REVOCON	- Remote Volume Control
RF	- Radio Frequency
	- Raised Face
	- Range Finder
	- Rapid Fire
	- Read Forward
	- Reconnaissance Fighter
	- Recovering Force
	- Register Finder
	- Replacement Factor
RFA	- Radio Frequency Amplifier
	- Request for Alteration (Request for Change preferred)
	- Request for Authority
	- Reserve Forces Act
RFC	- Radio Facility Chart
	- Radio Frequency Chart
	- Request for Change
RFCO	- Range Facilities Control Officer
RFD	- Reactor Flight Demonstration

RFE	- Radio Free Europe
	- Request for Estimate
RFECM	- Revised for Engineering Change Memorandum
RFEI	- Request for Engineering Information
RFFD	- Radio Frequency Fault Detection
RFG	- Receive Format Generator
RFI	- Radio Frequency Interference
	- Ready for Issue
	- Request for Information
	- Request for Inspection
RFIT	- Radio Frequency Interference Test
RFL	- Radio Frequency Laboratories
RFM	- Reactive Factor Meter
	- Refueling Mission
	- Reliability Figure of Merit
RFNA	- Red Fuming Nitric Acid
RFO	- Reason for Outage
	- Research Fiscal Office
RFP	- Request for Proposal
RFQ	- Request for Quotation
RFS	- Ready for Sea
	- Regardless of Feature Size
RFSS	- Reliability Failure Summary Support
RF/TFX	- Reconnaissance Fighter/Tactical Fighter Experimental
RFU	- Reliability Field Unit
RFWAR	- Requirements for Work and Resources
RG	- Rate Gyroscope
	- Reception Good
	- Recording Unit
	- Reset Gate
	- Reticulated Grating
	- Reverse Gate
R/G	- Radiation Guidance
RGA	- Range—Gemini to Agena
	- Residual Gas Analyzer
RGM	- Recorder Group Monitor
RGP	- Rate Gyro Package
RGS	- Radio Guidance System
RGZ	- Recommended Ground Zero
RH	- Relative Humidity
	- Right Hand
	- Rockwell Hardness
	- Round Head
RHA	- Records Holding Area
RHEO	- Rheostat
RHFEB	- Right Hand Forward Equipment Bay

■ RHI

RHI	- Radar Height Indicator
	- Range Height Indicator
	- Round Hill Installation
R/HR	- Roentgens per Hour
RHSC	- Right Hand Side Console
RHW	- Right Hand Word
RI	- Radar Input
	- Radar Intercept
	- Radio Inertial
	- Radio Interference
	- Range Instrumentation
	- Receiving Inspection
	- Reflective Insulation
	- Re-Initiate
	- Requires Identification
	- Rubber Insulation
R&I	- Receiving and Inspection
RIA	- Reset Indicators from Accumulator
	- Rock Island Arsenal
RIAA	- Recording Industry Association of America
RIAL	- Runway Identifiers and Approach Lighting
RIAS	- Research Institute for Advanced Studies
RIC	- Radar Input Control
	- Range Instrumentation Conference
RICASIP	- Research Information Center and Advisory Service on Information Processing
RICMO	- Radar Input Countermeasurers Officer
RICMT	- Radar Input Countermeasurers Technician
RID	- Radar Input Drum
	- Range Instrumentation Division
	- Reset Inhibit Drum
RIDD	- Range Instrumentation Development Division
RIE	- Refrigeration Installation Equipment
RIF	- Reduction in Force
	- Reliability Information File
RIFI	- Radio Interference Field Intensity
	- Radio Interference Free Instrument
RIFM	- Reliability Information File Method
RIFT	- Reactor In-Flight Test
RIFTS	- Reactor In-Flight Test System
RIGS	- Radio Inertial Guidance System
	- Runway Identifiers with Glide Slope
RIL	- Radio Interference Level
	- Reset Indicators of the Left Half
RILEM	- International Conference of Research and Testing Laboratories

RIM	- Radar Input Mapper
	- Radar Input Monitor
	- Radiant Intensity Measurement
	- Receipt, Inspection, and Maintenance
RIME	- Radio Inertial Monitoring Equipment
RIN	- Record Identification Number
	- Regulus Inertial Navigation
RINSMAT	- Resident Inspector of Naval Material
RIOMETER	- Relative Ionospheric Opacity Meter
RIOT	- Real-Time Input-Output Transducer
	- Real-Time Input-Output Translator
	- Resolution of Initial Operational Technique
RIP	- Recoverable Item Program
	- Relationship Improvement Program
RIPCO	- Receiving Inspection and Preparation for Checkout
RIPS	- Radar Impact Prediction System
	- Radio Isotope Power System
	- Range Instrumentation Planning Study
RIQAP	- Reduced Inspection Quality Assurance Program
RIR	- Receiving Inspection Report
	- Re-Inspection Report
	- Reliability Investigation Request
	- Reset Indicators of the Right Half
RIS	- Reset Indicators from Storage
RISE	- Research in Supersonic Environment
RISP	- Recoverable Interplanetary Space Probe
RISS	- Range Instrumentation and Support System
RIT	- Radar Inputs Test
	- Radio Network for Inter-American Telecommunications
	- Receiving and Inspection Test
RITA	- Reuseable Interplanetary Transport Approach Vehicle
RITE	- Rapid Information Technique for Evaluation
RIV	- Radio-Influence Voltage
R/J	- Ramjet
RJA	- Ramjet Addition
RJDS	- Reaction Jet Drivers
RJS	- Reaction Jet System
RKV	- Rose Knot Victory
RL	- Reference Line
	- Resistance Inductance
	- Right Label
	- Rocket Launcher
RLC	- Resistance-Inductance-Capacitance
	- Resistor-Inductor-Capacitor
RLCS	- Radio Launch Control System
RLE	- Research Laboratory for Electronics

■ RLI

RLI	- Red-Line Instrumentation
RLM	- Rearward Launched Missile
RLT	- Regional Landing Team
	- Rolling Liquid Transporter
RLTS	- Radio Linked Telemetry System
RM	- Radar Mapper
	- Radar Missile
	- Radiation Measurement
	- Radioman
	- Radio Monitor
	- Range Marks
	- Reaction Mass
	- Readout Material
	- Receiver, Mobile
	- Response Motivation
RMA	- Radio Manufacturers Association
RMAX	- Range Maximum
RMD	- Reaction Motors Division
RMG	- Radar Mapper Gap Filler
RMI	- Radio Magnetic Indicator
	- Reaction Motors, Incorporated
	- Reliability Maturity Index
RML	- Radar Mapper—Long-Range
RMM	- Radar Map Matching
RMP	- Rated Maximum Pressure
	- Resident Manufacturing Plan
	- Rocket Motor Propellant
RMPR	- Rated Mobilization and Professional Resource
RMR	- Regional Maintenance Representative
RMS	- Root Mean Square
RMSE	- Root Mean Square Error
RMU	- Remote Maneuvering Unit
RN	- Radio Navigation
	- Revision Notice
	- Reynolds Number
RNA	- Ribonucleic Acid
RNF	- Radio Noise Figure
RNIT	- Radio Noise Interference Test
RNV	- Radio Noise Voltage
RO	- Radio Operator
	- Range Operations
	- Receive Only
	- Recovery Operations
	- Reliability Office
	- Requisitioning Objective
	- Rough Opening
	- Route Order

R/O	- Receive Only
ROAD	- Reorganization of Objectives in Army Divisions
ROAMA	- Rome Air Materiel Area
ROAR	- Royal Optimizing Assembly Routine
ROB	- Radar Out of Battle
ROBIN	- Rocket Balloon Instrument
ROBO	- Rocket Bomber
ROC	- Range Operations Conference
	- Receiver Operating Characteristics
	- Required Operational Capability
ROCAT	- Rocket Catapult
ROCC	- Range Operations Conference Circuit
ROCKOON	- Rocket Launched from Balloon
ROCP	- Radar Out of Commission for Parts
ROD	- Range Operations Department
	- Range Operations Directive
ROD/AC	- Rotary Dual Input for Analog Computation
RODO	- Range Operations Duty Officer
ROE	- Reflector Orbital Equipment
ROFT	- Radar Off Target
ROG	- Rise Off Ground
ROIC	- Resident Officer In Charge
ROICC	- Resident Officer In Charge of Construction
ROINST	- Range Operations Instructions
ROJ	- Range of Jamming
ROK	- Republic of Korea
ROM	- Rough Order of Magnitude
ROMAC	- Range Operations Monitor Analysis Center
ROMACC	- Range Operational Monitoring and Control Center
ROMBUS	- Reusable Orbital Module-Booster and Utility Shuttle
ROMOTAR	- Range Only Measurement of Trajectory and Recording
ROO	- Range Operations Officer
ROOST	- Reusable One-Stage Orbital Space Truck
ROOT	- Relaxation Oscillator Optically Tuned
ROP	- Roster Option
	- Rotating Observation Platform
ROPS	- Range Operations Performance Summary
ROR	- Rate of Return
	- Rocket On Rotor
RO/RO	- Roll On—Roll Off
ROS	- Range Operations Supervisor
	- Reduced Operational Status
ROSE	- Rising Observational Sounding Equipment
ROSIE	- Reconnaissance by Orbiting Ship-Identification Equipment
ROT	- Radar on Target
ROTAB	- Rotary Table

■ ROTAC

ROTAC	- Rotary Oscillating Torque Actuators
ROTC	- Reserve Officers Training Corps
ROTI	- Recording Optical Tracking Instrument
ROTR	- Receive-Only Tape Reperforator
RP	- Real Property
	- Release Point
	- Reporting Post
	- Rocket Projectile
	- Rocket Propellant
RP-1	- Rocket Propellant No. 1 (kerosene)
RPA	- Radar Performance Analyzer
RPBG	- Revised Program and Budget Guidance
RPC	- Radar Planning Chart
	- Radar Processing Center
	- Remote Position Control
	- Resistance Products Company
RPD	- Radar Planning Device
	- Radar Prediction Device
	- Research Projects Division
RPDG	- RAND Program Development Group
RPE	- Range Planning Estimate
	- Related Production Equipment
	- Rocket Propulsion Establishment
RPEP	- Register of Planned Emergency Producers
RPF	- Radiometer Performance Factor
RPI	- Radar Precipitation Integrator
	- Rensselaer Polytechnic Institute
RPIE	- Real Property Installed Equipment
RPIO	- Registered Publication Issuing Office
RPL	- Running Program Language
RPM	- Reliability Performance Measure
	- Remote Performance Monitoring
	- Revolutions per Minute
RPMC	- Remote Performance Monitoring and Control
RPMI	- Revolutions per Minute Indicator
RPMIO	- Registered Publication Mobile Issuing Office
RPO	- Range Planning Office
	- Rotor Power Output
RPP	- Radar Power Programmer
	- Recurrence Prevention Program
	- Request for Proposal Preparation
	- Requisition Processing Point
RPPI	- Remote Plan Position Indicator
RPQ	- Request for Price Quotation
RPR	- Read Printer

R/S ■

RPS	- Revolutions per Second
	- Rotating Passing Scuttle
RPU	- Retention Pending Use
RPVT	- Relative Position Velocity Technique
RQS	- Ready Qualified for Standby
RR	- Rapid Rectilinear
	- Receiving Report
	- Recovery Room
	- Recurrence Rate
	- Rendezvous Radar
	- Repetition Rate
	- Respiration Rate
	- Round Robin
R/R	- Railroad
	- Readout and Relay
	- Record-Retransmit
	- Remove and Replace
R&R	- Refueling and Rearming
	- Remove and Replace
	- Repair and Retrofix
	- Routing and Recording
RRE	- Radar Research Establishment
RRF	- Retro-Rocket Fire
RRI	- Range Rate Indicator
RRR	- Rework Removal Rate
RRS	- Reaction Research Society
	- Restraint Release System
	- Retro-Rocket System
RRSTRAF	- Ready Reserve Strategic Army Forces
RRT	- Requirements Review Team
RR/T	- Rendezvous Radar/Transponder
RS	- Radar Simulator
	- Radar Start
	- Range Safety
	- Range Selector
	- Remote Station
	- Report of Survey
	- Reserve Stock
	- Resident School
	- Return to Situation
	- Reverberation Strength
	- Right Sign
R/S	- Range Safety
	- Range Surveillance
	- Regulating Station
	- Routing Slip

■ RSA

RSA	- Range Safety Approval
	- Rate Subsystem Analyst
	- Redstone Arsenal
RSBS	- Radar Safety Beacon System
RSC	- Range Safety Command
	- Range Safety Control
	- Release Schedule Code
	- Resident Shop Control
RS&C	- Reliability Surveillance and Control
RSCIE	- Remote Station Communication Interface Equipment
RSCS	- Rate Stabilization and Control System
RSCSS	- Range Safety Command Shutdown System
RSD	- Range Support Directive
	- Reliability Status Document
	- RUNCIBLE System Duplexer
RSDC	- Range Safety Data Coordinator
RSDP	- Remote Site Data Processor
RSDS	- Range Safety Destruct System
RSH	- Radar Status History
RSI	- Replacement Stream Input
	- Research Studies Institute
RSIUFL	- Release Suspension for Issue and Use of Following
RSL	- Reconnaissance and Security Line
RSLA	- Range Safety Launch Approval
RSM	- Radio Squadron, Mobile
	- Reconnaissance Strategic Missile
RSO	- Radiosonde Observation
	- Range Safety Officer
	- Reproduction Service Order
	- Research Satellite for Geophysics
RSO/MFSO	- Range Safety Officer/Missile Flight Safety Officer
RSOP	- Range Safety Operational Plan
RSP	- Radio Switch Planel
	- Range Support Plan
	- Reconnaissance and Security Positions
	- Route Selection Program
RSPP	- Radio Simulation Patch Panel
RSR	- Right Element Shift Right
RSS	- Range Safety Switch
	- Reactants Supply System
	- Reactive System Sensitivity
	- Root Sum Square
RST	- Read Symbol Table
	- Recording Specification Table
	- Reinforcing Steel
	- Request for Scheduling Test

RST	- Requirements for Scheduled Test
	- Right Store
	- Routine Sequence Table
RSV	- Ready Storage Vessel
RT	- Radio Telephone
	- Range Time
	- Range Tracking
	- Ratio Transfer
	- Reaction Time
	- Real Time
	- Receiver-Transmitter
	- Receive-Transmit
	- Reduction Table
	- Rejection Tag
	- Reperforator-Transmitter
	- Resistance Thermometer
	- Retro Table
	- Rocket Target
	- Room Temperature
	- Routine Tag
	- Rubber Tired
R/T	- Radio Telephone
	- Real Time
	- Rejection Tag
R-T	- Reperforator-Transmitter
R&T	- Research and Technology
RTA	- Reliability Test Assembly
	- Rise Time Analyzer
RTAG	- Range Technical Advisory Group
RTB	- Read Tape Binary
	- Resistance Temperature Bridge
	- Return to Base
RTC	- Radar Tracking Center
	- Radar Tracking Control
	- Real Time Command
	- Real Time Computer
	- Replacement Training Center
RTCA	- Radio Technical Commission for Aeronautics
RTCC	- Real Time Computer Complex
RTCS	- Real Time Computation System
RTCU	- Real Time Control Unit
RTD	- Range Time Decoder
	- Read Tape Decimal
	- Real Time Display
R&TD	- Research and Technology Division (of AFSC)

■ RTDD

RTDD	- Real Time Data Distribution
	- Remote Timing and Data Distribution
RTDDC	- Real Time Digital Data Correction
RTDHS	- Real Time Data Handling System
RTE	- Regenerative Turboprop Engine
RTEM	- Radar Tracking Error Measurement
RTF	- Rocket Test Facility
RTG	- Radioisotope Thermoelectric Generator
	- Reconnaissance Technical Group
RTGS	- Return to Government Stores
RTM	- Rapid Tuning Magnetron
	- Receiver-Transmitter-Modulator
	- Reconnaissance Tactical Missile
	- Running Time Meter
RTMA	- Radio and Television Manufacturers Association
RTN	- Return to Neuter
RTP	- Remote Transfer Point
RTQC	- Real Time Quality Control
RTS	- Radar Target Simulator
	- Radar Tracking Station
	- Range Time Signal
	- Remote Targeting System
RTSA	- Radio Tracking System Analyst
RTT	- Radiation Tracking Transducer
	- Radio Teletypewriter
RTTV	- Real Time Television
RTTY	- Radio Teletypewriter
RTU	- Recovery Task Unit
	- Replacement Training Unit
RTV	- Re-Entry Test Vehicle
	- Research Test Vehicle
	- Return to Vender
	- Rocket Test Vehicle
	- Room Temperature Vulcanizing
RTW	- Round the World
RTWS	- Raw Tape Write Submodule
RU	- Range User
	- Reproducing Unit
R&U	- Repairs and Utilities
RUG	- Recomp Users Group
RUM	- Remote Underwater Manipulator
RUMP	- Radio-Controlled Ultraviolet Measurement Program
	- Remote Underwater Marine Probe
RUNCIBLE	- Revised Unified New Compiler with IT Basic Language Extended
RUSDIC	- Russian Dictionary

RUSPAND	- Russian-Spanish Dictionary
RUSTAN	- Russian Text Analyzer
RV	- Rated Voltage
	- Re-Entry Vehicle
	- Relief Valve
R/V	- Re-Entry Vehicle
RVA	- Reactive Volt-Ampere Meter
RVB	- Rotation-Vibration Band
RVD	- Radius Vector Subroutine
RVDP	- Radar Video Data Processor
RVDT	- Rotary Variable Differential Transformer
RV/GC	- Re-Entry Vehicle and Ground Control
RVM	- Reactive Voltmeter
RVP	- Reid Vapor Pressure
	- Roll Vertical Pendulum
RVR	- Runway Visibility Range
RVS	- Re-Entry Vehicle Separation
RW	- Radiological Warfare
	- Read-Write
R/W	- Rework
RWASG	- Radiation Weapons Analysis Systems Group
RWC	- Read, Write, Compute
RWK	- Rework
RX	- Resolver-Transmitter
RZ	- Return to Zero
R&Z	- Range and Zero
RZL	- Return-to-Zero Level
RZM	- Return-to-Zero Mark

S

S	- Search
	- Secret
	- Side
	- Silk
	- Single
	- Slate
	- Soft
	- South
	- Stack
	- Sulfur
	- Switch
S-	- Saturn Configuration (Arabic Numerals)
	- Saturn Stage (Roman Numerals)
SA	- Saturn Apollo
	- Secretary of the Army
	- Semi-Automatic
	- Sense Amplifier
	- Sensible Atmosphere
	- Service Area
	- Shaft Alley
	- Shaft Angle
	- Small Arms
	- Spectrum Analyzer
	- Spin Axis
	- Springfield Armory
	- Standard Atmosphere
	- Subassembly
	- Supplementary Agreement
	- Support Activity
	- Support Agency
S/A	- Semi-Automatic
	- Spacecraft Adapter
	- Subaccount

S/A	- Subassembly
S&A	- Safe and Arm
	- Safety and Arming
	- Science and Applications
SAA	- Senior Army Advisor
	- Small Arms Ammunition
SAAB	- Svenska Aeroplan Aktiebolaget
SAAC	- Swiss American Aircraft Corporation
SAAD	- San Antonio Air Depot
SAAMA	- San Antonio Air Materiel Area
SAB	- Scientific Advisory Board
	- Society of American Bacteriologists
	- Solid Assembly Building
	- Systems Analysis Branch
SABCA	- Société Anonyme Belge de Constructions Aeronautiques (Belgium)
SABE	- Society for Automation in Business Education
SABL	- Serialized Assembly Breakdown List
SA-BO	- Sense Amplifier-Blocking Oscillator
SABOC	- SAGE BOMARC
SABOCC	- SAGE BOMARC Coordinating Committee
SABRE	- SAGE Battery Routing Equipment
	- Secure Airborne Radar Equipment
	- Self-Aligning Boost and Re-Entry System
SABRES	- Simulated ATABE Reply-Back System
SABU	- SAGE Backup (now BUIC)
SAC	- Semi-Automatic Coding
	- Strategic Air Command
	- Supplemental Air Carrier
	- Supreme Allied Commander
	- Synchronous Astro Compass
SACA	- Supreme Allied Command, Atlantic
SACAD	- Strategic Air Command Addressee Designator
SACC	- Supporting Arms Coordination Center
SACCEI	- Strategic Air Command Communications-Electronics Instruction
SACCOMNET	- Strategic Air Command Communications Network
SACCS	- Strategic Air Command Control System
SACEUR	- Supreme Allied Commander, Europe (NATO)
SACLANT	- Supreme Allied Commander, Atlantic (NATO)
SACM	- Strategic Air Command Manual
SAC/M	- Strategic Air Command Missiles
SACOM	- Ship's Advanced Communications
SACTO	- Sacramento Test Operations
SAD	- Site Activation Division
	- Situation Attention Display

■ SAD

SAD	- Special Adapter Device
	- Support Air Direction
	- Sympathetic Aerial Detonation
SADD	- Surface Anti-Submarine Development Detachment
SADGE	- SAGE Data Generator
SADIC	- Solid-State Analog-to-Digital Computer
SADIE	- Scanning Analog-to-Digital Input Equipment
	- Semi-Automatic Decentralized Intercept Environment
SADOPS	- Ships Angle-Tracking and Doppler System
SADR	- Six-Hundred-Megacycle Air Defense Radar
SADSAC	- Sample Data Simulator and Computer
SADTC	- SHAPE Air Defense Technical Center
SAE	- Shaft Angle Encoder
	- Shop Assisted Engineering
	- Society of Automotive Engineers
SAESIP	- Saturn-Apollo Electrical Systems Integration Panel
SAF	- Secretary of the Air Force
	- Spacecraft Assembly Facility
	- Strategic Air Force
SAFE	- Strategy and Force Evaluation
SAFGC	- Secretary of the Air Force General Council
SAFIS	- Secretary of the Air Force for Informational Services
SAFMA	- Secretary of the Air Force for Materiel
SAFMS	- Secretary of the Air Force for Missile and Satellite Systems
SAFO	- Safe Altitude Fuzing Option
SAFRD	- Secretary of the Air Force for Research and Development
SAFS	- Secondary Air Force Speciality
SAG	- System Application Group
SAGA	- System for Automatic Generation and Analysis
SAGCI	- Semi-Automatic Ground Control of Interceptors
SAGE	- Semi-Automatic Ground Environment
SA/GO	- Subaccount/General Order
SAHF	- Semi-Automatic Height Finder
SA&I	- Safety, Arming, and Initiation
SAIC	- Switch Action Interrupt Count
SAID	- Semi-Automatic Initiation and Deghosting
SAIM	- Semi-Automatic Inserting Machine
SAINT	- Satellite Inspector
SAIS	- South African Interplanetary Society
SAL	- San Salvador Island (remote site)
	- Sea-Animal Locomotion
	- Supersonic Aerophysics Laboratory
SALDRI	- Semi-Automatic Low-Data-Rate Input
SALE	- Simple Algebraic Language for Engineers
SALG	- Service Analysis and Liaison Group
SALON	- Satellite Balloon

SALTI	- Summary Accounting for Low-Dollar Turnover Items
SAM	- School of Aerospace Medicine
	- Selective Automatic Monitoring
	- Simulated Assignment Mode
	- Society for the Advancement of Management
	- Sort and Merge
	- Special Air Mission
	- Surface-to-Air Missile
	- Synchronous Amplitude Modulation
SAMA	- Sacramento Air Materiel Area (SMAMA preferred)
SAMAP	- Southern Air Materiel Area, Pacific
SAME	- Society of American Military Engineers
SAMOS	- Satellite-Missile Observation System
SAMPE	- Society of Aerospace Materials and Process Engineers
SAMS	- Satellite Automatic Monitoring System
SAM-SAC	- Specialized Aircraft Maintenance—Strategic Air Command
SANE	- Scientific Applications of Nuclear Explosions
SAO	- Smithsonian Astrophysical Observatory
	- Support Air Observation
SAOC	- Space and Aeronautics Orientation Class
SAP	- Semi-Armor-Piercing
	- Share Assembly Program
	- Symbolic Address Program
SAPIR	- System for the Automatic Processing and Indexing of Reports
SAPO	- Special Aircraft Project Office
SAR	- Search and Rescue
	- Semi-Automatic Rifle
	- Solar Aircraft Company
	- Special Aeronautical Requirement
	- Storage Address Register
	- Submarine Advance Reactor
	- System Analysis Report
SARA	- Ship Angle and Range
SARAC	- Steerable Array for Radar and Communications
SARAH	- Search and Rescue and Homing
	- Semi-Automatic Range, Azimuth, and Height
SARCAP	- Search and Rescue—Combat Air Patrol
SARCC	- Search and Rescue Control Center
SARD	- Simulated Aircraft Radar Data
	- Special Airlift Requirement Document
	- Support and Range Development
SARI	- Small Airport Runway Indicator
SARP	- Scramble and Recovery Procedure
SARUS	- Search and Rescue Using Satellites
SAS	- Scandinavian Airlines System
	- Secondary Alerting System

■ SAS

SAS	- South African Station
	- Stability Augmentation System
	- Supersonic Attack Seaplane
SASO	- Senior Air Surveillance Officer
SASP	- Special Ammunition Supply Point
SASS	- SAGE ATABE Simulation System
	- Strategic Alert Sound System
SAST	- Senior Air Surveillance Technician
SASTRO	- SAGE Strobe Training Operator
SAT	- Solar Atmospheric Tide
	- Squadron Allocation Table
SATAF	- Site Activation Task Force
SATAN	- Satellite Active Nullifier
	- Sensor for Airborne Terrain Analysis
SATAR	- Satellite for Aerospace Research
SAT-COM	- Satellite Communication
SATCOMA	- Satellite Communications Agency
SATE	- Semi-Automatic Test Equipment
SATELLAB	- Satellite Laboratory
SATF	- Strategic Area Task Force
SATIC	- Scientific and Technical Information Center
SATIF	- Scientific and Technical Information Facility
SATIN	- SAGE Air Traffic Integration
	- Satellite Inspection Technique
SATIRE	- Semi-Automatic Technical Information Retrieval
S-ATK	- Strike Attack
SATOBS	- Satellite Observations
SATP	- Single Aircraft Tracking Program
SATPATT	- Satellite Paper Tape Transfers
SATRAC	- Satellite Automatic Terminal Rendezvous and Coupling
SATS	- Small Airfields for Tactical Support
SATSA	- Signal Aviation Test and Support Activity
SAU	- Search Atlantic Unit
	- Surface Attack Unit
SAV	- Space-Air Vehicle
SAVDAT	- Save Data
SAVE	- System Analysis of Vulnerability and Effectiveness
SAW	- Simulate Antiaircraft Weapons
SAWC	- Special Air Warfare Center
SAWE	- Society of Aeronautical Weight Engineers
SAWG	- Schedule and Allocations Working Group
SAZO	- Seeker—Azimuth Orientation
SB	- Secondary Battery
	- Selection Board
	- Serial Binary
	- Sleeve Bearing

SB	- Space Booster
	- Splash Block
	- Status Board
	- Storage Building
	- Straight Binary
	- Submarine Base
	- Supply Bulletin
	- Surveillance Branch
	- Switchboard
S/B	- Serial Block
S&B	- Sterilization and Bath
SBA	- Small Business Administration
	- Standard Beam Approach
SBAMA	- San Bernardino Air Materiel Area
SBB	- Silicon-Borne Bonds
SBE	- Sub BIT Encoder
SBGP	- Strategic Bomber Group
SBI	- Santa Barbara Island
	- Satellite-Borne Instrumentation
SBIC	- Small Business Investment Company
SBM	- Subtract Magnitude
SBOS	- Silicon-Borne Oxygen System
SBP	- Simulated BOMARC Program
SBS	- Single Business Service
SBT	- Sheet, Bar, Tubing
	- Surface-Barrier Transistor
SBTC	- Search, Bomb and Terrain Clearance
S/BTCC	- SAGE/BOMARC Technical Coordination Commission
SBTM	- S-Band Telemetry Modification Kit
SBU	- Skirt Buildup
SBUE	- Switch—Backup Entry
SBX	- S-Band Transponder
SC	- Sandia Corporation
	- Sector Commander
	- Senior Controller
	- Separate Cover
	- Sequence Chart
	- Service Ceiling
	- Shaped Charge
	- Signal Conditioner
	- Signal Corps
	- Single Contact
	- Situation Console
	- Solar Cell
	- Solar Constant
	- Spacecraft

■ SC

SC	- Spacecraft Communicator
	- Specification Control
	- Statistical Control
	- Subcaliber
	- Superimposed Current
	- Supply Corps
S/C	- Spacecraft
	- Subcable
S&C	- Stabilization and Control
	- Strategic and Critical
SCA	- Sequence Control Area
	- Shipbuilders Council of America
	- Simulation Control Area
	- Spacecraft Adapter
S/CA	- Spacecraft Adapter
SCADAR	- Scatter Detection and Ranging
SCADS	- Sioux City Air Defense Sector
SCAMP	- Standard Configuration and Modification Program
	- Strategic Air Command Assembly Program
SCAN	- Self-Correcting Automatic Navigator
	- Switched Circuit Automatic Network
SCANS	- Scheduling and Control by Automated Network Systems
	- Systems Checkout Automatic Network Simulator
SCAP	- Silent, Compact Auxiliary Power
	- Supreme Commander, Allied Powers
SCAPE	- Self-Contained Atmospheric Protective Ensemble
SCAR	- Special Committee on Antarctic Research
	- Subcaliber Aircraft Rocket
	- Submarine Celestial Altitude Recorder
SCARF	- Side-Looking Coherent All-Range Focused
	- Special Committee for Adequacy of Range Facilities
SCAS	- Southwest Center of Advanced Studies
	- Spacecraft Adapter Simulator
SCASP	- Sequence of Coverage and Speed
SCAT	- Security Control of Air Traffic
	- Share Compiler-Assembler-Translator
	- Space Communications and Tracking
	- Speed Control Approach/Take-Off
	- Submarine Classification and Tracking
	- Supersonic Commercial Air Transport
SCATE	- Stromberg-Carlson Automatic Test Equipment
SCATER	- Security Control of Air Traffic and Electromagnetic Radiation
SCATS	- Sequentially-Controlled Automatic Transmitter Start
	- Simulation, Checkout, and Training System
SCC	- SAGE Control Center

SCC	- Satellite Control Center
	- Security Control Center
	- Sequence Control Chart
	- Simulation Control Center
	- Slice Control Central
	- Standard Commodity Classification
	- Super-Critical Cryogenics
	- Supply Control Center
SCCC	- SPASUR Command and Control Center
SCCN	- Subcontract Change Notice
SCCO	- Security Classification Control Officer
SCCPG	- Support and Combat Center Production Group
SCD	- Satellite Control Department
	- Source Control Drawing
	- Specification Control Drawing
	- Subcarrier Discriminator
SCE	- Schedule Compliance Evaluation
	- Selection Control Element
	- Service Checkout Equipment
	- Signal Conditioning Equipment
	- Stage Calibration Equipment
SCEL	- Signal Corps Engineering Laboratory
SCEO	- Satellite Control Engineering Office
SCEPTRON	- Spectral Comparative Pattern Recognizer
SCF	- Satellite Control Facility
	- Sequence Compatibility Firing
	- Support Carrier Force
	- System Chance Failure
SCFH	- Standard Cubic Feet per Hour
SCFM	- Standard Cubic Feet per Minute
SCFS	- Standard Cubic Feet per Second
SCG	- Specification Control Group
SCGD	- Specification Control Group Directive
SCGSS	- Super-Critical Gas Storage System
SCH	- Sector Command Headquarters
	- Store Channel
SCI	- Santa Cruz Island
	- Scientific Computers, Incorporated
	- Ship Controlled Intercept
SCICS	- Semiconductor Integrated Circuits
SCIEC	- Southern California Industry-Education Council
SCIP	- Self-Contained Instrument Package
SCIPP	- Silicon-Computing Instrument Patch-Programmed
SCL	- Selectivity Cross-Linked
	- Standard Classification List
	- Symbolic Correction Loader

■ SCM

SCM	- Simulated Command Module
	- Smith-Corona-Marchant, Incorporated
SCN	- Satellite Control Network
	- Sensitive Command Network
	- Specification Change Notice
SCNA	- Sudden Cosmic-Noise Absorption
SCNAWAF	- Special Category Navy with Air Force
SCN/SIN	- Sensitive Command Network/Sensitive Information Network
SCNT	- Sensitive Command Network Test
SCO	- Statistical Control Office
	- Subcarrier Oscillator
	- Sustainer Engine Cutoff (SECO preferred)
S/CO	- Spacecraft Observer
SCOB	- Scattered Clouds or Better
SCODA	- Scan Coherent Doppler Attachment
SCOOP	- Scientific Computation of Optimum Procurement
SCOOT	- SAC Cooperative Testing
SCOPE	- Sequential Customer Order Processing, Electronically
	- Stromberg-Carlson Operations Panel—Electrical
	- Subsystem for the Control of Operations and Plan Evaluation
	- System to Coordinate the Operation of Peripheral Equipment
SCOPT	- Subcommittee on Programming Technology
SCOR	- Special Committee on Oceanographic Research
SCORE	- Satellite Computer-Operated Readiness Equipment
	- Signal Communications by Orbiting Relay Equipment
	- Systematic Control of Range Effectiveness
SCOT	- Standby Compatible One Tape
SCP	- SAGE Change Proposal
	- SAGE Computer Project
	- Sector Command Post
	- Security Classification Procedure
	- Specification Change Proposal
	- Spherical Candlepower
	- Survey Control Point
SCPA	- SCS (Stabilization and Control System) Control Panel
SCPT	- SAGE Computer Programming Training
	- Security Control Point
SCR	- Signal Conditioner
	- Signal Corps Radio
	- Silicon Controlled Rectifier
	- Spares Coordination Record
	- Strip Chart Recorder
	- Subcontractor
SCRAM	- Selective Combat Range Artillery Missile
	- Several Compilers Reworked and Modified
	- Space Capsule Regulator and Monitor

SCRAM	- Supersonic Combustion Ramjet
SCRAMJET	- Supersonic Combustion Ramjet
SCRIPT	- Scientific and Commerical Subroutine Interpreter and Program Translator
SCS	- Secret Control Station
	- Security Control System
	- Silicon Controlled Switch
	- Single Change of Station
	- Single Channel Simplex
	- S/C (Spacecraft) Control System
	- Stabilization and Control System
SCSCO	- Secure Submarine Communications
SCSCP	- System Coordination for SAGE Computer Programming
SCSG	- SAGE Computer Support Group
SC/SM	- Spacecraft Systems Monitor
SCT	- Scanning Telescope
SC/TC	- Spacecraft Test Conductor
SCTCA	- SAC Channel and Traffic Control Agency
SCTF	- Santa Cruz Test Facility
SCU	- Scanner Control Unit
	- Signal Conditioning Unit
	- Static Checkout Unit
	- Statistical Control Unit
SCUBA	- Self-Contained Underwater Breathing Apparatus
SCV	- Sub-Clutter Visibility
SCVE	- Spacecraft Vicinity Equipment
SCWS	- Space Combat Weapon System
SD	- Secretary of Defense
	- Selenium Diode
	- Self-Destroying
	- Senior Director
	- Shell Destroying
	- Situation Display
	- Spaced Doublet
	- Special Duty
	- Subject Division
	- Sum of Digits
	- Sweep Driver
	- System Demonstration
S/D	- Secretary of Defense
	- Signal to Distortion
S&D	- Scheduling and Distribution
SDA	- Shaft Drive Axis
	- Source Data Automation
	- Supporting Data Analysis
	- Systems Dynamic Analyzer

■ SDAD

SDAD	- Satellite Digital and Analog Display
SDAS	- Scientific Data Automation System
SDAT	- Spacecraft Data Analysis Team
SDC	- SAGE Direction Center
	- SAGE Division Commander
	- Self-Destruct Circuit
	- Shipment Detail Card
	- Signal Data Converter
	- Situation Display Converter
	- Station Directory Control
	- System Development Corporation
SDCOI	- Sector Direction Center Operating Instruction
SDCP	- Supply Demand Control Points
SDCS	- Station Digital Command System
SDDU	- Simplex Data Distribution Unit
SDEG	- Special Doctrine and Equipment Group
SDF	- Single Degree of Freedom
SDFG	- Single-Degree-of-Freedom Gyroscope
SDG	- Simulated Data Generator
	- Situation Display Generator
	- Strap Down Gyroscope
SDGE	- Situation Display Generator Element
SDHE	- Spacecraft Data Handling Equipment
SDI	- Selective Dissemination of Information
	- Specially Designated Interceptor
SDIS	- Situation Display Indicator Section
SDL	- Security Devices Laboratory
	- Standard Distribution List
	- Systems Design Laboratory
SDM	- Situation Display Matrix
SDN	- SATIN Design Note
	- Separation Designation Number
SDO	- Senior Direction Officer
	- Staff Duty Officer
SDOP	- Ship Doppler
SDP	- Signal Data Processor
	- Site Data Processor
	- Station Directory Program
SDPO	- Space Defense Project Office
SDR	- Select Drum
	- Slow Death Response
	- Small Development Requirement
SDRP	- Simulated Data Reduction Program
SDRNG	- Sound Ranging
SDS	- Secret Delivery Station
	- Senior Direction Station

SDS	- Scientific Data Systems
	- Shop Direction Standard
	- Simulation Data Subsystem
	- Simulation Design Section
	- Splash Detection System
	- System Design Specification
	- Systems and Data Services
SDSS	- Self-Deploying Space Station
SDT	- Senior Director Technician
	- Shoot Down Test
	- Simulated Data Tape
	- Step Down Transformer
SDV	- Slowed Down Video
SE	- Single End
	- Single Engine
	- Southeast
	- Space Equivalent
	- Special Equipment
	- Systems Engineering
S/E	- Service Equipment
SEA	- Subterranean Exploration Agency
SEAC	- National Bureau of Standards Eastern Automatic Computer
SEADS	- Seattle Air Defense Sector
SEAL	- Sea-Air-Land
SEALS	- Sea-Air-Land System
SEAM	- Subset Extraction and Association
SEAT	- System Engineering Advisory Team
SEATAF	- Southern European Atomic Task Force
SEATO	- Southeast Asia Treaty Organization
SEC	- Scientific and Engineering Computation
	- Scientific Estimates Committee
	- Second
	- Securities and Exchange Commission
	- Southern Electronics Corporation
SECA	- Secretary of the Army
	- Segment Extended Coverage Altitude
SECAF	- Secretary of the Air Force
SECDEF	- Secretary of Defense
SECFLT	- Second Fleet
SECNAV	- Secretary of the Navy
SECO	- Self-Regulating Error Correcting Coder-Decoder
	- Sequential Control
	- Sustainer Engine Cutoff
SECOR	- Sequential Collation of Range
SECS	- Schedules and Equipment Coordination Section
SED	- Space Engineering Document

■ SED

SED	- Space Environment Division
	- Special Expanded Display
	- Student Educational Development
	- System Engineering Directorate
SEDD	- Systems Evaluation and Development Division
SEDGE	- SAGE Experimental Display Generator
	- Special Experimental Display Generation Program
SEDR	- Systems Engineering Department Report
SEE	- SAGE Evaluation Exercise
	- Situation Display Program
SEF	- Space Environmental Facility
SEFAR	- Sonic End Fire for Azimuth and Range
SEG	- Systems Engineering Group
SEIC	- Solar Energy Information Center
SELCAL	- Selective Calling
SELO	- Seeker—Elevation Orientation
SELS	- SELSYN
	- Severe Local Storm
SELSYN	- Self-Synchronous
SELT	- SAGE Evaluation Library Tape
SEM	- Subcontractor Engineering Memorandum
SEMC	- SAGE Evaluation Management Committee
SEMO	- Systems Engineering Management Organization
SENL	- Standard Equipment Nomenclature List
SENSE	- Sommers' Equivocation Network for Significant Expressions
SEODSE	- Special Explosive Ordnance Disposal Supplies and Equipment
SEP	- Scientific and Engineering Personnel
	- Space Electronic Package
	- Standard Electronic Package
SEPR	- Societe d'Etude de la Propulsion par Reaction
SEPS	- Service Module Electrical Power System
SER	- SNAP Experimental Reactor
SERB	- Study of Enhanced Radiation Belt
SEREB	- Societe pour l'Etude de la Realisation d'Engins Balistiques
SERF	- Sandia Engineering Reactor Facility
SERL	- Services Electronics Research Laboratory
SERLANT	- Service Force, Atlantic
SERPAC	- Service Force, Pacific
SERPS	- Service Propulsion System
SERT	- Space Electric Rocket Test
SERVO	- Servomechanism
SES	- Seaward Extension Simulator
	- Space Environment Simulator
	- Surface Effects Ship
	- Sylvania Electronic Systems

SESAME	- Service, Sort, and Merge
SESCO	- Secure Submarine Communications
SESE	- Secure Echo-Sounding Equipment
SET	- Sensory Evaluation Test
	- Solar Energy Thermionic
	- Systems Engineering Test
SETAF	- Southern European Task Force
SETD	- Systems Engineering and Technical Direction
SE&TD	- Systems Engineering and Technical Direction
SETE	- System Exercising for Training and Evaluation
SETEL	- Societe Europeene de Teleguidage
SETP	- Society of Experimental Test Pilots
	- Support Equipment Test Plan
	- Systems Engineering Test Program
SETS	- Solar Energy Thermionic Conversion System
SEVFLT	- Seventh Fleet
SEW	- SONAR Early Warning
SEWO	- Support Equipment Work Order
SF	- Safety Factor
	- Safety First
	- Semi-Finish
	- Shear Flow
	- Shock Front
	- Single Feed
	- Sound and Flash
	- Special Forces
	- Spot Face
	- Standard Form
	- Static Firing
SFA	- Sequential Functional Analysis
	- Simulated Flight—Automatic
	- Sun Finder Assembly
SFADS	- San Francisco Air Defense Sector
SFAR	- Sound Fixing and Ranging (SOFAR preferred)
SFB	- Structural Feedback
SFC	- Special Facilities Contract
	- Specific Fuel Consumption
SFCE	- Special Facility Contractor Equipment
SFC/OH	- Special Facilities Contract—Overhead
SFCP	- Shore Fire Control Party
SFE	- Society of Fire Engineers
SFEL	- Standard Facility Equipment List
SFF	- Straight Fixed Fee
SFG	- Special Facilities Group
SFIT	- Standard Field Installation Test
SFLZ	- Simulated Flight Size

■ SFM

SFM	- Simulated Flight—Manual
S&FM	- Space and Flight Missions
SFO	- Service Fuel Oil
SFOB	- Simulated Fuel Onboard
	- Special Forces Operational Base
SFOC	- Space Flight Operations Center
SFOD	- San Francisco Ordnance District
SFOF	- Space Flight Operations Facility
SFOM	- Space Flight Operations Memorandum
SFOP	- Space Flight Operations Plan
SFP	- Simulated Flight Plan
	- Space Flight Project
	- Sustainer Fuel Pressure
SFR	- System Failure Rate
SFS	- Saybolt Furol Seconds
	- Space Flight Systems
SFSD	- Star Field Scanning Device
SFX	- Sound Effects
SG	- Sawtooth Generator
	- Security Guard
	- Single Groove
	- Smoke Generator
	- Standing Group
	- Swamp Glider
SGA	- Standards of Grade Authorization
SGC	- Simulated Generation Control
	- Space-General Corporation
	- Sperry Gyroscope Company
SGITS	- Spacecraft/GOSS Interface Test System
SGLO	- Standing Group Liaison Officer
SGLS	- Space-to-Ground Link Subsystem
SGM	- Spark Gap Modulation
SGN	- Standing Group NATO
SGREP	- Standing Group Representative
SGS	- Secretary of the General Staff
SGSP	- Single Groove, Single Petticoat
SGST	- Society for the General Systems Theory
SH	- Scale Height
	- Scleroscope Hardness
	- Shipping
SHA	- Sidereal Hour Angle
SHACO	- Shorthand Coding System
SHAEF	- Supreme Headquarters Allied Expeditionary Force
SHAPE	- SAGE High-Altitude Prototype Environment
	- Supreme Headquarters Allied Powers, Europe
SHEP	- Solar High Energy Particles

SHF	- Super High Frequency (band 10)
SHIP	- Search-Height Integration Program
SHIPINSTL	- Ship Installation
SHIRAN	- S-Band High-Accuracy Ranging and Navigation
SHM	- Simple Harmonic Motion
SHOAP	- Symbolic Horribly Optimizing Assembly Program
SHOB	- Scale Height of Burst
SHOC	- SHAPE (SAGE High-Altitude Prototype Environment) Operations Center
SHODOP	- Short Range Doppler
SFOF	- Space Flight Operations Facility
SHORAN	- Short Range Navigation
SHP	- Shaft Horsepower
	- Sustainer Hydraulic Pressure
SHS	- Smoothing Heading Spot
	- Spherical Harmonic Series
SI	- Salinity Indicator
	- Scientific Instrumentation
	- Security Identification
	- Sense Indicator
	- Shipping Instruction
	- Spark Ignition
	- Switch Interpretation
	- System International
	- Systems Integration
S/I	- Systems Integration (SI preferred)
S&I	- Storage and Inspection
	- Surveillance and Inspection
SIA	- Strategic Industries Association
	- Systems Integration Area
SIAM	- Signal Information and Monitoring Service
	- Society of Industrial and Applied Mathematics
SIAP	- Standard Instrument Approach
SIBS	- Stellar Inertial Bombing System
SIC	- Satellite Information Center
	- Scientific Information Center
	- Survey Information Center
	- Systems Integration Contractor
SICBM	- Small Intercontinental Ballistic Missile
SID	- Situation Display
	- Space and Information Systems Division
	- Sudden Ionospheric Disturbance
S&ID	- Space and Information Systems Division
SIDL	- System Identification List
SIDOS	- Site Document Order Section
SIDS	- Stellar Inertial Doppler System

- **SIE**

SIE	- Science Information Exchange
	- Sensory Isolation Experiment
	- Society of Industrial Engineers
	- Southwestern Industrial Electronics Company
SIF	- Selective Identification Feature
	- Sound Intermediate Frequency
SIGBDP	- Special Interest Group for Business Data Processing
SIGPLAN	- Special Interest Group on Programming Languages
SIGSU	- Signal Corps Supply
SIITO	- Standard Installation Instruction Technical Order
SIL	- Set Indicators of the Left Half
	- Speech Interference Level
SILS	- Shipboard Impact Location System
SIM	- Simulate
SIMAC	- Sonic Instrument Measurement and Control
SIMCHE	- Simulation and Checkout Equipment
SIMCOM	- Simulator Compiler
SIMDC	- Simulated Direction Center
SIMFAC	- Simulation Facility
SIMICORE	- Simultaneous Multiple Image Correlation
SIM/L	- Simulated Launch
SIM M-R	- Simulation Monitor-Recorder
SIMPAC	- Simulation Package
SIMPU	- Simulator Punch
SIMSA	- Signal Material Support Agency
SIMSUP	- Simulation Supervisor
SIN	- Support Information Network
SIND	- Strobe Intersection Deghoster
SINS	- Ship's Inertial Navigation System
	- Stellar Inertial Navigation System
SIO	- Scripps Institute of Oceanography
	- Service Information Office
	- Staged in Orbit
SIOE	- Special Issue of Equipment
SIOP	- Single Integrated Operational Plan
SIP	- SAGE Improvement Program
	- Satellite Inspection Program
	- Satellite Interceptor Program
	- SCANS (Scheduling and Control by Automated Network Systems) Implementation Plan
	- Simulated Input Processor
	- Solar Instrument Probe
	- SONAR Instrumentation Probe
	- Standard Inspection Procedure
SIPRE	- Snow, Ice, and Permafrost Research Establishment
SIPS	- SAC Intelligence Data Processing System

SIPS	- Simulated Input Preparation System
SIR	- Set Indicators of the Right Half
	- Silo Installation Refurbish
	- Submarine Intermediate Reactor
SIRA	- Strategic Intelligence Research and Analysis
SIRC	- Spares Interim Report Control
SIS	- SAGE Interceptor Simulator
	- Simulation Interface System
	- Stage Interface Simulator
SI-SIC	- Siliconized Silicon Carbide
SISS	- Submarine Integrated SONAR System
SISTRS	- Simulator Strobe Tracking Study
SIT	- Simulation Input Tape
	- Statement of Inventory Transaction
	- Stevens Institute of Technology
SITE	- Search Information Tape Equipment
	- Spacecraft Instrumentation Test Equipment
SITP	- System Integration Test Plan
SITREP	- Situation Report
SIUFL	- Suspended Issue and Use of Following Lots
SIUSM	- Suspend from Issue and Use as Suspect Material
SIXATAF	- Sixth Allied Tactical Air Force
SIXFLT	- Sixth Fleet
SIZ	- Security Identification Zone
SKILL	- Satellite Kill
SL	- Scanning Slit
	- Sea Level
	- Soft Landing
	- Sound Locator
	- Support Line
	- Surface Launch
	- Star Line
	- Storage Location
S/L	- Space Laboratory
S&L	- Supply and Logistics
	- Systems and Logistics
SLA	- Special Libraries Association
SLAC	- Stanford Linear Accelerator Center
SLAE	- Society of Licensed Aircraft Engineers
SLAG	- Safe Launch Angle Gate
SLAM	- Select, Loft, and Monitor
	- Supersonic Low-Altitude Missile (now LASV)
SLAMS	- Successive Linear Approximation at Minimum Step
SLANG	- Systems Language
SLANT	- Simulator Landing Attachment for Night Landing Training
SLAR	- Side-Looking Aerial Radar

- **SLAR**

SLAR	- Slant Range
SLATE	- Small Lightweight Altitude Transmission Equipment
SLB	- Side Lobe Blanking
SLBM	- Sea Launched Ballistic Missile
	- Submarine Launched Ballistic Missile
SLC	- Side Lobe Cancellation
	- Simulated Linguistic Computer
SLCC	- Soft Launch Control Center
SLD	- Senior Load Distributor
SLE	- Stage Loose Equipment
SLEH	- Stage Loose Equipment Hardware
SLES	- Semi-Linear Erection System
SLEW	- Static Load Error Washout
SLF	- Sense Lights Off
SLG	- Site Liaison Group
SLICBM	- Sea Launched Intercontinental Ballistic Missile
SLIM	- Submarine Launched Inertial Missile
SLIRBM	- Sea Launched Intermediate Range Ballistic Missile
SLM	- Ship Launched Missile
	- Space Laboratory Module
SLN	- Sense Lights On
SLOE	- Special List of Equipment
SLOMAR	- Space Logistics, Maintenance, and Rescue Vehicle
SLOS	- Star Line of Sight
SLR	- Shift Left and Right
SLRA	- Soviet Long Range Aviation
SLRI	- Shipboard Long-Range Input
SLS	- Side Lobe Suppression
SLT	- Sense Light Test
SLTF	- Silo Launch Test Facility
SLU	- Structures Laboratory Unit
SLURP	- Servo Laboratory Utility Routine Program
SLV	- Satellite Launching Vehicle
	- Soft Landing Vehicle
	- Space Launch Vehicle
SLW	- Store Logical Word
	- Straight Line Wavelength
SM	- Service Module
	- Servomotor
	- Shear Modulus
	- Shipping Manifest
	- Space Medicine
	- Stabilized Member
	- Statute Mile
	- Stock Material
	- Strategic Missile

SMR

SM	- Supply Manual
	- Surface Missile
S/M	- Service Module (SM preferred)
	- Shipping Memorandum
S&M	- Sequencer and Monitor
SMA	- Squadron Maintenance Area
SMAC	- Sub-Micron Airosol Collector
SMACS	- Simulated Message Analysis and Conversion Subsystem
SMADS	- Sault Sainte Marie Air Defense Sector
SMAMA	- Sacramento Air Materiel Area
SMART	- Satellite Maintenance and Repair Technique
	- Sequential Mechanism for Automatic Recording and Testing
	- Supersonic Military Air Research Track
	- System Malfunction Analysis Reinforcement Trainer
SMASHT	- Simple Minded Approach to Squeezed Hollerith Text
SMC	- SAGE Maintenance Control
	- SNORT (SAGE Non-Real Time) Master Control
	- Spectrum Management Computer
SMCO	- SAGE Maintenance Control Office
SMCS	- SAGE Maintenance Control Section
SMD	- Shop Manufacturing Development
	- Submarine Mine Depot
	- System Measuring Device
SMDF	- SCATS (Simulation, Checkout, and Training System) Main Distributing Frame
SME	- Society of Military Engineers
	- Standard Medical Examination
SMEC	- Strategic Missiles Evaluation Committee
SMEK	- Summary Message Enable Keyboard
SMESS	- Santa Monica Experimental SAGE Sector
SMF	- Solar Magnetic Field
SMG	- Submachine Gun
SMGP	- Strategic Missile Group
SMIT	- Spin Motor Interruption Technique
SML	- Simulator Load
	- Support Material List
SMO	- So Much Of
	- Stabilized Master Oscillator
	- Stock Material Order
	- Systems Manger's Office
SMOC	- Simulation Mission Operation Computer
SMOG	- Smoke and Fog
SMOS	- Secondary Military Occupational Speciality
SMP	- STOD Management Program
SMPTE	- Society of Motion Picture and Television Engineers
SMR	- Systems Mangement Responsibility

■ SMRB

SMRB	- Strategic Missile Review Board
SMRD	- Spin Motor Rotation Detector
SMS	- Sequence Milestone System
	- Service Module Simulator
	- Standard Meteorological Station
	- Standard Modular System of Packaging
	- Strategic Missile Squadron
	- Strategic Missile System
	- Subject Matter Specialist
	- Surface Missile Systems
	- Synchronous Meteorological Satellite
SMSA	- Strategic Missile Support Agency
	- Strategic Missile Support Area
SMSB	- Strategic Missile Support Base
SMSD	- Ship's Magnet—Submarine Detector
SMSF	- Strategic Missile Support Facility
SMT	- Small Missile Telecamera
	- Square Mesh Tracking
	- Subject Matter Trainer
SMU	- Self-Maneuvering Unit
SN	- Saturn Nuclear
	- Secretary of the Navy
	- Sector Number
	- Serial Number
	- Service Number
	- Signal to Noise
	- Sortie Number
S/N	- Serial Number
	- Signal to Noise
SNACS	- Share News on Automatic Coding Systems
SNAFU	- Situation Normal—All Fouled Up
SNAL	- Site Number Assignment List
SNA&ME	- Society of Naval Architects and Marine Engineers
SNAP	- Systems for Nuclear Auxiliary Power
SNARK	- Snake-Shark
SNEC	- Saxton Nuclear Experimental Corporation
SNECMA	- Science Nationale d'Etude et de Construction de Moteurs d'Aviation
SNF	- System Noise Figure
SNI	- San Nicolas Island
SNL	- Standard Nomenclature List
SNORT	- SAGE Non-Real Time
	- Supersonic Naval Ordnance Research Track
SNPO	- Space Nuclear Propulsion Office
SNPRI	- Selected Nonpriority List Item
SNR	- Signal-to-Noise Ratio

SNSE	- Society of Nuclear Scientists and Engineers
SO	- Safety Officer
	- Sales Order
	- Send Only
	- Sequential Observer
	- Shipment Order
	- Shop Order
	- Slow Operate
	- Special Order
	- Stationary Orbit
	- Stockage Objective
	- Supply Officer
	- Surveillance Officer
S/O	- Send Only
	- Switchover
S-O	- Shut Off
SOA	- Speed of Advance
	- Special Operating Agency
	- Strap on Assembly
SOAP	- SAGE Operator Analysis Report
	- Self-Optimizing Automatic Pilot
	- Symbolic Optimum Assembly Program
SOAPD	- Southern Air Procurement District
SOAR	- SAGE Ocean Airborne Reporting
SOBLIN	- Self-Organizing Binary Logical Network
SOC	- Sector Operations Center
	- Simulation Operation Computer
	- Single Orbit Computation
	- Space Operations Center
	- Squadron Operations Center
SOCIR	- System of Corporate Information Retrieval
SOCKO	- Systems Operational Checkout
SOCOM	- Solar Communications System
SOCS	- Strategic Operational Communications System
SOD	- Small Object Detector
	- System Operational Description
	- System Operational Design
SODAR	- Sound Detection and Ranging
SOER	- Systems Office Evaluation Request
SOF	- Sound on Film
	- Status of Forces
SOFAR	- Sound Fixing and Ranging
SOFAR/BF	- Sound Fixing and Ranging Bomb Fuze
SOFCS	- Self-Organizing Flight Control System
SOF/MOP	- Status of Forces/Monitor Operational Plan
SOFPAC	- Special Operating Forces, Pacific

■ SOI

SOI	- Signal Operation Instructions
SOL	- Short Octal Load
	- Shut-Off Lights
	- System Oriented Language
SOLARIS	- Submerged Object Locating and Retrieving Identification System
SOLLAR	- Soft Lunar Landing and Return
SOLO	- Selective Optical Lock-On
SOLOMAN	- Simultaneous Operation Linked Ordinal Modular Network
SOM	- Ship Operations Manager
	- Start of Message
	- Sub-Orbital Mission
SOMP	- Start of Message Priority
SONAC	- SONAR Nacelle
SONAR	- Sound Navigation and Ranging
SONCM	- SONAR Countermeasurers and Deception
SO/NF	- Spin Orbit and Nuclear Force
SONOAN	- Sonic Noise Analyzer
SOOSE	- Sub-Orbital Offense Systems Group
SOP	- Senior Officer Present
	- Simulated Operating Procedure
	- Standing Operating Procedure
SOPA	- Senior Officer Present Afloat
SOPC	- SAGE Operational Problems Committee
SOPUSN	- Senior Officer Present, United States Navy
SOR	- Specific Operational Requirement
	- Start of Record
	- System Operational Recommendation
	- System Operational Requirement
SORAVAIL	- Sortie Availability
SORC	- Sound Ranging Control
SORO	- Special Operations Research Office
SORTE	- Summary of Radiation Tolerant Electronics
SORTI	- Satellite Orbital Track and Intercept
SORTIE	- Sub-Orbital Re-Entry Test Integrated Environment
SOS	- Share Operating System
	- Shop Order Shop
	- Shop Out of Stock
SOSS	- Strategic Orbital System Study
SOSUS	- Sound Surveillance System
SOT	- Systems Operations Test
SOTE	- System Operational Test Evaluation
SOTFE	- Special Operations Task Force, Europe
SOTIM	- Sonic Observation of the Trajectory and Impact of Missiles
SOV	- Shut-Off Valve
SOW	- Start of Word

SOX	- Solid Oxygen
SP	- Sample Part
	- Self-Propelled
	- Shear Plate
	- Short Persistance
	- Single Phase
	- Single Pole
	- Single Propellant
	- Single Purpose
	- Smokeless Powder
	- Smokeless Propellant
	- Solid Propellant
	- Space Probe
	- Spare
	- Spare Part
	- Special Project
	- Special Projectile
	- Special Purpose
	- Splashproof
	- Start Point
	- Static Pressure
	- Supine Position
	- Symbol Programmer
S/P	- Serial to Parallel
S-P	- Sequential-Phase
	- Serial to Parallel
S&P	- Security and Power
	- Stake and Platform
SPA	- SAGE Positional Analysis
	- S-Band Power Amplifier
	- Servo Power Assembly
	- Single Position Automatic Tester
	- Spectrum Analyzer
	- Sudden Phase Anomaly
	- System Problem Area
	- Systems and Procedures Association
SPACCS	- Space Command and Control System
SPACE	- Self-Programming Automatic Circuit Evaluator
	- Spacecraft Prelaunch Automatic Checkout Equipment
	- SPAR (Symbolic Program Assembly Routine) ALCOM COBOL Executive Program
	- Symbolic Programming Anyone Can Enjoy
SPACECOM	- Space Communications
SPACON	- Space Control
SPAD	- Satellite Position Prediction and Display
	- Satellite Protection for Area Defense

■ SPAD

SPAD	- Space Patrol for Air Defense
SPADATS	- Space Detection and Tracking System
SPADE	- Spare Parts Analysis, Documentation, and Evaluation
	- Sperry Air Data Equipment
SPADS	- Space Patrol Air Defense System
	- Spokane Air Defense Sector
SPAF	- Simulation Processor and Formatter
SPAN	- Space Communication Network
	- Stored Program Alpha-Numerics
SPANDAR	- Space and Range Radar
SPAR	- Satellite Position Adjusting Rocket
	- Seagoing Platform for Acoustic Research
	- Semper Paratus—Always Ready (Women's Coast Guard Reserve)
	- Super-Precision Approach Radar
	- Symbolic Program Assembly Routine
	- Synchronous Position Attitude Recorder
SPARC	- System of Programs for Analysis by RECOMP
SPARS	- Site Production and Reduction System
SPASUR	- Space Surveillance System
SPAT	- Silicon Precision Alloy Transistor
SPAWG	- Special Activity Wing
SPB	- Summary Plot Board
	- Systems Personnel Branch
SPC	- SAGE Problem Correction
	- SAGE Program Change
	- Sequence Parameter Checking
	- Single Prime Contractor
	- Special Common
	- Specific Propellant Consumption
	- Stored Program Command
	- Stored Program Control
SPCC	- SAGE Program Change Committee
	- Ship's Parts Control Center
SPCP	- Single Prime Contractor Policy
SPCR	- SAGE Program Change Request
SPD	- Situation Projected Display
	- System Program Director
SPDT	- Single-Pole Double-Throw
SPE	- Special Purpose Equipment
	- Stored Program Element
SPEARS	- Satellite Photo-Electronic Analog Rectification System
SPEC	- Specification
	- Stored Program Educational Computer
SPECDEVCEN	- Special Devices Center
SPECOMME	- Special Command Middle East

SPECTROL - Scheduling Planning Evaluation Cost Control
SPEDAC   - Solid-State, Parallel, Expandable, Differential Analyzer Computer
SPEED   - Self-Programmed Electronic Equation Delineator
         - Signal Processing in Evaluated Electronic Devices
         - Study and Performance Efficiency in Entry Design
SPEEL   - Shore Plant Electronic Equipment List
SPERT   - Schedule Program Evaluation Review Technique
         - Special Power Excursion Reactor Test
SPES   - Stored Program Element System
SPG   - Special Functions Group
SPGR   - Specific Gravity
SPH   - SAGE Positional Handbook
SPHT   - Specific Heat
SPI   - Society of Plastics Industry
         - Standard Practice Instructions
SPIA   - Solid Propellant Information Agency
SPIE   - Simulated Problem Input Evaluation
SPIN   - Special Purpose Individual Weapon
         - Standard Procedure Instructions
SPINVESWG - Special Investigation Wing
SPIRAL   - Sperry Inertial Radar Altimeter
SPIRE   - Spatial Inertial Reference Equipment
SPIS   - SAGE Program Identification Service
SPIW   - Special Purpose Infantry Weapon
SPL   - Sector Programming Leader
         - Sound Pressure Level
         - Speed-Phase Lock
SPM   - Self-Propelled Mount
         - Standard Parts Manual
SPMX   - Speed-O-Max
SPN   - Special Program Number
SPO   - Saturn Project Office
         - Special Projects Office (Navy)
         - System Project Office (Air Force)
SPOC   - Single Point Orbit Calculator
SPOOK   - Supervisor Program Over Other Kinds
SPP   - Solar Photometry Probe
         - System Package Program
SPPM   - Safe Passage Path Map
SPPO   - SAGE Performance Practices Operation
         - System Performance Practices Organization
SPR   - Sense Printer
         - Silicon Power Rectifier
         - Solid Propellant Rocket
         - Spare Parts Release

■ SPR

SPR	- Spare Parts Requirement
SPRC	- SAGE Program Review Committee
SPRCS	- Safe Passage Route Creation Sheet
SPRINT	- Selective Printing
	- Solid Propellant Rocket Intercept Missile
SPRTAP	- Specially Prepared Tape Program
SPS	- Samples per Second
	- Secondary Propulsion System
	- Service Propulsion System
	- Solar Probe Spacecraft
	- Solar Proton Stream
	- Space Power System
	- Standard Pressed Steel Company
	- Symbolic Programming System
SPST	- Single-Pole Single-Throw
SPT	- Sense Printer Test
	- Spare Parts Transmitter
SPTC	- Specified Period of Time Contract
SPTT	- Single-Pole Triple-Throw
SPTV	- Supersonic Parachute Test Vehicle
SPU	- Self-Propelled Underwater Missile
	- Sense Punch
SPUD	- Stored Program Universal Demonstrator
SPUR	- Separate Parts Usage Records
	- Single Precision Unpacked Rounded Floating Point Package
	- Space Power Unit Reactor
SPURT	- Spinning Unguided Rocket Trajectory
SPWG	- Spare Parts Working Group
SQ	- Square
	- Superquick
SQC	- Statistical Quality Control
SQ-CM	- Square Centimeter
SQ-DEL	- Superquick and Delay
SQDN	- Squadron
SQ-FT	- Square Foot
SQ-IN	- Square Inch
SQ-KM	- Square Kilometer
SQORD	- Separation, Quality Analysis of Radar Data
SR	- Saturable Reactor
	- Shift Register
	- Shipment Request
	- Ship Repair
	- Short Range
	- Slow Release
	- Soft Radiation
	- Solar Radiation

SR	- Solid Rocket
	- Sounding Rocket
	- Sound Ranging
	- Sound Rating
	- Special Regulations
	- Special Repair
	- Split Ring
	- Status Report
	- Study Requirement
	- System Requirement
	- Supporting Research
S/R	- Send and Receive
	- Shop Release
	- Slant Range
S&R	- Send and Receive
SRA	- Specialized Repair Activity
	- Special Receiving Area
	- Special Repair Area
	- Spin Reference Axis
	- Stores and Reclamation Area
	- Support Requirements Area
SRAM	- Short Range Attack Missile
SRAP	- Standard Range Approach
SRB	- System Requirements Branch
SRBM	- Short Range Ballistic Missile
SRC	- Shop Resident Control
	- Sound Ranging Control
	- Special Release Card
	- Specimen Return Container
	- Sperry Rand Corporation
	- Standard Requirements Code
SRCC	- Simplex Remote Communications Center
SRCP	- Special Reserve Components Program
SRCS	- Service Reaction Control System
SRD	- Shift Register Drive
	- Spacecraft Research Division (now Spacecraft Technology Division)
	- Standard Rate and Data
SRDL	- Signal Research and Development Laboratory
SRE	- Series Relay
	- Surveillance Radar Element
SRG	- Simulation and Reduction Group
SRGR	- Short Range Guided Rocket
SRI	- Stanford Research Institute
SRILTA	- Stanford Research Institute Lead Time Analysis
SRIP	- Short Range Impact Point
SRL	- Serial Reference List

■ SRL

SRL	- Systems Research Laboratory
SRM	- Strategic Reconnaissance Missile
SRN	- Simulation Reference Number
SRNC	- Severn River Naval Command
SRO	- Superintendent of Range Operations
SRP	- Seat Reference Point
SRR	- Supplier Rating Report
	- Surplus Review Record
SRS	- Selenium Rectifier Stacks
	- Simulated Remote Station
	- Solar Radiation Satellite
	- Statistical Reporting Service
SRSCC	- Simulated Remote Station Control Console
SRT	- Search Radar Terminal
	- Surface Craft Radio Transmitter
SRUF	- Source and Referent Unit Finder
SRW	- Strategic Reconnaissance Wing
SS	- Safety Supervisor
	- Safety Supplement
	- Samples per Second
	- Security Subsystem
	- Separation System
	- Set Screw
	- Signal Strength
	- Simulated Strike
	- Simulation Supervisor
	- Single Shot
	- Single Signal
	- Sliding Scale
	- Solar System
	- Sole Source
	- Solid State
	- Space Sciences
	- Space Simulator
	- Spin Stabilized
	- Stainless Steel
	- Stationary Satellite
	- Steamship
	- Stock Size
	- Storage Site
	- Strategic Support
	- Submarine
	- Subsector
	- Supersensitive
	- Surface to Surface
	- Surveillance Station

SSF

SS	- System Studies
S/S	- Samples per Second
	- Ship to Shore
	- Signal Strength
	- Subsystem
S&S	- Situation and Status
SSA	- Silo Subassembly
	- Small Search Area
	- Soaring Society of America
	- Space Suit Assembly
S&SA	- Staging and Support Area
SSB	- Signal Sight Back
	- Single Sideband
	- Soft Service Building
	- Space Science Board
SSC	- SAGE Sector Commander
	- Sensor Signal Conditioner
	- Solid State Circuit
	- Stock Shortage Control
SSCBM	- Shipping and Storage Container—Ballistic Missile
SSCC	- Space Surveillance Control Center
SSCDS	- Shall Ship Combat Data System
SSCE	- Squadron Supervisory and Control Equipment
SS/CF	- Signal Strength, Center Frequency
SSCU	- Special Signal Conditioning Unit
SSD	- Solid State Detectors
	- Space Systems Division (of AFSC)
	- Special Systems Department
	- Stabilized Ship Detector
	- Status Source Document
	- Surveillance Situation Display
	- System Status Display
	- System Study Director
SS/D	- Synchronization Separator and Digitizer
SS&D	- Synchronization Separator and Digitizer
SSDF	- Space Science Development Facility
SSDF-NAA	- Space Science Development Facility—North American Aviation
SSDS	- Small Ship Data System
SSE	- Solid State Electrolytes
	- Spacecraft Simulation Equipment
	- Special Support Equipment
SSEC	- Selective Sequence Electronic Calculator
	- Subsystem Executive Control Program
SSECO	- Second Stage Engine Cutoff
SSF	- Saybolt Seconds Furol

349

■ SSF

SSF	- Service Storage Facility
SSFL	- Steady-State Fermi Level
SS/FM	- Single-Sideband Frequency Modulation
SSG	- Search Signal Generator
	- Security Service Guide
	- Submarine, Guided Missile
SSG(N)	- Submarine, Guided Missile, Nuclear
SSGS	- Standard Space Guidance System
SSHP	- Single Shot Hit Probability
SSI	- San Salvador Island
	- Second Stage Ignition
	- Sector Scan Indicator
	- Standing Signal Instructions
SSIP	- Subsystem Integration Plan
SSKP	- Single-Shot Kill Probability
SSLORAN	- Sky-Wave Synchronized Long Range Navigation
SSLS	- Standard Space Launch System
SSM	- Set Sign Minus
	- Single Sideband Modulation
	- Solid State Materials
	- Spacecraft Systems Monitor
	- Surface-to-Surface Missile
SSMD	- Silicon Stud-Mounted Diode
SSN	- Specification Serial Number
SSO	- Saturn Systems Office
	- Spacecraft Systems Officer
	- Statistical Service Office
	- Steady-State Oscillation
	- Submarine Supply Office
	- System Staff Office
SSOC	- Space Surveillance Operations Center
SSP	- Set Sign Plus
	- Silo Support Plan
	- Single-Shot Probability
	- Sustained Superior Performance
SSPO	- Support System Project Office
SSR	- Spin-Stabilized Rocket
	- Support Staff Rooms
	- System Subroutines
SSRL	- Systems Simulation Research Laboratory
SSS	- Simulation Study Series
	- Squadron Supervisory Station
	- Strategic Support Squadron
	- System Study Section
SSSB	- System Source Selection Board
SSSBP	- System Source Selection Board Procedure

SSSR	- SAGE System Status Report
	- Union of Soviet Socialist Republics (USSR)
SST	- Simulated Structural Test
	- Spacecraft Systems Test
	- Stainless Steel
	- Supersonic Transport
SSTL	- Sector System Training Leader
SSTM	- SAGE System Training Mission
SSTO	- Second Stage Tail Off
SSTP	- SAGE System Training Program
SSTU	- SAGE System Training Unit
SSU	- Saybolt Seconds Universal
	- Statistical Service Unit
SSV	- Ship-to-Surface Vessel
SSVN	- Subsystem and Vehicle Number
SSW	- Scramble Status and Weather
ST	- Schuler Tuning
	- Service Test
	- Shock Tube
	- Shock Tunnel
	- Single Throw
	- Skin Track
	- Sounding Tube
	- Spaced Triplet
	- Special Test
	- Stores Transfer
	- Systems Test
S/T	- Search/Track
	- Sonic Telegraphy
STA	- Satellite Test Annex
	- Spark Thrust Augmentor
	- Store Address
STAB	- Supersonic Test of Aerodynamic Bomb
STABS	- SACCS Training and Battle Simulation
STACAP	- Status and Capability
STADAN	- Space Tracking and Data Acquisition Network (formerly MINITRACK)
STAE	- Second Time Around Echo
STAF	- Standard Test and Administrative Form
STAFF	- Stellar Acquisition Flight Feasibility Program
STAG	- Strategy and Tactics Analysis Group
STAI	- Simulation Tape Alarm Indicator
STAIR	- Structural Analysis Interpretative Routine
STALO	- Stabilized Local Oscillator
STAMIC	- Set Theoretic Analysis and Measure of Information Characteristics

■ STANAG

STANAG	- Standardization Agreement
STAP	- Standard Test and Administrative Procedure
STAPP	- Standard Tape Print Program
	- Single Thread All Purpose Program
STAR	- Scientific and Technical Aerospace Reports
	- Selective Training and Retention
	- Shield Test Air Facility Reactor
	- Ship-Tended Acoustic Relay
	- Space Thermionic Auxiliary Reactor
	- Standard Test and Administrative Report
	- Statistical Treatment of Radar Returns
	- System Training Analysis Report
STARCOM	- Strategic Army Communications System
STARE	- Steerable Telemetry Antenna Receiving Equipment
STARS	- Satellite Telemetry Automatic Reduction System
START	- Spacecraft Technology and Advanced Re-Entry Test
STAS	- Safe-to-Arm System
STAT	- Seabee Technical Assistance Team
STATSVS	- Statistical Services
STC	- SAGE Test Committee
	- Satellite Test Center
	- Sensitivity Time Control
	- Short Time Constant
	- Silicon Transistor Corporation
	- Simulation Tape Conversion
	- Spacecraft Test Conductor
	- Standard Transmission Code
	- Systems Test Complex
	- System Technical Center
	- System Training Command
STCO	- System Technical Coordinator
STCT	- System Technical Coordinator Technician
STCW	- System Time Code Word
STD	- Set Driver
	- Sodium Thermionic Detector
	- Spacecraft Technology Division (formerly SRD)
	- Standard
	- Store Decrement
	- Systems Test Division
	- System Technology Division
STDG	- System Training Design Group
STE	- Special Test Equipment
	- System Test Engineer
	- System Training Exercise
STEADY	- Simulation Tables—Environment and Dynamic
STEDI	- Space Thrust Evaluation and Disposal Investigation

STEEL	- Simulation Test Environment to Evaluate Team Load
STEIN	- System Test Environment Input
STEM	- Stay Time Extension Module
	- Stellar Tracker Evaluation Missile
	- Storable Tubular Extendible Member
STEP	- Simple Transition to Electronic Processing
	- Simulated Tracking Evaluation Program
	- Space Terminal Evaluation Program
	- Supervisory Tape Executive Program
	- System Training and Exercising Program
STEPS	- Solar Thermionic Electric Power System
	- Staff Training Exercise for Programming Supervisor
STF	- SACCS Test Facility
	- S-Band Temperature, Fahrenheit
	- System Test Facility
STG	- Space Task Group
STG-AR	- Staging Area
STGT	- Secondary Target
S-TGT	- Secondary Target
STI	- Store Indicators
S&TI	- Scientific and Technical Information
STIC	- Simulated Time in Climb
STICTION	- Static Friction
STIG	- System Test and Integration Group
STINCO	- SAGE-Talos Integration Committee
STINGS	- Stellar Inertial Guidance System
STIT	- Simulated Time in Turn
STK	- Situation Track Display
STL	- Simulated Table Load
	- Space Technology Laboratories, Incorporated
	- Status and Telling
	- Stockage List
	- Studio-to-Transmitter Link
	- Support Table Load
STLI	- Stockage List Item
STLO	- Scientific and Technical Liaison Office
STM	- Senior Track Monitor
	- Silo Test Missile
	- Structural Test Model
	- System Master Tape
	- System Training Mission
STMLOAD	- System Master Tape Load
STMP	- Single Track MORT (Master Operational Recording Tape) Processing
STMU	- Special Test and Maintenance Unit
STN	- SAC Telephone Network

- **STO**

STO	- SAGE Test Officer
	- Standing Tool Order
STOD	- System Training Operations Department
STOL	- Short Take-Off and Landing
STON	- Short Ton
STOP	- Search Terminal Option
	- System Test Operating Instruction
STORC	- Self-Ferrying Trans-Ocean Rotory-Wing Crane
STP	- Seattle Test Program
	- Selective Tape Print
	- Simultaneous Track Processor
	- System Temperature and Pressure
	- System Test Procedure
	- System Training Program
STPD	- System Training Production Department
STPS	- Systems Test Planning Section
STPX	- System Training Program Exercise
STR	- SAC Target Register
	- Sea Test Range
	- Submarine Thermal Reactor
STRAC	- Strategic Army Corps
STRADAP	- Storm Radar Data Processor
STRAF	- Strategic Army Forces
STRANGE	- SAGE Tracking and Guidance Evaluation System
STRAP	- Strategic Planner
STRATAD	- Strategic Aerospace Division
STRATCOM	- Strategic Communication System
STRAW	- Simultaneous Tape Read and Write
STREP	- Simulation Tape Reconstruction Program
STRICOM	- Strike Command
STRIP	- Specification Technical Review and Improvement Program
	- Standard Taped Routines for Image Processing
STROBE	- Satellite Tracking of Balloons and Emergencies
STRS	- SAGE Training Requirements Section
STS	- SAGE Training Specialist
	- Satellite to Satellite
	- Senior Training Specialist
	- Ship to Shore
	- Stockpile-to-Target Sequence
	- System Test Station
	- System Training Section
	- System Training Specialist
	- System Trouble Survey
STSV	- Satellite to Space Vehicle
STT	- Store Tag
STU	- Special Test Unit

STU	- Static Test Unit
	- Submersible Test Unit
	- Systems Test Unit
	- System Training Unit
STV	- Separation Test Vehicle
	- Space Test Vehicle
	- Special Test Vehicle
	- Standard Test Vehicle
	- Structural Test Vehicle
	- Supersonic Test Vehicle
STVC	- Space Thermal Vacuum Chamber
STWE	- Society of Technical Writers and Editors
STWP	- Society of Technical Writers and Publishers
STZ	- Store Zero
SU	- Set Up
	- Supercommutation
SUB	- Submarine
SUBIC	- Submarine Integrated Control
SUBLANT	- Submarine Force, Atlantic
SUBNO	- Substitutes Not Desirable
SUBOK	- Substitution OK
SUBPAC	- Submarine Force, Pacific
SUBROC	- Submarine Rocket
SUEL	- Sperry Utah Engineering Laboratory
SUM	- Surface-to-Underwater Missile
	- System Check and Utility Master
SUMS	- Sperry UNIVAC Material System
SUN	- Symbols, Units, and Nomenclature
SUNFED	- Special United Nations Fund for Economic Development
SUP	- System Utilization Procedure
SUPARCO	- Space and Upper Atmospheric Research Committee
SUPG	- System Utilization Procedural Guide
SURANO	- Surface Radar and Navigation Operation
SURC	- Syracuse University Research Corporation
SURCAL	- Surveillance Calibration Satellite
SURE	- Symbolic Utility Revenue Environment
SURF	- Surface Force
SURGE	- Sorting, Updating, Report Generating, Etc.
SURIC	- Surface Ship Integrated Control
SURVAL	- Simulator Universal Radar Variability Library
SURWAC	- Surface Water Automatic Computer
SUS	- Saybolt Universal Seconds
SUSMOP	- Senior United States Military Observer, Palestine
SUT	- Supporting Utility
	- System Under Test
SV	- Safety Valve

SV

SV	- Self-Verification
	- Simulated Video
	- Slowed-Down Video
	- Space Vehicle
S/V	- Space Vehicle
	- Surface Vessel
SVE	- Space Vehicle Electronics
	- Sun Vehicle Earth
	- Swept Volume Efficiency
SVS	- Space Vehicle Simulator
	- Space Vehicle System
SVTP	- Sound, Velocity, Temperature, Pressure
SVTSV	- Space Vehicle to Space Vehicle
SW	- Salt Water
	- Sea Water
	- Short Wave
	- Single Weight
	- Southwest
	- Special Weapon
	- Spot Weld
	- Standby Weight
	- Switchband Wound
S/W	- Specification of Wiring
S&W	- Smith and Wesson
SWA	- Support Work Authorization
SWAC	- National Bureau of Standards Western Automatic Computer
SWACS	- Space Warning and Control System
SWAG	- Systems Work Assignment Group
SWAMI	- Standing Wave Area Monitor Indicator
SWAT	- Sidewinder Acquisition Track
SWB	- Sandia Wind Balloon
	- Short Wheelbase
SWC	- Special Weapons Command
SWD	- Senior Weapons Director
SWDB	- Special Weapons Development Board
SWDT	- Senior Weapons Director Technician
SWE	- Society of Women Engineers
SWEEP	- Structures with Error Expurgation Program
SWEL	- Special Weapons Equipment List
SWELSTRA	- Special Weapons Equipment List Single Theater Requisitioning Agency
SWET	- Simulated Ward Experimental Test
	- Special Weapons Equipment Test
SWF	- Short Wave Fadeout
SWFC	- Surface Weapons Fire Control
SWG	- Special Working Group

SWG	- Standard Wire Gauge (British)
S&WG	- Surveillance and Weapons Group
SWIG	- SAGE Weapon Integration Group
SWIS	- Special Weapons Integration Subcommittee
SWM	- Special Warfare Mission
SWO	- Staff Weather Officer
SWP	- Safe Working Pressure
	- Standard Work Procedure
SWPAN	- Special Weapons Project Analysis
SWPC	- Small War Plants Corporation
SWR	- Standing Wave Ratio
SWR^2	- Scaled Weapons Radius Squared
SWRI	- Southwest Research Institute
SWS	- Space Weapon System
	- Special Weapon System
SWSD	- Special Weapons Supply Depot
SWSIL	- Seattle Weapon System Integration Laboratory
SWT	- Supersonic Wind Tunnel
SWTA	- Special Weapon Training Allowance
SWTL	- Surface-Wave Transmission Line
SWX	- Status and Weather
SWXO	- Staff Weather Officer
SXA	- Store Index in Address
SXD	- Store Index in Decrement
SXT	- Sextant
	- Space Sextant
SYADS	- Syracuse Air Defense Sector
SYC	- Symbolic Corrector
SYCATE	- Symptom-Cause-Test
SYM	- Symmetrical
SYMPAC	- Symbolic Program for Automatic Control
SYN	- Synchronous
	- Synthetic
SYNC	- Synchronize
	- Synchronizer
SYNCOM	- Synchronous Communications Satellite
SYNSCP	- Synchroscope
SYS	- System
SYSTO	- Systems Staff Office
SYSTOS	- Systems Offices
SZVR	- Silicon Zener Voltage Regulator

T

T	- Target
	- T-Bar
	- Tee
	- Temperature
	- Tera
	- Thrust
	- Time
	- Timer
	- Ton
	- Tooth
	- Trainer
	- Training
	- Truss
T-	- Time Prior to Launch
T-O	- Time of Launch
T+	- Time after Launch
TA	- Table of Allowance
	- Target Aircraft
	- Target Area
	- Test Announcer
	- Training Analyst
	- Travel Authorization
	- True Altitude
	- Trunnion Angle
	- Type Approval
T/A	- Table of Allowance
T-A	- Thymine-Adenine
T&A	- Transfer and Accountability
TAAR	- Target Area Analysis—Radar
TAB	- Tactical Action Board
	- Target Acquisition Battalion
	- Technical Abstract Bulletin
	- Technical Advisory Board

TAB	- Technical Assistance Board
	- Title Abstract Bulletin
TABSIM	- Table Simulation
TABSOL	- Tabular Systems-Oriented Language
TABSTONE	- Target and Background Signal-to-Noise Evaluation
TAC	- Tactical Air Command
	- Tactical Air Coordinator
	- Technical Activities Committee
	- Technical Area Coordinator
	- Test Advisory Committee
	- Traceability and Configuration
	- TRANSAC Assembler Compiler
	- Transistorized Automatic Control
	- Translator, Assembler, Compiler
	- Transmitter, Assembler, Compiler
TACA	- Tactical Air Coordinator, Airborne
TACAN	- Tactical Air Navigation
TACAN-DME	- Tactical Air Navigation Distance Measuring Equipment
TACC	- Tactical Air Control Center
TACCAR	- Time Averaged Clutter Coherent Airborne Radar
TACCS	- Tactical Air Command Control System
TACDEN	- Tactical Data Entry
TACG	- Tactical Air Control Group
TACH	- Tachometer
TACNAV	- Tactical Navigation
TACP	- Tactical Air Control Party
TACRON	- Tactical Air Control Squadron
TACS	- Tactical Air Control System
	- Theater Area Communications Systems
	- Transtage Attitude Control System
TACT	- Transistor and Component Tester
TAD	- Tactical Action Display
	- Tactical Air Direction
	- Target Acquisition Data
	- Target Area Designation
	- Target Area Designator
	- Technical Acceptance Demonstration
	- Temporary Additional Duty
	- Thrust-Augmented Delta
	- Twin and Add
TADC	- Tactical Air Direction Center
TADOR	- Table Data Organization and Reduction
TADP	- Tactical Air Direction Post
TADS	- Teletypewriter Automatic Dispatch System
TAF	- Tactical Air Force
TAFAD	- Task Force Air Defense

■ TAFCSD

TAFCSD	- Total Active Federal Commissioned Service to Date
TAFMSD	- Total Active Federal Military Service to Date
TAFUBAR	- Things Are Fouled Up Beyond All Recognition
TAG	- Technical Advisory Group
	- The Acronym Generator
	- The Adjutant General
TAGO	- The Adjutant General's Office
TAHA	- Tapered Aperture Horn Antenna
TAHQ	- Theater Army Headquarters
TAIP	- Terminal Area Impact Point
TAIR	- Test Assembly Inspection Record
TALANT	- Thiokol Nuclear Development Center; Allison Division, General Motors; Linde Division, Union Carbide; and Nuclear Development Corporation Team
TALOG	- Theater Army Logistical Command
TALT	- Tracking Altitude
TAM	- Tactical Air Missile
	- Technical Ammunition
	- Technical Area Management
	- Technical Area Manager
TAMP	- Tactical Anti-Missile Measurement Program
TANS	- Terminal Area Navigation System
TAOC	- Tactical Air Operations Center
	- The Army Operations Center
TAP	- Terrestrial Auxiliary Power
TAPAC	- Tape Automatic Positioning and Control
TAPE	- Tape Automatic Preparation Equipment
	- Technical Advisory Panel for Electronics
TAPEX	- Tape Executive Program
TAPPI	- Technical Association of the Pulp and Paper Industry
TAPRE	- Tracking in an Active and Passive Radar Environment
TAR	- Team Acceptance Review
	- Technical Action Request
	- Terrain Avoidance Radar
	- Thrust Augmented Rocket
	- Track Address Reigster
	- Training Analyst Report
TARC	- Theater Army Replacement Command
	- Thru Axis Rotational Control
TARE	- Telemetry Automatic Reduction Equipment
TAREF	- TAG (The Acronym Generator) Reference
TARFU	- Things Are Really Fouled Up
TARN	- Team Acceptance Review Notice
TARP	- Tactical Airborne Recording Package
TARS	- Tactical Air Research and Survey Office
	- Terrain Analog Radar Simulator
	- Three-Axis Reference System

TARTC	- Theater Army Replacement and Training Command
TAS	- Target Acquisition System
	- The Army Staff
	- True Air Speed
TASC	- Terminal Area Sequencing and Control
TASCON	- Television Automatic Sequence Control
TASD	- Tactical Action Situation Display
TASI	- Time Assignment Speech Interpolation
TASO	- Television Allocations Study Organization
TASR	- Terminal Area Surveillance Radar
TASS	- Tactical Air Support Section
	- Tech (Carnegie Institute of Technology) Assembly System
TASSA	- The Army Signal Supply Agency
TAT	- Target Aircraft Transmitter
	- Technical Acceptance Team
	- Thrust-Augmented Thor
	- Turn Around Time
	- Type Approval Test
TATC	- Terminal Air Traffic Control
TATCS	- Terminal Air Traffic Control System
TATSA	- Transportation Aircraft Test and Support Activity
TAVE	- Thor-Agena Vibration Experiment
TAVET	- Temperature, Acceleration, Vibration Environmental Tester
TAW	- Tactical Assault Weapon
TAWCS	- Tactical Air Weapons Control System
TAWS	- Total Airborne Weapon System
	- Transonabuoy Automatic Weather System
TAX	- Training Assessment Exercise
TB	- Technical Bulletin
	- Terminal Board
	- Torpedo Bombing
T/B	- Talk Back
	- Title Block
T&B	- Top and Bottom
TBA	- Tires, Batteries, and Accessories
	- To Be Activated
	- To Be Added
TBC	- The Boeing Company (formerly BAC)
TBD	- Target Bearing Designator
TBF	- Tail Bomb Fuze
	- Training Base Facility
TBGP	- Tactical Bomb Group
TBI	- Target Bearing Indicator
	- To Be Initiated
T-BIRD	- Terrestrial Ballistic Infrared Development
TBL	- Terminal Ballistics Laboratory

■ TBM

TBM	- Tactical Ballistic Missile
	- Theater Ballistic Missile
TBMAA	- Travel by Military Aircraft Authorized
TBMX	- Tactical Ballistic Missile Experiment
TBO	- Time Between Overhauls
TBR	- To Be Released
TBS	- Tape and Buffer System
	- Training and Battle Simulation
TBT	- Terminal Ballistic Track
TBX	- Tactical Bomber, Experimental
TC	- Tape Core
	- Technical Characteristics
	- Technical Circular
	- Technical Communications
	- Telecomputing Corporation
	- Temperature Coefficient
	- Test Conductor
	- Test Console
	- Test Controller
	- Thermocouple
	- Thrust Chamber
	- Time Chapter
	- Time Close
	- Timing Channel
	- Tracking Camera
	- Training Center
	- Training Circular
	- Training Concepts
	- Transaction Code
	- Transfer Control
	- Transitional Control
	- Trip Cell
	- Trip Coil
	- Troop Carrier
	- True Course
	- Type Certificate
T/C	- Telecommunications
	- Test Conductor
	- Thermocouple
	- Thrust Chamber
	- Time-Cost
	- Time-Cycle
T-C	- Turbine-Compressor
T&C	- Turbine and Controls
TCA	- Temperature Control Amplifier
	- Thrust Chamber Assembly
	- Transfer Control A Register

TCAP	- Tactical Channel Assignment Panel
TCB	- Technical Coordination Bulletin
TCBM	- Transcontinental Ballistic Missile
TCBV	- Temperature Coefficient of Breakdown Voltage
TCC	- Telecommunications Coordinating Committee
	- Temporary Council Committee (NATO)
	- Test Conductor Console
	- Thiokol Chemical Corporation
	- Time Compression Coding
	- Tracking and Control Center
	- Traffic Control Center
	- Traffic Control Complex
	- Transit Control Center
	- Transport Control Center
TCD	- Tentative Classification of Defects
	- Test Communications Division
	- Thyratron Core Driver
TCE	- Talker Communication Error
	- Time Critical Equipment
	- Total Composit Error
TCEA	- Training Center for Experimental Aerodynamics (NATO)
TCF	- Technical Certification Form
TCFP	- Thrust Chamber Fuel Purge
TCG	- Transponder Control Group
	- Tune-Controlled Gain
TCH	- Transfer in Channel
TCI	- Terrain Clearance Indicator
TCL	- Transistor Coupled Logic
TCM	- Telemetry Code Modulation
	- Temperature Control Model
TCMS	- Track Combat Status
TCN	- Track Channel Number
	- Transfer on Channel Not in Operation
TCO	- Termination Contracting Office
	- Test Control Officer
	- Transfer on Channel in Operation
	- Translational Control
TCOA	- Translational Control A
TCOB	- Translational Control B
TCOP	- Thrust Chamber Oxidizer Purge
TCP	- TAG (The Acronym Generator) Converter Program
	- Thrust Chamber Pressure
	- Traffic Control Post
	- Tricalcium Phosphate
TCR	- Telemetry Compression Routine
	- Transfer Control Register

- **TCS**

TCS	- Tactical Call Sign
	- Target Cost System
	- Technical Change Summary
	- Temperature Control System
	- Temporary Change of Station
	- Transportable Communications System
TCSC	- Trainer Control and Simulation Computer
TCSP	- Traffic Control Simulation Project
TCSS	- Track Coordination Subsection
TCTM	- Time Compliance Technical Manual
TCTO	- Time Compliance Technical Order
TCU	- Tape Control Unit
	- Transmission Control Unit
TCV	- Thrust Chamber Valve
	- Troop Carrying Vehicle
TCVC	- Tape Control Via Console
TCVR	- Transceiver
TCWG	- Tele-Communications Working Group
TD	- Table of Distribution
	- Tabular Data
	- Tactical Decision
	- Tank Destroyer
	- Technical Data
	- Technical Direction
	- Technical Directive
	- Technical Director
	- Test Data
	- Test Directive
	- Test Director
	- Time Delay
	- Time Difference
	- Time Division
	- Touchdown
	- Track Data
	- Track Display
	- Transmitter-Distributor
	- Turbine Driven
T/D	- Telemetry Data
TDA	- Table of Distribution-Augmentation
	- Trunnion Drive Axis
	- Tunnel-Diode Amplifier
T&DA	- Tracking and Data Acquisition
TDAS	- Tracking and Data Acquisition System
TDB	- Training Development Branch
TDC	- Technical Data Center
	- Technical Development Center

TDC	- Time Delay Closing
	- Top Dead Center
	- Track Data Central
TDCM	- Transistor Driver Core Memory
TDCO	- Torpedo Data Computer Operator
TDCT	- Track Data Central Tables
TDD	- Target Detecting Device
	- Target Detection Device
	- Telemetry Data Digitizer
	- Test Data Division
	- Timing Data Distributor
	- Tracy Defense Depot
TDDL	- Time Division Data Link
TDEC	- Technical Development Evaluation Center
	- Telephone Line Digital Error Checking
TDFG	- Three-Degree-of-Freedom Gyroscope
	- Two-Degree-of-Freedom Gyroscope
TDH	- Total Dynamic Head
TDI	- Target Data Inventory
	- Tear Down Inspection
TDIO	- Timing Data Input-Output
TDIPRE	- Target Data Inventory Master Tape Preparation
TDM	- Technical Direction Meeting
	- Telecommunications Data-Link Monitor
	- Telemetric Data Monitor
	- Test Data Memorandum
	- Time Division Multiplex
	- Tooling Design Manual
	- Trouble Detection and Monitoring
TDMS	- Telegraph Distortion Measuring System
TDN	- Target Doppler Nullifier
TDO	- Time Delay Opening
TDP	- Target Direction Post
	- Technical Development Plan
	- Tracking Data Processor
TDPF	- Target Data Planning File
TDPFO	- Temporary Duty Pending Further Orders
TDR	- Technical Data Report
	- Technical Documentary Report
	- Tool Design Request
	- Torque-Differential Receiver
TDS	- Tank Desiccant System
	- Tape Data Selector
	- Test Data Specification
	- Training Development Staff

■ TDS

TDS	- Track Data Simulation
	- Tract Data Storage
TDT	- Tower Disconnect Technician
TDTG	- True Date-Time Group
TDU	- Talos Defense Unit
TDV	- Tool Design Variance
	- Touchdown Velocity
	- Tumbleweed Diagnostic Vehicle
	- Twin and Divide
TDX	- Torque-Differential Transmitter
TDY	- Temporary Duty
TE	- Task Element
	- Terminal Equipment
	- Testing
	- Thermal Element
	- Threat Entry
	- Trailing Edge
	- Transearth
	- Transporter-Erector
	- Transverse Electric
	- Transverse Electrostatic
	- Twin Engine
T/E	- Telemetry Event
	- Test Equipment
	- Transporter-Erector
T-E	- Transporter-Erector
T&E	- Test and Evaluation
TEAR	- Traffic Equipment Assignment Requirement
TEAS	- Threat Evaluation and Action Selection
TEB	- Technical Evaluation Board
	- Tone Encoded Burst
	- Triethylborane
TEC	- Test, Evaluation, and Control
	- Turn Error Compensation
TECHTAF	- Technical Training Air Force
TECR	- Test Equipment Calibration Record
TED	- Test, Evaluation, and Development
	- Training Equipment Development
TEDS	- Test Equipment Design Section
TEEZI	- Threat Evaluation Equipment Zone of Interior
TEF	- Transfer on End of File
TEFLON	- Tetrafluoroethylene Resin
TEI	- Trans-Earth Injection
TEJ	- Transverse Expansion Joint
TEL	- Training Equipment List
	- Transporter-Erector-Launcher

TELCO	-	Telephone Communications
TELECOM	-	Telecommunications
TELECON	-	Telephone Conversation
	-	Teletypewriter Conference
TELEDAC	-	Telemetric Data Converter
TELEPAK	-	Telemetering Package
TELERAN	-	Television and Radar Navigation
TELESAT	-	Telecommunications Satellite
TELETYPE	-	Teletypewriter
TELEX	-	International Telephone Exchange
	-	Teletypewriter Exchange Service, Automatic (Western Union)
TELSCOM	-	Telemetry-Surveillance-Communications
TEM	-	Test Excursion Module
	-	Transverse Electromagnetic
TEMP	-	The Expanded Memory Print Program
TEMPO	-	Technical Military Planning Operation
	-	Total Evaluation of Management and Production Output
TENOC	-	Ten-Year Oceanographic Research Program
TEPI	-	Training Equipment Planning Information
TEPU	-	Technical Publication
TER	-	Transmitted Engineering Requirement
	-	Travel Expense Report
TERASCA	-	Terrier-ASROC-Cajuna
TERCOM	-	Terrain Contour Mapping
TEREL	-	Terrain Elevation at Interception Point
TERG	-	Training Equipment Requirements Guide
TERREL	-	Terrain Elevation
TES	-	Test Evaluation Support
	-	Training Evaluation Section
TETRA	-	Terminal Tracking Telescope
TEVROC	-	Tailored-Exhaust-Velocity Rocket
TEW	-	Tactical Early Warning
TEWA	-	Threat Evaluation and Weapon Assignment
TEWR	-	Thrust-to-Earth Weight Ratio
TEX	-	Teleprinter Exchange Service, Automatic (Western Union)
TF	-	Tactical Fighter
	-	Task Force
	-	Test Facility
	-	Test Fixture
	-	Test Flight
	-	Test to Failure
	-	Tool Fabrication
	-	Training Film
	-	Type of Flight
TFA	-	Transfer Function Analyzer
TFAT	-	Theoretical First Appearance

■ TFCS

TFCS	- Torpedo Fire Control System
TFCSD	- Total Federal Commissioned Service to Date
TFE	- Time From Event
	- Test Facilities and Equipment
TFG	- Transmit Format Generator
TFO	- Test Facility Operation
TFOL	- Tape File Octal Load
TFR	- Trouble and Failure Report
TFT	- Temporary Facility Tool
TFX	- Tactical Fighter, Experimental
TG	- Tail Gear
	- Task Group
	- Terminal Guidance
	- Torpedo Group
	- Tracking and Guidance
T&G	- Tongue and Groove
	- Tracking and Guidance
TGA	- Thermogravimetric Analysis
T-GAM	- Training-Guided Aircraft Missile
TGI	- Target Intensifier
TGL	- Toggle
TGS	- Telemetry Ground Station
	- Triglycine Sulfate
	- True Ground Speed
TGSE	- Test Ground Support Equipment
TGT	- Target
TH	- Terrain Height
	- True Heading
T&H	- Transportation and Handling
THAS	- Tumbleweed High Altitude Sample
THERM	- Thermometer
THERMISTOR	- Thermal Resistor
THERMO	- Thermostat
THESIS	- The Honeywell Engineering Status Information System
THI	- Temperature-Humidity Index
THOR	- Transistorized High-Speed Operations Recorder
THORAD	- Thor Advanced
THP	- Thrust Horsepower
THQ	- Theater Headquarters
THUD	- Thorium, Uranium, Deuterium
TI	- Tape Inverter
	- Target Identification
	- Technical Inspection
	- Technical Interchange
	- Texas Instruments, Incorporated
	- Time Index

TI	- Track Identity
	- Track Initiation
	- Track Initiator
	- Traffic Identification
TIA	- Training Integration Area
TIARA	- Target Illumination and Rescue Aircraft
TIC	- Tape Identification Card
	- Target Intercept Computer
	- Technical Information Center
	- Technical Integration Committee
	- Trainer Instructor Console
	- Transducer Information Center
TICTAC	- Time Compression Tactical Communications
TID	- Technical Information Department
TIDE	- Tactical International Data Exchange
TIDOS	- Table and Item Documentation System
TIDR	- Tool Investigation and Disposition Report
TIDY	- Track Identification
TIF	- Technical Information File
	- Transfer If Indicators Off
TIG	- Teletype Input Generator
	- The Inspector General
	- Transearth Injection Geometry
	- Tungsten Inert Gas
TIH	- Time In Hold
TII	- Table and Item Inventory
TIM	- Time Indicator, Miniature
	- Time Initiator Monitor
	- Time Meter
	- Tracking Information Memorandum
	- Tracking Instrument Mount
TIMM	- Thermionic Integrated Micro-Module
TIMOT	- Track, Initiation, Monitoring Overlap Technician
TIMS	- The Institute of Management Sciences
TIO	- Technical Implementation Office
	- Transfer If Indicators On
TIP	- Teletype Input Processing
	- Test In Process
	- Track Initiation and Prediction
TIPS	- Technical Interest Profiles
	- Telemetry Impact Prediction System
	- TIC (Technical Information Center) Intelligence Processing System
TIPSY	- Task Input Parameter Synthesizer
TIR	- Technical Information Release
	- Technical Intelligence Report

■ TIR

TIR	- Total Indicator Reading
	- Total Internal Reflection
TIREC	- TIROS Ice Reconnaissance
TIROS	- Television Infrared Observation Satellite
TIS	- Target Information System
	- Termination Inventory Schedule
	- Tern Island Station
	- Test Instrumentation System
	- Track Initiation Supervisor
TISE	- Technical Information Service
TIU	- Tape Identification Unit
TIVI	- Transearth Injection Velocity Increment
TIX	- Transfer on Index
TJ	- Turbojet
TJAG	- The Judge Advocate General
TJC	- Tower Jettison Command
	- Trajectory Chart
TJD	- Trajectory Diagram
TJM	- Tower Jettison Motor
TJOC	- Theater Joint Operations Center
TJP	- Turbojet Propulsion
TK	- Turbocompressor (USSR)
TL	- Talk-Listen
	- Test Link
	- Thrust Line
	- Tie Line
	- Time Length
	- Total Load
	- Trunckload
TLC	- Thin-Layer Chromatography
TLCC	- Training Launch Control Center
TLE	- Temperature Limited Emission
	- Thin-Layer Electrophoresis
TLF	- Terminal Launch Facility
TLG	- Tail Landing Gear
	- Telegraph
TLI	- Translunar Injection
TLM	- Telemetry
TLO	- Technical Liaison Officer
TLP	- Tabular List of Parts
TLRP	- Track Last Reference Position
TLS	- Telemetry Listing Submodule
	- Telescope
TLT	- Target Liaison Talker
	- Transportable Link Terminal

TLU	- Table Look Up
	- Threshold Logic Unit
TM	- Tactical Missile
	- Take-Off Mass
	- Task Memorandum
	- Technical Manual
	- Technical Memorandum
	- Telemetry
	- Temperature Meter
	- Time Modulation
	- Tone Modulation
	- Tooling Memorandum
	- Track Monitor
	- Training Manual
	- Transverse Magnetic
	- Trial Modification
T/M	- Telemetry
T&M	- Time and Material
	- Tooling and Material
TMAS	- Taylor Manifest Anxiety Scale
TMC	- Temporary Minor Change
	- The Marquardt Corporation
	- The Martin Company
	- Transportation Materiel Command
T/MC	- Talker per Megacycle
TMCA	- Titanium Metals Corporation of America
TMCOMP	- Telemetry Computation
TMCP	- Technical Manual Control Panel
TMD	- Theoretical Maximum Density
	- Time Division Multiplex
	- Total Mean Downtime
TMDT	- Total Mean Downtime
TMG	- Tactical Missile Group
TMGE	- Thermo-Magneto-Galvanic Effects
TMHF	- Transit Missile Hold Facility
TMI	- Tracking Merit Interception
	- Transfer on Minus
TMIS	- Technical Meetings Information Service
TML	- Titanium Metallurgical Laboratory
TMM	- Test Message Monitor
TMMD	- Tactical Moving Map Display
TMO	- Total Materiel Objective
TMOS	- Two Main Orbiting Spacecraft
TMP	- Theodolite Measuring Point
TMPROC	- Telemetry Processing

■ **TMR**

TMR	- Technical Memorandum Report
	- Tooling Material Request
	- Total Material Requirement
TMRBM	- Transportable Medium Range Ballistic Missile
TMS	- Track Monitor Special
	- Track Monitor Supervisor
	- Transport Monitor System
	- Type, Model, and Series
T/M/S	- Type, Model, and Series
TMSD	- Total Military Service to Date
TMT	- Transonic Model Tunnel
TMU	- Twin and Multiply
TMW	- Tactical Missile Wing
	- Textile Machine Works
TMX	- Tactical Missile, Experimental
TN	- Technical Note
	- Thermonuclear
	- Track Number
TNA	- The National Archives
TNC	- Track Number Conversion
TNO	- Transfer on No Overflow
TNS	- Track Number Sorted Table
TNT	- Trinitrotoluene
TNTBP	- Trinitrotoluene and Black Powder
TNX	- Transfer on No Index
TNZ	- Transfer on No Zero
TO	- Table of Organization
	- Tail Gear
	- Take-Off
	- Technical Order
	- Theater of Operations
	- Time Open
	- Time Over
	- Tooling Order
	- Tracking Officer
	- Transportation Officer
	- Turnout
T-O	- Take-Off
T/O	- Table of Organization
T&O	- Test and Operation
TOA	- Total Obligatory Authority
	- Transferred on Assembly
TOB	- Test One BIT
TOC	- Tactical Operations Center
	- Technical Order Compliance
	- Theater of Operations Command

372

TOC	- Timing Operations Center
	- Traffic Order Change
	- Trainer Operator Console
TOD	- Technical Objectives Document
	- Time of Delivery
TOE	- Tables of Organization and Equipment
	- Talker Omission Error
TO&E	- Tables of Organization and Equipment
TOES	- Tables of Organization and Equipment
	- Trade Off Evaluation System
TOF	- Time of Filing
	- Time of Flight
TOGO	- Time to Go
TOJ	- Track on Jamming
TOLO	- Time of Lockout
TOM	- Test Set, Overall, Missile
	- Tracking Operations Memorandum
	- Translator Octal Mnemonic
TOMCAT	- Theater of Operations Missile Continuous-Wave Anti-Tank Weapon
TOMCIS	- Test of Multiple Corridor Identification System
TOO	- Time of Origin
TOP	- Test Operations Project
	- Time of Penetration
	- Torque Oil Pressure
TOPHAT	- Terrier Operation Proof High Altitude Target
TOPS	- The Operational PERT System
TOPSI	- Topside Sounder, Ionosphere
TOR	- Tactical Operations Report
	- Technical Operating Report
	- Test Operations Resources
	- Time of Receipt
	- Tool Order Release
	- Training Observer-Recorder
	- Training Operations Report
TORACCS	- Tool Order-Reporting and Cost Control System
TORES	- Toxicological Research
TORF	- Time of Retro-Fire
TORPCM	- Torpedo Countermeasures and Deception
TORS	- Tool Order Reporting System
TOS	- Term of Service
	- TIROS Operational System
	- Training Operations Staff
TOSS	- TIROS Operational Satellite System
	- Training Operations Support Section

■ TOT

TOT	- Time of Transmission
	- Time on Tape
	- Time on Target (artillery support)
	- Time over Target (air support)
	- Tracking Officer Technician
TOTO	- Tongue of the Ocean
TOV	- Transfer on Overflow
TOW	- Take-Off Weight
	- Tube-Launched, Optically-Tracked, Wire-Guided Anti-Tank Missile
TOWA	- Terrain and Obstacle Warning and Avoidance
TP	- Target Practice
	- Technical Publication
	- Telemetry Processor
	- Temperature Profile
	- Test Plan
	- Test Point
	- Test Procedure
	- Tie Plate
	- Time Pulse
	- Tracking Program
	- Training Plan
	- Triple Pole
	- True Position
	- True Profile
	- Turbopump
	- Turn Point
T&P	- Target and Penetration
TPA	- Tape Pulse Amplifier
	- Technical Publications Announcements
	- Test Plans and Analysis
	- Test Point Access
	- Test Preparation Area
	- Track Producing Area
	- Turbopump Assembly
TPC	- Telecommunications Planning Committee
	- Time Polarity Control
TPCPR	- Tool and Production Change Planning Record
TPD	- Time Pulse Distributor
TPDT	- Triple-Pole Double-Throw
TPE	- Test Project Engineer
TP&E	- Test Planning and Evaluation
TPG	- Timing Pulse Generator
TPI	- Tape Phase Inverter
	- Target Position Indicator
	- Task Parameter Interpretation

TR

TPI	- Technical Proficiency Inspection
	- Teeth per Inch
	- Threads per Inch
TPL	- Test Parts List
	- Training Parts List
	- Transfer on Plus
TPM	- Tape Preventive Maintenance
	- Telemetry Processor Module
	- Transmission and Processing Model
TPMG	- The Provost Marshal General
TPOCP	- Turbopump Oxidizer Cavity Purge
TPP	- Test Point PACE (Prelaunch Automatic Checkout Equipment)
	- Tool and Production Planning
TPPCR	- Tool and Production Planning Change Record
TPPM	- Tool and Production Planning Manual
TPR	- Technical Performance Report
	- Teleprinter
	- Telescopic Photographic Recorder
TPS	- Tangent Plane System
	- Task Parameter Synthesizer
	- Technical Publishing Society
	- Test Preparation Sheet
	- Thermal Protection System
	- Track Processing Special
TPSI	- Torque Pressure in Pounds per Square Inch
TPSS	- Training Program Subsystem
TPST	- Triple-Pole Single-Throw
TPSY	- Task Parameter Synthesizer
TPT	- Time Priority Table
TP-T	- Target Practice with Tracer
TPTG	- Tuned Plane Tuned Grid
T&QA	- Test and Quality Assurance
T/Q/C	- Time-Quality-Cost
TQE	- Technical Quality Evaluation
TR	- Tape Recorder
	- Target Recognition
	- Technical Report
	- Test Request
	- Test Requirement
	- Time to Retro-Fire
	- Tooling Request
	- Torque Synchro Receiver
	- Tracking Radar
	- Transformer-Rectifier
	- Transmit-Receive
	- Transmitter-Receiver

■ **TR**

TR	- Transom
	- Transportation Request
	- Trouble Report
T/R	- Transformer-Rectifier
	- Transmit-Receive
	- Transmitter-Receiver
T&R	- Test and Return
	- Transmit and Receive
TRAAC	- Transit Research and Attitude Control Satellite
TRABOT	- Terrier Radar and Beacon Orientation Test
TRAC	- Text Reckoning and Compiling
	- Total Record Access Control
TRACALS	- Traffic Control, Approach, and Landing System
TRACE	- Task Reporting and Current Evaluation
	- Taxiing and Routing of Aircraft Coordinating Equipment
	- Tracking and Communication, Extra-Terrestrial
	- Transistor Radio Automatic Circuit Evaluator
TRACOMP	- Tracking Comparison
TRADAD	- Trace to Destination and Advise
TRADAT	- Transit Data Transmission System
TRADCOM	- Transporation Research and Development Command
TRADEX	- Target Resolution and Discrimination Experiment
	- Tracking Radar Experiment
TRADIC	- Transistorized Airborne Digital Computer
TRAJ	- Trajectory
TRAJ/HE	- Trajectory Heating Envelope
TRAJ/PS	- Trajectory Parametric Study
TRALANT	- Training Command, Atlantic
TRAMP	- Target Radiation Measurement Program
	- Test Retrieval and Memory Print
TRAMPS	- Temperature Regulator and Missile Power Supply
TRANET	- Tracking Network
	- Transit Network
TRANS	- Transformer
	- Transmittance
TRANSAC	- Transistorized Automatic Computer
TRANSDIV	- Transport Division
TRANSEC	- Transmission Security
TRANSRON	- Transport Squadron
TRAP	- Terminal Radiation Airborne Program
TRAPAC	- Training Command, Pacific
TRAWL	- Tape Read and Write Library
TRC	- Tracking, Radar-Inputs, and Correlation
	- Type Requisition Code
TRCC	- Tripartite Research Coordination Committee
TRCCC	- Tracking Radar Central Control Console

TRW ∎

TRCVR	- Transceiver
TRD	- Test Requirements Document
TR&D	- Training Research and Development
TRDA	- Three-Axis Rotational Control—Direct A
TRDB	- Three-Axis Rotational Control—Direct B
TRDTO	- Tracking Radar Data Take-Off
TREAT	- Transient Reactor Test Facility
TRECOM	- Transportation Research and Engineering Command
TRER	- Transient Radiation Effect on Radiation
TRF	- Test Requirements Handbook
	- Thermal Radiation at Microwave Frequencies
	- Tuned Radio Frequency
TRH	- Track History
TRICE	- Transistorized Real-Time Incremental Computer
TRID	- Track Identity
TRIDOP	- Tri-Doppler
TRIPOLD	- Transit Injector Polaris-Derived
TRIZON	- Three-Ton RAZON
TRLFSW	- Tactical Range Landing Force Support Weapon
TRM	- Trial Run Model
TRN	- Tooling Requirement Notice
	- Track Reference Number
TRNA	- Three-Axis Rotational Control—Normal A
TRNB	- Three-Axis Rotational Control—Normal B
TROO	- Transponder On-Off
TRP	- Traffic Regulation Point
TRPO	- Track Reference Printout
TRQ	- Torque
TRR	- Tactical Range Recorder
	- Target Ranging Radar
TR/R	- Technical Request/Release
TRRR	- Tool Rejection Rework Release
TRRT	- Tool Rejection Rework Tag
TRS	- Technical Reference Service
	- Test Requirements Summary
	- Tetrahedral Research Satellite
	- Time Reference System
TRSR	- Taxi and Runway Surveillance Radar
TRSSGM	- Tactical Range Surface-to-Surface Guided Missile
TRSSM	- Tactical Range Surface-to-Surface Missile
TRU	- Transportable Radio Unit
TRUD	- Time Remaining Until Dive
TRUMP	- Target Radiometric Ultraviolet Measuring Program
TRUT	- Time Remaining Until Transition
TRV	- Two-Rotor Vertical
TRW	- Thompson-Ramo-Wooldridge, Incorporated

■ TS

TS	- Taper Shank
	- Temperature Switch
	- Tensile Strength
	- Terminal Service
	- Test Set
	- Top Secret
	- Tough Situation
	- Tracking Supervisor
	- Transit Storage
	- Trouble Shoot
T/S	- Target Seeker
TSA	- Track Subsystem Analyst
	- Transfer System A
TSAD	- Trajectory-Sensitive Arming Device
TsAGI	- Central Aero-Hydrodynamics Institute (USSR)
T/S-AZ	- Target Selector—Azimuth
TSB	- Twin Sideband
TSC	- Tactical Support Center
	- Test Set Connection
	- Transfer System C
	- Transportable Communications System
TS-C	- Tooling Supplement to Contract
TSCA	- Target Satellite Controlled Approach
TSCLT	- Transportable Satellite Communications Link Terminal
TSCO	- Top Secret Control Officer
TSD	- Technical Support Directorate
	- Test Sequence Document
	- Theater Shipping Document
	- Track Situation Display
	- Traffic Status Display
TSDF	- Target System Data File
TSDU	- Target System Data Update
TSE	- Test Support Equipment
T/S-EL	- Target Selector—Elevation
TSF	- Tower Shield Facility
	- Turbulent Shear Flow
TSFC	- Thrust Specific Fuel Consumption
TSG	- Technical Services Group
	- Technical Specialty Group
	- The Surgeon General
	- Time Signal Generator
TSI	- Technical Standardization Inspection
	- Technical Systems, Incorporated
TsIAM	- P. I. Baranov Central Scientific-Research Institute of Aviation-Engine Construction (USSR)

TsIATIM	- Central Scientific Research Institute of Aviation Fuels and Lubricants (USSR)
TSIMS	- Telemetry Simulation Submodule
TSM	- Table Simulation
TSMC	- Transportation Supply Maintenance Command
TsNIIMF	- Central Scientific-Research Institute of the Maritime Fleet (USSR)
TSO	- Technical Standard Order
TSOP	- Tactical Standing Operating Procedure
TSOR	- Tentative Specific Operational Requirement
TSQ	- Time and Super Quick
TSR	- Tactical-Strike-Reconnaissance
	- Test Schedule Request
TSS	- Technical Services Section
	- Telecommunications Switching System
TST	- Test Support Table
	- Trouble Shooting Time
TSTM	- Tactical Subsystem Training Mission
TSU	- Technical Service Unit
	- Time Standard Unit
	- Twin and Subtract
TsUMB	- Central Administration of Weights and Measurements (USSR)
TSW	- Test Switch
TSX	- Telephone Satellite, Experimental
	- Transfer and Set Index
TT	- Solid Fuel (USSR)
	- Tactical Training
	- Technical Test
	- Texas Tower
	- Thrust Termination
	- Tooling Tag
	- Tracking Technician
	- Tracking Telescope
T/T	- Timing and Telemetry
TTA	- Travel Time Authorized
TTAF	- Technical Training Air Force
TTB	- Test Two BITS
TTC	- Tactical Training Center
	- Tape to Card
	- Target Track Central
	- Test and Training Center
	- Tracking, Telemetry, and Command
	- Translunar Trajectory Characteristics
TTDT	- Tactical Test Data Translator
TTE	- Telephone Terminal Equipment
	- Tentative Table of Equipment
	- Time to Event

■ TTET

TTET	- Turbine Transport Evaluation Team
TTF	- Test to Failure
	- Thoriated-Tungsten Filament
TTG	- Time to Go
TTGR	- Time-to-Go Rating
TTL	- Teletype Telling
	- Transistor-Transistor Logic
TTM	- Two-Tone Modulation
TTO	- Transmitter Turn-Off
TTP	- Tape to Print
TTPC	- Tripartite Technical Procedures Committee
TTR	- Target Tracking Radar
	- Teletype Translator
	- TRAP Translator
TTSR	- Temporary Threshold Shift Reduction
TTTT	- Tartar-Talos-Terrier-Typhon
TTY	- Teletype
	- Teletypewriter
TTY Q/R SS	- Teletype Query-Reply Subsystem
TU	- Remote Control (USSR)
	- Task Unit
TUC	- Time of Useful Consciousness
TUG	- TRANSAC Users Group
TUSLOG	- The United States Logistics Group
TV	- Television
	- Terminal Velocity
	- Test Vehicle
	- Time Variation of Gain
TVA	- Temporary Variance Authority
TVC	- Thrust Vector Control
TVCS	- Thrust Vector Control System
TVD	- Turboprop Engine (USSR)
TVDP	- Terminal Vector Display Unit
TVDS	- Toxic Vapor Detection System
TVEL	- Track Velocity
TVG	- Time Varied Gain
TVI	- Television Interference
TVM	- Thrust Vectoring Motor
TVOR	- Terminal VHF Omnidirectional Range
TVR	- Test Verification Report
TVSO	- Television Space Observatory
TW	- Tail Wind
	- Take-Off Weight
	- Test Wing
	- Time Word
	- Traveling Wave

TWA	- Trans World Airlines
	- Traveling Wave Amplifier
TWG	- Telemetry Working Group
	- Test Working Group
TWM	- Traveling Wave MASER
TWNP	- Tape-Wound Nylon Phenolic
TWP	- Technical War Plan
TWR	- Thrust-to-Weight Ratio
	- Tower
TWS	- Tactical Weather Station
	- Track-While-Scanning
TWSO	- Tactical Weapon System Operation
TWT	- Transonic Wind Tunnel
	- Traveling Wave Tube
	- Travel With Troops
TWTS	- Traveling Wave Tubes
TWX	- Teletypewriter Exchange (AT&T)
	- Time Wire Transmission
TX	- Torque Synchro Transmitter
	- Transmit
TXH	- Transfer on Index High
TXI	- Transfer with Index Incremented
TXL	- Transfer on Index Low
TXRX	- Transmitter-Receiver
TYDAC	- Typical Digital Automatic Computer
TYS	- Tensile Yield Strength
TYSD	- Total Years Service Date
TZE	- Transfer on Zero

U

U	- Unclassified
	- Unit
	- Unknown
	- Uranium
	- Utility
UA	- Microampere
	- Unit First Appearance
	- Urinalysis
UAC	- United Aircraft Corporation
	- Utility Airplane Council
UACTE	- Universal Automatic Control and Test Equipment
UAIDE	- Users of Automatic Information Display Equipment
UAL	- Unit Allowance List
	- Unit Authorization List
	- Upper Acceptance Level
UAM	- Underwater-to-Air Missile
	- Unnormalized Aid Magnitude
UAMC	- Utility Assemble Master COMPOOL
UAP	- United Aircraft Products, Incorporated
	- Upper Atmosphere Phenomena
UAR	- Upper Atmosphere Research
UATP	- Universal Air Travel Plan
UAUM	- Underwater-to-Air-to-Underwater Missile
UB	- Underwater Battery
	- Unit Bond
	- Utica-Bend
U/B	- Usage Block
UBC	- Universal Buffer Controller
UBD	- Utility Binary Dump
UBITRON	- Undulating Beam Interaction Electron Tube
U-Bomb	- Uranium-cased Atomic or Hydrogen Bomb
UC	- Unit Cooler
	- Upper Case

UCAL	- Upper Conformance Altitude
UCC	- Union Carbide Corporation (formerly UC&CC)
	- Utility Control Center
UC&CC	- Union Carbide & Carbon Corporation (now UCC)
UCDP	- Uncorrected Data Processor
UCE	- Unit Checkout Equipment
	- Unit Correction Entry
UCI	- Utility Card Input
UCIS	- Uprange Computer Input System
UCK	- Utility Checker
UCL	- Upper Control Limit
UCLA	- University of California at Los Angeles
UCLRL	- University of California Lawrence Radiation Laboratory
UCMJ	- Uniform Code of Military Justice
UCN	- Uniform Control Number
UCO	- Utility Compiler
UCON	- Utility Control
UCOS	- Uprange Computer Output System
UCP	- Utility Control Program
UCRL	- University of California Radiation Laboratory (see UCLRL)
UCSD	- Universal Communications Switching Device
UDC	- Universal Decimal Classification
UDEC	- Unitized Digital Electronic Calculator
UDIT	- Utility Data Insert Task
UDL	- Uniform Data Link
	- Up Data Link
UDMH	- Unsymmetrical Dimethyl Hydrazine
UDOFT	- Universal Digital Operational Flight Trainer
UDOP	- UHF Doppler System
UDR	- Utility Data Reduction
UDRC	- Utility Data Reduction Control
UDRO	- Utility Data Reduction Output
UDT	- Underwater Demolition Team
	- Universal Data Transcriber
UDTI	- Universal Digital Transducer Indicator
UDU	- Underwater Demolition Unit
UE	- Unit Entry
	- Unit Equipment
	- Until Exhausted
	- Update and Ephemeris
UEAC	- Unit Equipment Aircraft
UEE	- Unit Essential Equipment
UER	- Unplanned Event Record
	- Unsatisfactory Equipment Report
UET	- Universal Engineer Tractor
UETA	- Universal Engineer Tractor, Armored

■ **UETRT**

UETRT	- Universal Engineer Tractor, Rubber-Tired
UF	- Microfarad
	- Unit Final Fade
	- Unit of Fire
UFA	- Unnormalized Floating Add
UFCS	- Underwater Fire Control System
UFM	- Unnormalized Floating Multiply
UFO	- Unidentified Flying Object
UFS	- Unnormalized Floating Subtract
UFSPC	- Updated Field-Site Production Capability
UG	- Underground
UGL	- Utility General
UGLIAC	- United Gas Laboratory Internally-Programmed Automatic Computer
UH	- Microhenry
	- Unit Heater
UHA	- Ultra High Altitude
UHF	- Ultra High Frequency
UHF/DF	- Ultra High Frequency/Direction Finding
UHF/HF	- Ultra High Frequency/High Frequency
UHR	- Ultra High Resistance
UHT	- Umbilical Handling Technician
UHV	- Ultra High Vacuum
UI	- Unit Issue
	- Unit Item
UIT	- Utility Interim Tape
UK	- United Kingdom
	- Unknown
UL	- Underwriters' Laboratory, Incorporated (also ULI)
ULF	- Ultra Low Frequency
ULI	- Underwriters' Laboratory, Incorporated (also UL)
	- Universal Logic Implementer
UL-LL	- Upper-Limit, Lower-Limit
UL-LLC	- Upper-Limit, Lower-Limit Comparator
ULLV	- Unmanned Lunar Logistics Vehicle
ULM	- Utility Library Merge
ULN	- Unlaunchable
ULO	- Utility Library Output
UM	- Unit Manufacture
UMA	- Ultrasonic Manufacturers Association
UMC	- Universal Match Corporation
UMD	- Unit Manning Document
UME	- Unit Mission Equipment
	- Unit Mobility Equipment
UMR	- Upper Maximum Range
UMT	- Universal Military Training

UMTS	- Universal Military Training and Service
UMWS	- Unmanned Weapons Simulator
UN	- Unified Coarse Thread
	- United Nations
UNAAF	- Unified Action Armed Forces
UNACOM	- Universal Army Communication System
UNASGD	- Unassigned
UNAUTH	- Unauthorized
UNB	- Universal Navigation Beacon
UNC	- United Nations Command
	- United Nuclear Corporation
UNCDRP	- Universal Card Read-In Program
UNCL	- Unclassified
UNCOL	- Universal Computer-Oriented Language
UNCOPUOS	- United Nations Committee on Peaceful Uses of Outer Space
UNCR	- United Nations Command (Rear)
UNE	- Universal Nonlinear Element
UNEF	- Unified Extra Fine Thread
	- United Nations Emergency Forces
UNESCO	- United Nations Educational, Scientific, and Cultural Organization
UNF	- Unified Fine Thread
UNGA	- United Nations General Assembly
UNIADI	- Ukranian Aerodynamics Scientific Research Institute (USSR)
UNICODE	- UNIVAC Automatic Coding System
UNICOM	- Universal Integrated Communication System
UNIF COEF	- Uniformity Coefficient
UNIPLOT	- UNIVAC Plotting Routine
UNIPOL	- Universal Procedures-Oriented Language
UNISAP	- UNIVAC Share Assembly Program
UNIVAC	- Universal Automatic Computer
UNOPAR	- Universal Operator Performance Analyzer and Recorder
UNPS	- Universal Power Supply
UNRRA	- United Nations Relief and Rehabilitation Administration
UNS	- Unified Special Thread
UNSC	- United Nations Security Council
UNSECNAV	- Under Secretary of the Navy
UNTT	- United Nations Trust Territory
U/O	- Used On
UOA	- Used On Assembly
UOC	- Ultimate Operational Capability
UOL	- Underwater Object Locator
	- Utility Octal Load
UOV	- Unit of Variance
UP	- Unrotative Projectile
	- Users Project

■ UPAP

UPAP	-	Utility Parameter Assembly Program
UPE	-	Unit Proficiency Exercise
UPI	-	Unit Primarily Intended
UPP	-	Utility Print Punch
UPREAL	-	Unit Property Record and Equipment Authorization List
UPREL	-	Unit Property Record and Equipment List
UPS	-	United Proficiency System
	-	Universal Polar Stereographic
UPT	-	Utility Parameter Test
UPU	-	Universal Postal Union
UPWT	-	Unitary Plan Wind Tunnel
UQL	-	Unacceptable Quality Level
UR	-	Unsatisfactory Report
U/R	-	Uprange
URBM	-	Ultimate Range Ballistic Missile
URG	-	Universal Radio Group
URGR	-	Underway Replenishment Group
URI	-	Unintentional Radar Interference
	-	Utility Read-In Program
URLTR	-	Your Letter
URMSG	-	Your Message
URQ	-	Unsatisfactory Report Questionnaire
URSI	-	Union Radio Scientifique Internationale (also ISRU)
URT	-	Unit Run Time
US	-	Undersize
	-	United States
USA	-	United States Army
	-	United States of America
USAADCDA	-	United States Army Air Defense Combat Developments Agency
USAADCEN	-	United States Army Air Defense Center
USAADS	-	United States Army Air Defense School
USAAESWBD	-	United States Army Airborne, Electronics and Special Warfare Board
USAAFIO	-	United States Army Aviation Flight Information Office
USAAGCDA	-	United States Army Adjutant General Combat Developments Agency
USAAMC	-	United States Army Artillery and Missile Center
USAAMS	-	United States Army Artillery and Missile School
USAARMBD	-	United States Army Armor Board
USAARMC	-	United States Army Armor Center
USAARMCDA	-	United States Army Armor Combat Developments Agency
USAARMS	-	United States Army Armor School
USAARTYBD	-	United States Army Artillery Board
USAARTYCDA	-	United States Army Artillery Combat Developments Agency

USAATBD - United States Army Arctic Test Board
USAATC - United States Army Arctic Test Center
USAATCO - United States Army Air Traffic Coordinating Officer
USAAVNBD - United States Army Aviation Board
USAAVNC - United States Army Aviation Center
USAAVNCDA - United States Army Aviation Combat Developments Agency
USAAVNS - United States Army Aviation School
USABAAR - United States Army Board for Aviation Accident Research
USACACDA - United States Army Civil Affairs Combat Developments Agency
USACAG - United States Army Combined Arms Group
USACARMSCDA - United States Army Combined Arms Combat Developments Agency
USACBRCDA - United States Army CBR (Chemical, Biological, Radiological) Combat Developments Agency
USACDC - United States Army Combat Developments Command
USACDEC - United States Army Combat Developments Experimentation Center
USACECDA - United States Army Communications-Electronics Combat Developments Agency
USACGSC - United States Army Command and General Staff College
USACMLCSCH - United States Army Chemical Corps School
USACOMZEUR - United States Army Communications Zone, Europe
USACRAPAC - United States Army Command Reconnaissance Activities, Pacific Command
USACRF - United States Army Counterintelligence Records Facility
USACSSG - United States Army Combat Service Supply Group
USADSC - United States Army Data Services and Administrative Systems Command
USAECDA - United States Army Engineer Combat Developments Agency
USAECOM - United States Army Electronics Command
USAEMC - United States Army Engineer Maintenance Center
USAEPG - United States Army Electronic Proving Ground
USAERDL - United States Army Engineer Research and Development Laboratories
USAES   - United States Army Engineer School
USAF   - United States Air Force
USAFA   - United States Air Force Academy
USAFACS - United States Air Force Air Crew School
USAFBS - United States Air Force Bombardment School
USAFE   - United States Air Forces in Europe
USAFI   - United States Armed Forces Institute
USAFIT - United States Air Force Institute of Technology
USAFNS - United States Air Force Navigation School
USAFPS - United States Air Force Pilot School
USAFR   - United States Air Force Reserve

■ USAFSS

USAFSS   - United States Air Force Security Service
USAFTS   - United States Air Force Technical School
USAIAS   - United States Army Institute of Advanced Studies
USAIC   - United States Army Infantry Center
USAICDA   - United States Army Infantry Combat Developments Agency
USAINTC   - United States Army Intelligence Center
USAINTCA - United States Army Intelligence Corps Agency
USAINTCDA - United States Army Intelligence Combat Developments Agency
USAINTS   - United States Army Intelligence School
USAIPSG   - United States Army Industrial and Personnel Security Group
USAIS   - United States Army Infantry School
USALDJ   - United States Army Logistics Depot, Japan
USALMC   - United States Army Logistics Management Center
USALS   - United States Army Language School
USAMAPLA - United States Army Military Assistance Program Logistics Agency
USAMC   - United States Army Materiel Command
USAMDLC   - United States Army Materiel Development and Logistic Command
USAMICOM - United States Army Missile Command
USAMOCOM - United States Army Mobility Command
USAMUCOM - United States Army Munitions Command
USANWSG   - United States Army Nuclear Weapon Systems Safety Group
USAOCDA   - United States Army Ordnance Combat Developments Agency
USAOGMS   - United States Army Ordnance Guided Missile School
USAORRF   - United States Army Ordnance Rocket Research Facility
USAOSA   - United States Army Overseas Supply Agency
USAOSWD   - United States Army Office Special Weapons Developments
USAPHS   - United States Army Primary Helicopter School
USAPRDC   - United States Army Polar Research and Development Center
USAQMCDA - United States Army Quartermaster Combat Developments Agency
USAQMS   - United States Army Quartermaster School
USARADBD - United States Army Air Defense Board
USARAL   - United States Army, Alaska
USARCARIB - United States Army, Caribbean
USARCEN   - United States Army Records Center
USARDL   - United States Army Research and Development Laboratories
USAREUR   - United States Army, Europe
USARFT   - United States Army Forces, Taiwan
USARHAW   - United States Army, Hawaii
USARIS   - United States Army Information School
USARJ   - United States Army, Japan
USARMIS   - United States Army Mission
USARMLO   - United States Army Liaison Officer

USARPAC - United States Army, Pacific
USARYIS - United States Army, Ryukyu Islands
USASA - United States Army Security Agency
USASADEA - United States Army Signal Air Defense Engineering Agency
USASATSA - United States Army Signal Aviation Test Support Activity
USASC - United States Army Subsistence Center
USASCC - United States Army Strategic Communications Command
USASCS - United States Army Signal Center and School
USASETAF - United States Army Southern European Task Force
USASGV - United States Army Support Group, Vietnam
USASIMSA - United States Army Signal Materiel Support Agency
USASMCOM - United States Army Supply and Maintenance Command
USASRDL - United States Army Signal Research and Development Laboratory
USASSA - United States Army Signal Supply Agency
USASSD - United States Army Special Security Detachment
USASTC - United States Army Signal Training Center
USASTCFM - United States Army Signal Training Center and Fort Monmouth
USASWCDA - United States Army Special Warfare Combat Developments Agency
USASWS - United States Army Special Warfare School
USATATSA - United States Army Transportation Aircraft Test and Support Activity
USATC - United States Army Training Center
USATCAD - United States Army Training Center, Air Defense
USATCDA - United States Army Transportation Combat Developments Agency
USATCFA - United States Army Training Center, Field Artillery
USATECOM - United States Army Test and Evaluation Command
USATMC - United States Army Transportation Materiel Command
USATRECOM - United States Army Transportation Engineering Command
USAW - Underwater Security Advance Warning
USAWC - United States Army War College
USAWECOM - United States Army Weapons Command
USB - Upper Sideband
USBE - Unified S-Band Equipment
USC - Under Separate Cover
 - United States Code
USCC - United States Chamber of Commerce
U/SCC - Utility and Support Control Committee
USCG - United States Coast Guard
USC&G - United States Coast and Geodetic Survey (USCGS preferred)
USCGA - United States Coast Guard Academy
USCGAD - United States Coast Guard Air Detachment

■ USCGAS

USCGAS	- United States Coast Guard Air Station
USCGC	- United States Coast Guard Cutter
USCGS	- United States Coast and Geodetic Survey
USCIB	- United States Communications Intelligence Board
USCINCEUR	- United States Commander-In-Chief, Europe
USCM	- Unit Simulated Combat Mission
USCONARC	- United States Continental Army Command
USD	- United States Drone
USDA	- United States Department of Agriculture
USDC	- Underwater Search, Detection, Classification
USE	- United States Equipment
	- Unit Support Equipment
	- UNIVAC Scientific Exchange
USEC	- Microsecond
USER	- Unique-to-Site Equipment Review
USEUCOM	- United States European Command
	- United States European Communications
USFA	- United States Forces, Austria
USFS	- United States Forest Service
	- United States Frequency Standard
USG	- United States Gage
	- United States Government
USGS	- United States Geological Survey
USIA	- United States Information Agency
USIS	- United States Information Service
USLO	- United States Liaison Office
USM	- Underwater-to-Surface Missile
	- Unnormalized Subtract Magnitude
USMA	- United States Military Academy
USMC	- United States Marine Corps
USMCR	- United States Marine Corps Reserve
USMS	- United States Maritime Service
USN	- United States Navy
USNA	- United States Naval Academy
USNAF	- United States Naval Avionics Facility
USNAVEUR	- United States Navy, Europe
USNB	- United States Naval Base
USNC	- United States National Committee
USNCB	- United States Naval Construction Battalion
USNEES	- United States Naval Engineering Experiment Station
USNFR	- United States Naval Fleet Reserve
USNI	- United States Naval Institute
USNMDL	- United States Navy Mine Defense Laboratory
USNMR	- United States National Military Representative
USNMTC	- United States Naval Missile Test Center
USNPGS	- United States Navy Postgraduate School

USNR	- United States Naval Reserve
USNRDL	- United States Navy Radiological Defense Laboratory
USNS	- United States Naval Station
	- United States Navy Ship
USNSMSES	- United States Naval Ship Missile Systems Engineering Station (also NSMSES)
USNUSL	- United States Navy Underwater Sound Laboratory
USOFA	- Under Secretary of the Army
USOFAF	- Under Secretary of the Air Force
USP	- United States Pharmacopoeia
	- Utility Storage Print
USPCC	- Utility and Support Programming Control Committee
USPFO	- United States Property and Fiscal Officer
USPHS	- United States Public Health Service
USS	- United States Ship
	- United States Standard
	- United States Steel
USSBS	- United States Strategic Bombing Survey
USSR	- Union of Soviet Socialist Republics
USSTAF	- United States Strategic Air Force
USSTRICOM	- United States Strike Command
UST	- United States Testing Company
USW	- Under Sea Warfare
USWB	- United States Weather Bureau
UT	- Unit Time
	- Universal Time
	- Users Test
UTAL	- Upper Transition Altitude
UTC	- United Technology Center
	- United Technology Corporation
	- Unit Time Coding
	- Utility Tape Copy
UTIL	- Utility
	- Utilization
UTILIDOR	- Utility Corridor
UTLM	- Up Telemetry
UTM	- Universal Test Message
	- Universal Transverse Mercator
UTO	- Uncorrected Universal Time
UTP	- Utility Tape Processor
UTRON	- Utility Squadron
UTRS	- Universal Trouble Reporting System
UTS	- Unit Training Standard
	- Utility Interim Table Simulation
UTWG	- Utility Wing
UUF	- Micromicrofarad

■ **UUM**

UUM	- Underwater-to-Underwater Missile
UUT	- Unit Under Test
UV	- Microvolt
	- Ultraviolet
	- Under Voltage
UVASER	- Ultraviolet Amplification by Stimulated Emission of Radiation
UVD	- Under Voltage Device
UW	- Ultrasonic Wave
	- Unconventional Warfare
U/W	- Used With
UWATU	- Underwater Training Unit
UWR	- Underwater Range
UXB	- Unexploded Bomb
UXO	- Unexploded Ordnance
UXPLD	- Unexploded Bomb

V	- Valve
	- Vanadium
	- Varnish
	- Velocity
	- Verbal
	- Vertical Lift
	- Video
	- Voice
	- Volt
V-2	- Vergeltungswaffen Zwei
VA	- Video Amplifier
	- Volt-Ampere
V-A	- Verlort-Azimuth
	- Viper-Arrow
VAB	- Vertical Assembly Building
VAC	- Vector Analog Computer
	- Volts Alternating Current
VACTL	- Vertical Assembly Component Test Laboratory
VAF	- Vendor Approval Form
VAFB	- Vandenberg Air Force Base
VAM	- Voltammeter
VANT	- Vibration and Noise Tester
VAOR	- VHF Aural Omnirange
VAR	- Vertical Aircraft Rocket
	- Video-Audio Range
	- Visual-Aural Range
	- Volt-Ampere Reactive
VARISTOR	- Variable Resistor
VAST	- Versatile Automatic Specification Tester
VATE	- Versatile Automatic Test Equipment
VB	- Valve Box
	- Vertical Bomb
VC	- Varnish Cambric

- **VC**

VC	- Versatility Code
	- Vertical Center
	- Voice Ciphony
	- Voice Coil
VCA	- Visual Course Adapter
VCB	- Vertical Location of the Center of Buoyancy
VCC	- Vandenberg Control Center
	- Vehicle Crew Chief
VCG	- Vertical Location of the Center of Gravity
VCL	- Vertical Centerline
VCM	- Vibrating Coil Magnetometer
VCNO	- Vice Chief of Naval Operations
VCO	- Vernier Engine Cutoff
	- Voltage-Controlled Oscillator
VCofS	- Vice Chief of Staff
VCOS	- Voltage-Controlled Oscillators
VCP	- Vector Correction Program
	- Vehicle Collecting Point
VCS	- Vice Chief of Staff
	- Voltage Calibration Set
VCSL	- Voice Call Sign List
VCSR	- Voltage-Controlled Shift Register
VCT	- Voltage-Controlled Transfer
VCXO	- Voltage-Controlled Crystal Oscillator
VD	- Vertical Drive
	- Voltage Detector
V&D	- Visual and Dimensional
VDC	- Volts Direct Current
VDD	- Visual Display Data
	- Voice Digital Display
VDE	- Variable Display Equipment
VDF	- VHF Direction Finding
	- Video Frequency
VDFG	- Variable Diode Function Generator
VDH	- Variable Length Divide or Halt
VDP	- Variable Length Divide or Proceed
	- Vehicle Deadlined for Parts
	- Vertical Data Processing
VDPI	- Vehicle Direction and Position Indicator
VDS	- Variable Depth SONAR
	- Visual Docking Simulator
VE	- Value Engineering
VEB	- Variable Elevation Beam
VECO	- Vernier Engine Cutoff
VECOS	- Vertical Checkout Set
VEDAR	- Visible Energy Detection and Ranging

VEDS	- Vehicle Emergency Detection System
VEFCO	- Vertical Functional Checkout
VEH	- Vehicle
VEP	- Value Engineering Program
VERA	- Vision Electronic Recording Apparatus
VERC	- Vehicle Effectiveness Remaining Converter
VERIS	- Veterans Record Integration System
VERDAN	- Versatile Differential Analyzer
VERLORT	- Very Long Range Tracking
VERNITRAC	- Vernier Tracking by Automatic Correlation
VERTIJET	- Vertical Take-Off and Landing Jet
VEST	- Vertical Earth Scanning Test
VETP	- Vandenberg Engineering Test Program
VEV	- Voice-Excited Vocoder
VEWS	- Very Early Warning System
VF	- Vector Field
	- Verification of Function
	- Video Frequency
	- Voice Frequency
VFC	- Video Frequency Carrier
	- Video Frequency Channel
	- Voice Frequency Carrier
	- Voice Frequency Channel
VFCT	- Voice Frequency Carrier Teletype
VFO	- Variable Frequency Oscillator
VFR	- Visual Flight Rules
VFTG	- Voice Frequency Telegraph
VG	- Velocity Gravity
	- Vertical Gyroscope
V-G	- Velocity-Gravity
VGA	- Variable Gain Amplifier
VGI	- Vertical Gyroscope Indicator
VGP	- Vertical Ground Point
VGPI	- Visual Glide Path Indicator
VGSI	- Visual Glide Slope Indicator
VH	- Very High
	- Vickers Hardness
VHA	- Very High Altitude
VHAA	- Very High Altitude Abort
VHB	- Very High Bombardment
VHF	- Very High Frequency (band 8)
VHF/DF	- Very High Frequency Direction Finding
VHO	- Very High Output
VHP	- Very High Performance
	- Very High Pressure
VI	- Velocity, Inertial

■ VI

VI	- Viscosity Index
VIAM	- All-Union Scientific-Research Institute of Aviation Materials (USSR)
VIB	- Vertical Integration Building
VIBRECON	- Vibration-Recording Console
VICI	- Velocity Indicating Coherent Integrator
VIDAC	- Visual Information Display and Control
VIG	- Video Integrating Group
VIGS	- Visual Glide Slope
VIL	- Very Important Launch
VILP	- Vector Impedance Locus Plotter
VINS	- Velocity Inertial Navigation System
VIOC	- Variable Input-Output Code
VIP	- Variable Information Processing
	- Very Important Person
VIPERSCAN	- Viper Rocket with Scanner
VIPP	- Variable Information Processing Package
VIPS	- Voice Interruption Priority System
VIRNS	- Velocity Inertial Radar Navigation System
VIS	- Visual Instrumentation Subsystem
VISC	- Viscosity
VITAL	- Verification of Interceptor Tactics Logic
VJST	- Voice Jamming Simulator, TBS
VJSW	- Voice Jamming Simulator, Weapons
VKF	- Von Karman Gas Dynamics Facility
VL	- Vertical Lower
VLA	- Very Low Altitude
VLCR	- Variable Length Cavity Resonance
VLF	- Vertical Launch Facility
	- Very Low Frequency (band 4)
VLFS	- Variable Low Frequency Standard
VLM	- Variable Length Multiply
VLR	- Vector Length Subroutine
	- Very Long Range
	- Very Low Range
VLS	- Vapor-Liquid-Solid
VM	- Vector Message
	- Velocity Meter
	- Velocity Modulation
	- Voltmeter
V/M	- Volts per Meter
VMA	- Vehicle Maintenance Area
VMC	- Velocity, Minimum Control
	- Visual Meteorological Conditions
V/MIL	- Volts per Mil
VMO	- World Meteorological Organization (USSR)

VN	- Vestibular Neurons
VNAF	- Vietnamese Air Force
VNII	- All-Union Instrument Research Institute (USSR)
VNIIM	- All-Union Scientific Research Institute of Meteorology (USSR)
VNIITI	- All-Union Institute of Scientific and Technical Information (USSR)
VNITOPRIBOV	- All-Union Scientific Engineering and Technical Society of Instrument Making (USSR)
VNZMAE	- All-Union Institute of Terrestrial Magnetism and Atmospheric Electricity (USSR)
VO	- Verbal Orders
	- Voice
V/O	- Volume Percent
VOC	- Voice-Operated Coder
VOCM	- Vehicle Out of Commission for Maintenance
VOCODER	- Voice Coder
VOCP	- Vehicle Out of Commission for Parts
VODARO	- Vertical Ozone Distribution from the Absorption and Radiation of Ozone
VODAS	- Voice-Operated Device Anti-Singing
VODAT	- Voice-Operated Device for Automatic Transmission
VODER	- Voice Coder
VOGAD	- Voice Ground Air Defense
VOIS	- Visual Observation Instrumentation Subsystem
VOL	- Volume
VOM	- Volt-Ohm-Milliammeter
VOR	- VHF Omnidirectional Range
VORDME	- VHF Omnidirectional Range Distance Measuring Equipment
VORTAC	- VHF Omnidirectional Range/Tactical Air Navigation
VOSE	- Vacuum Operation of Spacecraft Equipment
VOT	- VOR Test Signal
VOX	- Voice-Operated Transmission
VP	- Vapor Pressure
	- Variable Pitch Propeller
	- Vent Pipe
	- Vertical Polarization
	- Vulnerable Point
V/P	- Vertical Polarization
VPC	- Vapor-Phase Chromatography
VPLCC	- Vehicle Propellant Loading Control Center
VPM	- Vehicles per Mile
	- Volts per Meter
	- Volts per Mil
VPS	- Vibrations per Second
VR	- Vendor Rating
	- Voltage Regulator

■ VR

VR	- Voltage Relay
VRB	- VHF Recovery Beacon
VRC	- Visible Record Computer
VRD	- Vacuum Tube Relay Driver
VRL	- Vertical Reference Line
VRMS	- Voltage Root Mean Square
VRPS	- Voltage Regulated Power Supply
VRSA	- Voice Reporting Signal Assembly
VRSD	- Vought Range Systems Division
VRU	- Voice Readout Unit
VS	- Vascular System
	- Vapor Seal
	- Vent Stack
	- Vector Space
	- Versus
	- Vibration Spectra
VSB	- Vestigial Sideband
VSC	- Vibration Safety Cutoff
	- Voltage Saturated Capacitor
VSCF	- Variable-Speed Constant-Frequency
VSD	- Vendor's Shipping Document
VSE	- Vehicle Systems Engineer
VSFR	- Visibility Forecast
VSM	- Vestigial Sideband Modulation
VSMF	- Vendor Specifications Microfilm File
VSMS	- Video Switching Matrix System
VSO	- Very Stable Oscillator
VSR	- Very Short Range
VSTOL	- Vertical and/or Short Take-Off and Landing
V/STOL	- Vertical and/or Short Take-Off and Landing (VSTOL preferred)
VSTP	- Visual Satellite Tracking Program
V-S-V	- Vehicle-Surface-Vehicle
VSWR	- Voltage Standing Wave Ratio
VT	- Proximity Fuze Device
	- Vacuum Tube
	- Variable Thrust
	- Variable Time
	- Voice Tube
V-T	- Velocity-Time
VTA	- Variable Transfer Address
VTF	- Vertical Test Facility
	- Vertical Test Fixture
VTM	- Voltage Tunable Magnetron
VTO	- Vertical Take-Off
	- Voltage Tunable Oscillator

VTOHL	- Vertical Take-Off, Horizontal Landing
VTOL	- Vertical Take-Off and Landing
VTP	- Vandenberg Test Program
VTR	- Vehicle Time Reproducer
	- Video Tape Recorder
VTS	- Vandenberg Tracking Station
	- Vertical Test Site
	- Vertical Test Stand
VTTI	- All-Union Precision Instrument Industry Trust (USSR)
VTU	- Volunteer Reserve Training Unit
VTVM	- Vacuum Tube Voltmeter
VU	- Vertical Upper
	- Volume Unit
VUP	- Vela Uniform Program
VVIA	- N. E. Zhukovskii Air Force Engineering Academy (USSR)
VVR	- Variable Voltage Rectifier
VVS	- Vertical Velocity Sensor
VVSS	- Vertical Volute Spring Suspension
VWL	- Variable Word Length
VWP	- Variable Width Pulse
VWSS	- Vertical Wire Sky Screen

W	- Water
	- Watt
	- Waveguide
	- Weight
	- West
	- White
	- Wide
	- Wire
	- Word
WA	- Watertown Arsenal
	- Waveform Analyzer
	- Weapon Assignment
	- Word Add
WAAB	- War Alert Action Book
WAACS	- Western Airways and Air Communications Service
WAADS	- Washington Air Defense Sector
WAB	- Weapons Allocation Branch
	- When Authorized By
WAC	- Wake Analysis and Control
	- Women's Army Corps
	- Worked All Continents
	- World Aeronautical Chart
WACO	- Written Advice of Contracting Officer
WAD	- Weapon Assignment Display
WADC	- Western Air Defense Command
	- Wright Air Development Center
WADD	- Wright Air Development Division
WADF	- Western Air Defense Force
WADS	- Wide-Area Data Service
WAF	- Women in the Air Force
WAFB	- Whiteman Air Force Base
WAFFLE	- Wide Angle Fixed Field Locating Equipment
WAIS	- Wechsler Adult Intelligence Scale

WAL	- Watertown Arsenal Laboratory
WALDO	- Wichita Automatic Linear Data Output
WALP	- Weapons Assignment Linear Programming
WAMOSCOPE	- Wave Modulated Oscilloscope
WAO	- Weapons Allocation Officer
WAP	- Weapon Assignment Program
	- Work Assignment Procedure
WARM	- Weapons Assignment Research Model
WARP	- WIN Adaptation Reduction Program
WAS	- Weapons Application Study
WASP	- Weather Atmospheric Sounding Projectile
	- Westinghouse Advanced Systems Planning
	- Women's Air Force Service Pilots
WATS	- Wide Area Telephone Service
	- Wide Area Transmission System
WAVES	- Women Accepted for Volunteer Emergency Service
WAX	- Weapon Assignment and Target Extermination
WB	- Weather Bureau
	- Wheelbase
	- Wideband
	- Will Be
W/B	- Wideband
	- Wire Bundle
WBAN	- Weather Bureau Air Force/Navy
WBCT	- Wideband Current Transformer
WBD	- Wideband Data
WBDL	- Wideband Data Link
WBI	- Will Be Issued
WBNL	- Wideband Noise Limiting
WBNS	- Water Boiler Neutron Source
WBP	- Wartime Basic Plan
WBR	- Whole Body Radiation
WBTE	- Weapons Battery Terminal Equipment
WC	- War College
	- Weapon Carrier
	- Wet Countdown
	- Work Card
WCCMORS	- West Coast Classified Military Operations Research Symposium
WCD	- Weather Card Data
WCDB	- Wing Control During Boost
WCE	- Weapon Control Equipment
WCEMA	- West Coast Electronic Manufacturers Association
WCF	- White Cathode Follower
WCM	- Wired-Core Matrix
	- Word Combine and Multiplexer

■ WCMR

WCMR	- Western Contract Management Region
WCP	- Wartime Capabilities Plan
	- Wing Command Post
WD	- Weapons Director
	- When Directed
	- Wind Direction
	- Word
W/D	- Withdrawn
WDA	- Weapons Defended Area
WDC	- Western Defense Command
	- World Data Centers
WDD	- Western Development Division
WDE	- Weapons Direction Evaluation
WDF	- Wall Distribution Frame
WDI	- Warhead Detection Indicator
WDL	- Western Development Laboratory
WDP	- Weapons Direction Program
WDPC	- Western Data Processing Center
WDR	- Write Drum
WDS	- Weapons Delivery System
	- Weapons Direction Section
	- Weapons Direction Simulator
	- Weapons Direction Supervisor
WDT	- Weapons Director Technician
WEACAP	- Weapon Capability
WEADES	- Western Electric Air Defense Engineering Service
WEAPD	- Western Air Procurement District
WEARCON	- Weather Observation and Forecasting Control
WEASEL	- Weapon Selection
WEC	- Westinghouse Electric Company
WECO	- Western Electric Company
WECS	- Water-Glycol Evaporator Control System
WEDAC	- Westinghouse Digital Airborne Computer
WEDGE	- Waterless Electrical Data Generating Effortless
	- Weapon Development Glide Entry
WEF	- Write End of File
WEMA	- Western Electronics Manufacturers Association
WESCOM	- Weapon System Cost Model
WESCON	- Western Electronic Show and Convention
WEST	- Weapons Evaluation and Subsystem Training
	- Western Electric System Test
WESTAF	- Western Transport Air Force
WESTAR	- Waterways Experiment Station Terrain Analyzer Radar
WESVAL	- Weapon System Evaluation
WETAC	- Westinghouse Electronic Tubeless Analog Computer
WEU	- Western European Union

WEXVAL	- Weapons Examination-Evaluation
WFNA	- White Fuming Nitric Acid
WG	- Waveguide
	- Western Gear Corporation
	- Wing
W/G	- Water-Glycol
W-G	- Water-Glycol
WGAI	- Working Group Agenda Item
WGL	- Weapon Guidance Laboratory
WGM	- Working Group Meeting
WGR	- Western GEEIA Region
WGS	- World Geodetic System
WGT	- Water-Glycol Cooling Unit Technician
	- Weapons Guidance and Tracking
WH	- Warhead
	- Watt-Hour
	- Western Hemisphere
W/H	- Warhead
	- Watt-Hour
W-H	- Watt-Hour
WHASA	- Washington Headquarters Army Security Agency
WHD	- Western Hemisphere Defense
	- Write Head Driver
WHIP	- Wideband High-Intercept Probability
WHM	- Watt-Hour Meter
WHO	- World Health Organization
WHOI	- Woods Hole Oceanographic Institute
WHR	- Watt-Hour
WHS	- White Sands, New Mexico (remote site)
WIFNA	- Inhibited White Fuming Nitric Acid
WILCO	- Will Comply
WIN	- Weapon Interception
WINA	- Witton Network Analyzer
WIND	- Weather Information Network and Display
WINDS	- Weather Information Network and Display System
WIPS	- Washington Intelligence Data Processing System
WIR	- Weekly Intelligence Review
W&IR	- Work and Inspection Record
WIS	- Wallops Island Station
WISE	- Whirlwind I SAGE Evaluation
	- Women in Space Earliest
WITS	- Weather Information Telemetry System
WJCC	- Western Joint Computer Conference
WK	- Week
WL	- Warning Light
	- Water Line

■ WL

WL	- Wind Load
W/L	- Water Line
	- Weapon Loading
WLAC	- Watson Laboratories Air Materiel Command
WLBM	- Wings-Level Bombing System
WLD	- Warning Light Relay Driver
	- West Longitude Date
WLI	- Water Landing Impact
WM	- Wattmeter
	- Woman Marine
W/M	- Words per Minute
WMC	- Weapons Monitoring Center
WMO	- World Meteorological Organization
WMPA	- Wet Maximum Power Available
WMR	- Woomera (Australia) Missle Range
WMS	- Waste Management System
	- World Magnetic Survey
WN	- Will Not
WNB	- Will Not Be
WNP	- Will Not Proceed
WNR	- Western NORAD Region
WO	- Warning Order
	- Work Order
W/O	- Weight Percent
	- Without
WOC	- Without Compensation
WOCP	- Weapon Out of Commission for Parts
WOE	- Without Equipment
W/OE&SP	- Without Equipment and Spare Parts
WOG	- Water, Oil, and Gas
WOLF	- Flow spelled backward designating reverse flow solid rocket
WOM	- Weapons Output Makeup
	- Woomera, Australia (remote site)
WOO	- Western Operations Office
WOPE	- Without Personnel and Equipment
WOR	- Wear-Out Rate
	- Work Order Request
WORTAC	- Westinghouse Overall Radar Tester and Calibrator
WOSAC	- World-Wide Synchronization of Atomic Clocks
WOSD	- Weapons Operational Systems Development
WOTCU	- Wave-Off and Transition Control Unit
WOW	- Word on Way
WP	- Waterproof
	- Weatherproof
	- White Phosphorus
	- Will Proceed

WP	- Working Point
	- Working Pressure
	- Work Package
WPAFB	- Wright-Patterson Air Force Base
WPB	- War Plan, Basic
	- War Production Board
	- Write Printer Binary
WPBC	- Western Pacific Base Commander
WPC	- Watts per Candle
WPD	- Write Printer Decimal
WPI	- Worcester Polytechnic Institute
WPM	- Words per Minute
WPNSTA	- Weapons Station
WPO	- West Pacific Ocean
WPP	- Weapons Production Program
WPS	- Wartime Capability Play, Short Range
	- Words per Second
WPU	- Write Punch
WPWOD	- Will Proceed Without Delay
WR	- Wall Receptacle
	- Weapon Radius
	- Weapon Requirements
	- Write
WRA	- Wind Restraint Area
WRAIR	- Walter Reed Army Institute of Research
WRAMA	- Warner Robins Air Materiel Area
WRAP	- Weapons Readiness Achievement Program
WRBC	- Weather Relay Broadcast Center
WRC	- Weather Relay Center
	- Winston Research Corporation
WRE	- Weapons Research Establishment
WREDAC	- Weapons Research Establishment Digital Automatic Computer
WRL	- Willow Run Laboratory
	- Wing Reference Line
WRM	- War Readiness Materiel
WRR	- WARM Run Recording
WRRC	- Willow Run Research Center
WRRS	- Wire Relay Radio System
WRS	- Weather Relay Squadron
	- Write Select
WRUSS	- Western Reserve University Relay Searching Selector
WS	- Wallops Station
	- Water Supply
	- Weapon System
	- Wind Speed

- **WS**

WS	– Work Statement
W/S	– Will Slip
WSA	– Working Standard Area
WSC	– Weapon System Contractor
WSCE	– Weapon System Calibration Equipment
WSCP	– Weapon System Stock Control Plan
	– Weapon System Stock Control Point
WSD	– Weapon System Directive
	– Weapon System Director
WSDC	– Weapon System Design Criteria
WSDD	– Weapon Status Digital Display
WSDP	– Weapon System Development Plan
WSE	– Weapon Support Equipment
	– Weapon System Equipment
	– Western Society of Engineers
	– Work Shop Equipment
WSECL	– Weapon System Equipment Component List
WSED	– Weapon System Evaluation Division
WSEG	– Weapon System Evaluation Group
WSEM	– Weapon Support Equipment Maintenance
	– Weapon System Evaluator Missile
WSIL	– Weapon System Integration Laboratory
WSM	– Weapon System Manager
WSMO	– Weapons System Materiel Officer
WSMR	– White Sands Missile Range
WSO	– Washington Standardization Officer
WSOE	– Weapon System Operational Equipment
WSP	– Weapon System Program
WSPACS	– Weapon System Planning and Control System
WSPG	– Weapon System Phasing Group
	– White Sands Proving Ground
WSPO	– Weapon System Project Office
WSPR	– Weapon System Problem Report
WSR	– Weapon System Review
WSS	– Warfare Systems School
	– Weapon Support System
	– Weapon System Storage
	– Wind Shear Spike
WSSA	– Weapon System Support Activities
WSSC	– Weapon System Support Center
WSSCA	– White Sands Signal Corps Agency
WSSCL	– Weapon System Stock Control List
WSSL	– Weapon System Stock List
WSSM	– Weapon System Supply Manager
WSSR	– Weekly Sector Status Report
WSSS	– Weapon System Storage Site

WST	- Weapon System Test
	- Weapon System Training
	- Write Symbol Table
WT	- Watertight
	- Weight
	- Wind Tunnel
WTB	- Write Tape Binary
WTD	- Write Tape Decimal
WTG	- Wind Tape Generation
WTL	- Wind Tunnel (WT preferred)
WTR	- Western Test Range (AFWTR preferred)
	- Work Transfer Record
WTS	- War Training Service
WU	- Western Union
WUTC	- Western Union Telegraph Company
WUX	- Western Union Message
WW	- Weather Wing
	- Wire-Wound
	- World War
	- World-Wide
WW I	- Whirlwind I
WW II	- Whirlwind II
WWCC	- World-Wide Coordinating Committee
WWMCCS	- World-Wide Military Command and Control System
WWV	- World Wide Time (U.S. National Bureau of Standards)
WWVB	- World Wide Time Boulder, Colorado (U. S. National Bureau of Standards)
WWVH	- World Wide Time Hawaii (U. S. National Bureau of Standards)
WX	- Weather
WX-AM	- Weather and Air Movements

X	- Experimental
XB	- Experimental Bomber
XCONN	- Cross Connection
XCVR	- Transceiver
XDP	- X-ray Density Probe
XEG	- X-ray Emission Gauge
XEQ	- Executive
XFER	- Transfer
XFMR	- Transformer
XGAM	- Experimental Guided Aircraft Missile
XHV	- Extreme High Vacuum
XLR	- Experimental Liquid Rocket
XLWB	- Extra-Long Wheelbase
XM	- Experimental Missile
XMAS	- Extended Mission Apollo Simulation
XMFR	- Transformer
XMIT	- Transmit
XMITS	- Transmits
XMSN	- Transmission
XMT	- Transmit
XMTG	- Transmitting
XMTL	- Transmittal
XMTR	- Transmitter
XO	- Executive Officer
XPDR	- Transponder
XPNDR	- Transponder
X-RAY	- Energetic High-Frequency Electromagnetic Radiation
XSECT	- Cross Section
XSM	- Experimental Strategic Missile
XSPV	- Experimental Solid-Propellant Vehicle
XSTR	- Transistor
XT	- Crosstalk
XTASI	- Exchange of Technical Apollo Simulation Information

XTEL	- Crosstell
XTI	- Crosstell Input
XTLO	- Crystal Oscillator
XTRAN	- Experimental Language
XTS	- Crosstell Simulator
XUV	- Extreme Ultraviolet
XW	- Experimental Warhead
XWS	- Experimental Weapon Specification
	- Experimental Weapon System

Y

Y	- Prototype
	- Yaw
	- Yellow
YAK	- Yakovlev (Russian aircraft designer)
YAP	- Yaw and Pitch
YD	- Yard
YDSO	- Yards and Docks Supply Office
YIG	- Yttrium Iron Garnet
YM	- Prototype Missile
	- Yawing Moment
YP	- Yellow Phone
	- Yield Point
YR	- Year
YS	- Yield Strength
YSM	- Prototype Strategic Missile
YTS	- Yuma Test Station

Z

Z	- Zebra
	- Zinc
	- Zone
	- Zulu
ZA	- Zero and Add
ZAR	- Zeus Acquisition Radar
ZDC	- Zeus Defense Center
ZDCTBS	- Zeus Defense Center Tape and Buffer System
ZDP	- Zero Delivery Pressure
ZDR	- Zeus Discrimination Radar
ZEBRA	- Zero Energy Breeder Reactor Assembly
ZEEP	- Zero Energy Experimental Pile
ZEL	- Zero-Length Launcher
ZETA	- Zero Energy Thermonuclear Apparatus
ZETR	- Zero Energy Thermal Reactor
ZF	- Zone of Fire
ZFS	- Zero-Field Splitting
ZI	- Zone of Interior
ZIP	- Zinc-Impurity Photodetector
	- Zone Improvement Plan
ZK	- Nike-Zeus at Kwajalein
ZLC	- Zero-Lift Cord
ZM	- Nike-Zeus at Point Mugu
ZMAR	- Zeus Multifunction Array Radar
ZMMD	- Zurich, Mainz, Munich, Darmstadt
ZODIAC	- Zone Defense Integrated Active Capability
ZPA	- Zeus Program Analysis
ZPEN	- Zeus Project Engineer Network
ZPO	- Zeus Project Office
ZPPR	- Zero Power Plutonium Reactor
ZPR	- Zero Power Reactor
ZS	- Zero and Subtract
ZT	- Zone Time

- ZTO

ZTO - Zone Transportation Office
ZURF - Zeus Uprange Facility
ZW - Nike-Zeus at White Sands
ZZB - Zanzibar, Tanganyika (remote site)

MISSILE, ROCKET, PROBE DESIGNATION SYSTEM*

Status

J	- Special Test, Temporary	Vehicles on special test programs by authorized organizations and vehicles on bailment contract having a special configuration to accommodate the test. At completion of the test, the vehicles will be either returned to their original configuration or returned to standard operational configuration.
N	- Special Test, Permanent	Vehicles on special test programs by authorized activities and vehicles on bailment contract, whose configurations are so drastically changed that return of the vehicles to their original configurations or conversion to standard operational configurations is beyond practicable or economical limits.
X	- Experimental	Vehicles in a developmental or experimental stage but not established as standard vehicles for service use.
Y	- Prototype	Preproduction vehicles procured for evaluation and test of a specific design.
Z	- Planning	Vehicles in the planning or predevelopment stage.

---

*Accepted for joint service usage 27 June 1963 as per Air Force Regulation 66-20, Army Regulation 705-36, and BUWEPS Instruction 8800.2. These regulations implement DOD Directive 4000.20 issued 11 December 1962. The Air Force is the authority for the maintenance of the designation system and for the assignment of new designations. This system does not cover the space vehicles, space boosters, rocket systems designated for line-of-sight ground fire against ground targets, and naval torpedoes.

MISSILES, ROCKETS, AND PROBES

Launch Environment*

A	- Air	Air launched.
B	- Multiple	Capable of being launched from more than one environment.
C	- Coffin	Horizontally stored in a protective enclosure and launched from the ground.
H	- Silo Stored	Vertically stored below ground level and launched from the ground.
L	- Silo Launched	Vertically stored and launched from below ground level.
M	- Mobile	Launched from a ground vehicle or movable platform.
P	- Soft Pad	Partially or nonprotected in storage and launched from the ground.
R	- Ship	Launched from a surface vessel such as ship, barge, etc.
U	- Underwater	Launched from a submarine or other underwater device.

Mission

D	- Decoy	Vehicles designed or modified to confuse, deceive, or divert enemy defenses by simulating an attack vehicle.
E	- Special Electronic	Vehicles designed or modified with electronic equipment for communications, countermeasures, electronic radiation sounding, or other electronic recording or relay missions.
G	- Surface Attack	Vehicles designed to destroy enemy land or sea targets.
I	- Intercept-Aerial	Vehicles designed to intercept aerial targets in defensive or offensive roles.
Q	- Drone	Vehicles designed for target, reconnaissance, or surveillance purposes.
T	- Training	Vehicles designed or permanently modified for training purposes.

*The launch environment symbol is optional when a status symbol is applicable.

MISSILES, ROCKETS, AND PROBES

U	- Underwater Attack	Vehicles designed to destroy enemy submarines or other underwater targets or to detonate underwater.
W	- Weather	Vehicles designed to observe, record, or relay data pertaining to meteorological phenomena.

Vehicle Type

M	- Guided Missile	Unmanned, self-propelled vehicles designed to move in a trajectory or flight path all or partially above the earth's surface and whose trajectory of course, while the vehicle is in motion, is capable of being controlled remotely or by homing systems, or by inertial and/or programmed guidance from within. This term does not include space vehicles, space boosters, or naval torpedoes, but does include target and reconnaissance drones.
N	- Probe	Non-orbital instrumented vehicles not involved in space missions that are used to penetrate the aerospace environment and transmit or report back information.
R	- Rocket	Self-propelled vehicles without installed or remote control guidance mechanisms, whose trajectory or flight path cannot be altered after launch. Rocket systems designed for line-of-sight ground fire against ground targets are not included.

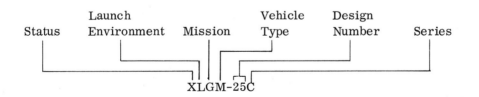

Example:

XLGM-25C - Experimental, Silo Launched, Surface Attack, Missile, design number 25, and "C" series (third modification).

AIRCRAFT DESIGNATION SYSTEM*

Special Use

G	- Permanently Grounded (For Instruction and ground training purposes.)
J	- Special Test, Temporary (Modified for special testing. Upon completion of tests, plane will be restored to its original design.)
N	- Special Test, Permanent (Permanently modified for testing.)
X	- Experimental (Not yet adapted for service use.)
Y	- Prototype (Purchased in limited numbers for complete testing of design.)
Z	- Planning (Indicates the aircraft is in the early stages of planning or development.)

Mission Modification

A	- Attack
C	- Cargo/Transport
D	- Director (For controlling drone aircraft or missiles.)
E	- Special Electronic Installation (For airborne early warning etc.)
H	- Search/Rescue
K	- Tanker
L	- Cold Weather (For Arctic or Antarctic operations.)
M	- Missile Carrier
Q	- Drone
R	- Reconnaissance
S	- Anti-Submarine
T	- Trainer
U	- Utility
V	- Staff
W	- Weather

---

*Accepted for joint service usage September 1962 as per Air Force Regulation 66-11, Army Regulation 700-26, and BUWEPS Instruction 13100.7. Previously the Army, Air Force, and Navy each had its own aircraft designating system.

AIRCRAFT

Basic Mission

A	- Attack
B	- Bomber
C	- Cargo/Transport
E	- Special Electronic Installation
F	- Fighter
H	- Helicopter
K	- Tanker
O	- Observation
P	- Patrol
S	- Anti-Submarine
T	- Trainer
U	- Utility
V	- VTOL and STOL (Planes designed for vertical take-off and landing. Also aircraft capable of taking off and landing in a minimum prescribed distance.)
X	- Research
Z	- Airship

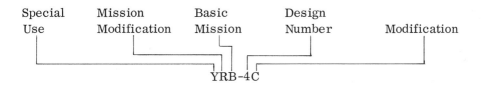

Special Use | Mission Modification | Basic Mission | Design Number | Modification

YRB-4C

Example:

YRB-4C - Prototype, Reconnaissance, Bomber, design number 4, and third modification.

Initially, an aircraft has only the basic mission letter and design number. Modification letters are added according to design changes. When the original, basic mission is changed, a mission modification letter is added. If the aircraft is designated for special use, a specific, special use letter is added. In most cases, an aircraft will not have both the special use and mission modification letters, but to eliminate confusion in cases when this does happen, the special-use letters are different from the mission modification letters.

SHIP DESIGNATIONS*

AD	- Destroyer Tender
ADG	- Degaussing Vessel
AE	- Ammunition Ship
AF	- Store Ship
AFDB	- Large Auxiliary Floating Dry Dock
AFDL	- Small Auxiliary Floating Dry Dock
AFDM	- Medium Auxiliary Floating Dry Dock
AG	- Miscellaneous
AGB	- Icebreaker
AGC	- Amphibious Force Flagship
AGP	- Motor Torpedo Boat Tender
AGR	- Radar Picket Ship
AGS	- Surveying Ship
AGSC	- Coastal Surveying Ship
AG(SS)	- Auxiliary Submarine
AH	- Hospital Ship
AK	- Cargo Ship
AK(SS)	- Cargo Submarine
AKA	- Attack Cargo Ship
AKD	- Cargo Ship, Dock
AKL	- Light Cargo Ship
AKN	- Net Cargo Ship
AKS	- General Stores Issue Ship
AKV	- Cargo Ship and Aircraft Ferry
AN	- Net Laying Ship
AO	- Oiler
AOE	- Fast Combat Support Ship
AOG	- Gasoline Tanker
AOR	- Replenishment Fleet Tanker
AO(SS)	- Submarine Oiler
AP	- Transport
AP(SS)	- Transport Submarine

---

*Addition of the suffix "(N)" to the identifying classification of a naval vessel or service craft indicates that the particular vessel or craft has nuclear propulsion.

SHIPS

APA	- Attack Transport
APB	- Self-Propelled Barracks Ship
APC	- Small Coastal Transport
APD	- High Speed Transport
APL	- Barracks Ship (non-self-propelled)
AR	- Repair Ship
ARB	- Battle Damage Repair Ship
ARC	- Cable Repairing or Laying Ship
ARD	- Auxiliary Floating Dry Dock
ARG	- Internal Combustion Engine Repair Ship
ARL	- Landing Craft Repair Ship
ARS	- Salvage Vessel
ARSD	- Salvage Lifting Vessel
ARST	- Salvage Craft Tender
ARV	- Aircraft Repair Ship
ARVA	- Aircraft Repair Ship (Aircraft)
ARVE	- Aircraft Repair Ship (Engine)
AS	- Submarine Tender
ASR	- Submarine Rescue Vessel
ATA	- Auxiliary Ocean Tug
ATF	- Fleet Ocean Tug
ATR	- Rescue Ocean Tug
AV	- Seaplane Tender
AVB	- Advanced Aviation Base Ship
AVM	- Guided Missile Ship
AVP	- Small Seaplane Tender
AVS	- Aviation Supply Ship
AW	- Distilling Ship
BB	- Battleship
BBG	- Guided Missile Capital Ship
CA	- Heavy Cruiser
CAG	- Guided Missile Heavy Cruiser
CB	- Large Cruiser
CBC	- Large Tactical Command Ship
CG	- Guided Missile Cruiser
CL	- Light Cruiser
CLAA	- Antiaircraft Light Cruiser
CLC	- Tactical Command Ship
CLG	- Guided Missile Light Cruiser
CVA	- Attack Aircraft Carrier
CVE	- Escort Aircraft Carrier
CVHA	- Assault Helicopter Aircraft Carrier
CVHE	- Escort Helicopter Aircraft Carrier
CVL	- Small Aircraft Carrier
CVS	- ASW Support Aircraft Carrier
CVU	- Utility Aircraft Carrier

■ SHIPS

DD	- Destroyer
DDC	- Corvette
DDE	- Escort Destroyer
DDG	- Guided Missile Destroyer
DDR	- Radar Picket Destroyer
DE	- Escort Vessel
DEC	- Control Escort Vessel
DER	- Radar Picket Escort Vessel
DL	- Frigate
DLG	- Guided Missile Frigate
DM	- Minelayer, Destroyer
DMS	- Minesweeper, Destroyer
IFS	- Inshore Fire Support Ship
IX	- Unclassified Miscellaneous
LCU	- Utility Landing Craft
LPH	- Amphibious Assault Ship
LSD	- Dock Landing Ship
LSFF	- Flotilla Flagship Landing Ship
LSIL	- Infantry Landing Ship (Large)
LSM	- Medium Landing Ship
LSMR	- Medium Landing Ship (Rocket)
LSSL	- Support Landing Ship (Large) Mk III
LST	- Tank Landing Ship
MCS	- Mine Countermeasures Support Ship
MHC	- Minehunter, Coastal
MMA	- Minelayer, Auxiliary
MMC	- Minelayer, Coastal
MMF	- Minelayer, Fleet
MSB	- Mine Sweeping Boat
MSC	- Minesweeper, Coastal (nonmagnetic)
MSC(O)	- Minesweeper, Coastal (Old)
MSF	- Minesweeper, Fleet (steel hulled)
MSI	- Minesweeper, Inshore
MSO	- Minesweeper, Ocean (nonmagnetic)
MSS	- Minesweeper, Special
PC	- Submarine Chaser (173-foot)
PCE	- Escort (180-foot)
PCER	- Rescue Escort (180-foot)
PCS	- Submarine Chaser (136-foot)
PF	- Patrol Escort
PGM	- Motor Gunboat
PR	- River Gunboat
PT	- Motor Torpedo Boat
PY	- Yacht
SC	- Submarine Chaser (110-foot)
SS	- Submarine

SHIPS

SSB	- Fleet Ballistic Missile Submarine
SSG	- Guided Missile Submarine
SSK	- Anti-Submarine Submarine
SSR	- Radar Picket Submarine
SST	- Target and Training Submarine
X	- Submersible Craft
YAG	- Miscellaneous Auxiliary
YC	- Open Lighter
YCF	- Car Float
YCK	- Open Cargo Lighter
YCV	- Aircraft Transportation Lighter
YD	- Floating Derrick
YDT	- Diving Tender
YF	- Covered Lighter (self-propelled)
YFB	- Ferryboat or Launch
YFD	- Yard Floating Dry Dock
YFN	- Covered Lighter (non-self-propelled)
YFNB	- Large Covered Lighter
YFND	- Covered Lighter (for use with dry dock)
YFNG	- Covered Lighter (special purpose)
YFNX	- Lighter (special purpose)
YFP	- Floating Power Barge
YFR	- Refrigerated Covered Lighter (self-propelled)
YFRN	- Refrigerated Covered Lighter (non-self-propelled)
YFRT	- Covered Lighter (range tender)
YFT	- Torpedo Transportation Lighter
YFU	- Harbor Utility Craft
YG	- Garbage Lighter (self-propelled)
YGN	- Garbage Lighter (non-self-propelled)
YHB	- House Boat
YM	- Dredge
YMP	- Motor Mine Planter
YMS	- Auxiliary Motor Mine Sweeper
YNG	- Gate Vessel
YO	- Fuel Oil Barge (self-propelled)
YOG	- Gasoline Barge (self-propelled)
YOGN	- Gasoline Barge (non-self-propelled)
YON	- Fuel Oil Barge (non-self-propelled)
YOS	- Oil Storage Barge
YP	- Patrol Vessel
YPD	- Floating Pile Driver
YR	- Floating Workshop
YRD	- Submarine Repair and Berthing Barge
YRBM	- Submarine Repair, Berthing and Messing Barge
YRDH	- Floating Dry Dock Workshop (Hull)
YRDM	- Floating Dry Dock Workshop (Machine)

■ SHIPS

YRL	- Covered Lighter (repair)
YSD	- Seaplane Wrecking Derrick
YSR	- Sludge Removal Barge
YTB	- Large Harbor Tug
YTL	- Small Harbor Tug
YTM	- Medium Harbor Tug
YTT	- Torpedo Testing Barge
YV	- Drone Aircraft Catapult Control Craft
YW	- Water Barge (self-propelled)
YWN	- Water Barge (non-self-propelled)

COMMUNICATION ELECTRONIC EQUIPMENT DESIGNATION SYSTEM

(Joint AN Nomenclature System)

SET OR EQUIPMENT DESIGNATION SYSTEM

Installation

A	- Airborne (installed and operated in aircraft)
B	- Underwater mobile, submarine
C	- Air transportable (inactivated, do not use)
D	- Pilotless carrier
F	- Fixed
G	- Ground, general ground use (includes two or more ground installations)
K	- Amphibious
M	- Ground, mobile (installed as operating unit in a vehicle which has no function other than transporting the equipment)
P	- Pack or portable (animal or man)
S	- Water surface craft
T	- Ground, transportable
U	- General utility (includes two or more general installation classes, airborne, shipboard, and ground)
V	- Ground, vehicular (installed in vehicle designed for functions other than carrying electronic equipment, etc., such as tanks)
W	- Water surface and underwater

Type of Equipment

A	- Invisible light, heat radiation
B	- Pigeon
C	- Carrier
D	- RADIAC
E	- Nupac
F	- Photographic
G	- Telegraph or teletype
I	- Interphone and public address
J	- Electromechanical (not otherwise covered)
K	- Telemetering
L	- Countermeasures
M	- Meteorological
N	- Sound in air

■ COMMUNICATION ELECTRONIC EQUIPMENT

P	- Radar
Q	- SONAR and underwater sound
R	- Radio
S	- Special types, magnetic, etc. or combinations of types
T	- Telephone (wire)
V	- Visual and visible light
W	- Armament (peculiar to armament, not otherwise covered)
X	- Facsimile or television

Purpose

A	- Auxiliary assemblies (not complete operating sets)
B	- Bombing
C	- Communications (receiving and transmitting)
D	- Direction finder
E	- Ejection and/or release
G	- Fire control or search light directing
H	- Recording and/or reproducing (graphic meteorological and sound)
L	- Searchlight control (inactivated, use "G")
M	- Maintenance and test assemblies (including tools)
N	- Navigational aids (including altimeters, beacons, compasses, RACONS, depth sounding, approach and landing)
P	- Reproducing (inactivated, do not use)
Q	- Special, or combination of purposes
R	- Receiving, passive detecting
S	- Detecting and/or range and bearing
T	- Transmitting
W	- Control
X	- Identification and recognition

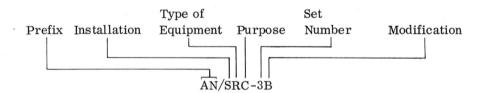

Example:

AN/SRC-3B - Water surface craft, radio, communications, set number three, second modification.

COMMUNICATION ELECTRONIC EQUIPMENT ∎

COMPONENT DESIGNATION SYSTEM

Component Indicator

AB	- Supports, antenna
AM	- Amplifiers
AS	- Antennae, complex
AT	- Antennae, simple
BA	- Battery, primary type
BB	- Battery, secondary
BZ	- Signal devices, audible
C	- Controls
CA	- Commutator assembly, SONAR
CB	- Capacitor bank
CG	- Cable assemblies, radio frequency
CK	- Crystal kits
CM	- Comparators
CN	- Compensators
CP	- Computers
CR	- Crystals
CU	- Couplers
CV	- Converters (electronic)
CW	- Covers
CX	- Cable assemblies, non-radio frequency
CY	- Cases and cabinets
D	- Dispensers
DA	- Load, dummy
DT	- Detecting heads
DY	- Dynamotors
E	- Hoists
F	- Filters
FN	- Furniture
FR	- Frequency measuring devices
G	- Generators, power
GO	- Goniometers
GP	- Ground rods
H	- Head, hand and chest sets
HC	- Crystal holder
HD	- Air-conditioning apparatus
ID	- Indicators, non-cathode ray tube
IL	- Insulators
IM	- Intensity measuring devices

■ COMMUNICATION ELECTRONIC EQUIPMENT

IP	- Indicators, cathode ray tube
J	- Junction devices
KY	- Keying devices
LC	- Tools, line construction
LS	- Loudspeakers
M	- Microphones
MA	- Magazines
MD	- Modulators
ME	- Meters
MF	- Magnets or magnetic field generators
MK	- Miscellaneous kits
ML	- Meteorological devices
MT	- Mountings
MX	- Miscellaneous
MU	- Memory units
O	- Oscillators
OA	- Operating assemblies
OC	- Oceanographic devices
OS	- Oscilloscope, test
PD	- Prime drivers
PF	- Fittings, pole
PG	- Pigeon articles
PH	- Photographic articles
PP	- Power supplies
PT	- Plotting equipment
PU	- Power equipments
R	- Receivers
RC	- Reels
RD	- Recorder-reproducers
RE	- Relay assemblies
RF	- Radio frequency component
RG	- Cables, radio frequency, bulk
RL	- Reeling machines
RO	- Recorders
RP	- Reproducers
RR	- Reflectors
RT	- Receiver and transmitter
S	- Shelters
SA	- Switching devices
SB	- Switchboards
SG	- Generals, signal
SM	- Simulators
SN	- Synchronizers
ST	- Straps
SU	- Optical device
T	- Transmitters

COMMUNICATION ELECTRONIC EQUIPMENT

TA	- Telephone apparatus
TB	- Towed body
TC	- Towed cable
TD	- Timing devices
TF	- Transformers
TG	- Positioning devices
TH	- Telegraph apparatus
TK	- Tool kits
TL	- Tools
TN	- Tuning units
TR	- Transducers
TS	- Test items
TT	- Teletypewriter and facsimile apparatus
TV	- Tester, tube
TW	- Tapes and recording wires
U	- Connectors, audio and power
UG	- Connectors, radio frequency
V	- Vehicles
VS	- Signaling equipment, visual
WD	- Cables, two conductor
WF	- Cables, four conductor
WM	- Cables, multiple conductor
WS	- Cables, single conductor
WT	- Cables, three conductor
ZM	- Impedance measuring devices

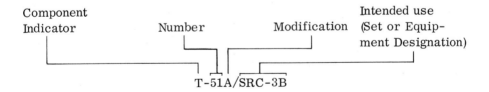

```
Component                                    Intended use
Indicator          Number       Modification  (Set or Equip-
                                              ment Designation)
                       T-51A/SRC-3B
```

Example:

T-51A/SRC-3B - Transmitter, number 51, first modification, part of or used with water surface craft, radio, communications, set number 3, second modification.